LONDON MATHEMATICAL SOCIETY LECTURE NOTE SERIES

Managing Editor: Professor N.J. Hitchin, Mathematics Institute,
University of Oxford, 24–29 St Giles, Oxford OX1 3LB, United Kingdom

The titles below are available from booksellers, or, in case of difficulty, from Cambridge University Press.

46	*p*-adic Analysis: a short course on recent work, N. KOBLITZ
59	Applicable differential geometry, M. CRAMPIN & F.A.E. PIRANI
66	Several complex variables and complex manifolds II, M.J. FIELD
86	Topological topics, I.M. JAMES (ed)
87	Surveys in set theory, A.R.D. MATHIAS (ed)
88	FPF ring theory, C. FAITH & S. PAGE
89	An F-space sampler, N.J. KALTON, N.T. PECK & J.W. ROBERTS
90	Polytopes and symmetry, S.A. ROBERTSON
92	Representations of rings over skew fields, A.H. SCHOFIELD
93	Aspects of topology, I.M. JAMES & E.H. KRONHEIMER (eds)
96	Diophantine equations over function fields, R.C. MASON
97	Varieties of constructive mathematics, D.S. BRIDGES & F. RICHMAN
98	Localization in Noetherian rings, A.V. JATEGAONKAR
99	Methods of differential geometry in algebraic topology, M. KAROUBI & C. LERUSTE
100	Stopping time techniques for analysts and probabilists, L. EGGHE
104	Elliptic structures on 3-manifolds, C.B. THOMAS
105	A local spectral theory for closed operators, I. ERDELYI & WANG SHENGWANG
107	Compactification of Siegel moduli schemes, C.-L. CHAI
109	Diophantine analysis, J. LOXTON & A. VAN DER POORTEN (eds)
113	Lectures on the asymptotic theory of ideals, D. REES
114	Lectures on Bochner-Riesz means, K.M. DAVIS & Y.-C. CHANG
116	Representations of algebras, P.J. WEBB (ed)
119	Triangulated categories in the representation theory of finite-dimensional algebras, D. HAPPEL
121	Proceedings of *Groups - St Andrews 1985*, E. ROBERTSON & C. CAMPBELL (eds)
128	Descriptive set theory and the structure of sets of uniqueness, A.S. KECHRIS & A. LOUVEAU
130	Model theory and modules, M. PREST
131	Algebraic, extremal & metric combinatorics, M.-M. DEZA, P. FRANKL & I.G. ROSENBERG (eds)
132	Whitehead groups of finite groups, ROBERT OLIVER
133	Linear algebraic monoids, MOHAN S. PUTCHA
134	Number theory and dynamical systems, M. DODSON & J. VICKERS (eds)
137	Analysis at Urbana, I, E. BERKSON, T. PECK, & J. UHL (eds)
138	Analysis at Urbana, II, E. BERKSON, T. PECK, & J. UHL (eds)
139	Advances in homotopy theory, S. SALAMON, B. STEER & W. SUTHERLAND (eds)
140	Geometric aspects of Banach spaces, E.M. PEINADOR & A. RODES (eds)
141	Surveys in combinatorics 1989, J. SIEMONS (ed)
144	Introduction to uniform spaces, I.M. JAMES
146	Cohen-Macaulay modules over Cohen-Macaulay rings, Y. YOSHINO
148	Helices and vector bundles, A.N. RUDAKOV *et al*
149	Solitons, nonlinear evolution equations and inverse scattering, M. ABLOWITZ & P. CLARKSON
150	Geometry of low-dimensional manifolds 1, S. DONALDSON & C.B. THOMAS (eds)
151	Geometry of low-dimensional manifolds 2, S. DONALDSON & C.B. THOMAS (eds)
152	Oligomorphic permutation groups, P. CAMERON
153	L-functions and arithmetic, J. COATES & M.J. TAYLOR (eds)
155	Classification theories of polarized varieties, TAKAO FUJITA
156	Twistors in mathematics and physics, T.N. BAILEY & R.J. BASTON (eds)
158	Geometry of Banach spaces, P.F.X. MÜLLER & W. SCHACHERMAYER (eds)
159	Groups St Andrews 1989 volume 1, C.M. CAMPBELL & E.F. ROBERTSON (eds)
160	Groups St Andrews 1989 volume 2, C.M. CAMPBELL & E.F. ROBERTSON (eds)
161	Lectures on block theory, BURKHARD KÜLSHAMMER
162	Harmonic analysis and representation theory, A. FIGA-TALAMANCA & C. NEBBIA
163	Topics in varieties of group representations, S.M. VOVSI
164	Quasi-symmetric designs, M.S. SHRIKANDE & S.S. SANE
166	Surveys in combinatorics, 1991, A.D. KEEDWELL (ed)
168	Representations of algebras, H. TACHIKAWA & S. BRENNER (eds)
169	Boolean function complexity, M.S. PATERSON (ed)
170	Manifolds with singularities and the Adams-Novikov spectral sequence, B. BOTVINNIK
171	Squares, A.R. RAJWADE
172	Algebraic varieties, GEORGE R. KEMPF
173	Discrete groups and geometry, W.J. HARVEY & C. MACLACHLAN (eds)
174	Lectures on mechanics, J.E. MARSDEN
175	Adams memorial symposium on algebraic topology 1, N. RAY & G. WALKER (eds)
176	Adams memorial symposium on algebraic topology 2, N. RAY & G. WALKER (eds)
177	Applications of categories in computer science, M. FOURMAN, P. JOHNSTONE & A. PITTS (eds)
178	Lower K- and L-theory, A. RANICKI
179	Complex projective geometry, G. ELLINGSRUD *et al*
180	Lectures on ergodic theory and Pesin theory on compact manifolds, M. POLLICOTT
181	Geometric group theory I, G.A. NIBLO & M.A. ROLLER (eds)
182	Geometric group theory II, G.A. NIBLO & M.A. ROLLER (eds)
183	Shintani zeta functions, A. YUKIE

184 Arithmetical functions, W. SCHWARZ & J. SPILKER
185 Representations of solvable groups, O. MANZ & T.R. WOLF
186 Complexity: knots, colourings and counting, D.J.A. WELSH
187 Surveys in combinatorics, 1993, K. WALKER (ed)
188 Local analysis for the odd order theorem, H. BENDER & G. GLAUBERMAN
189 Locally presentable and accessible categories, J. ADAMEK & J. ROSICKY
190 Polynomial invariants of finite groups, D.J. BENSON
191 Finite geometry and combinatorics, F. DE CLERCK et al
192 Symplectic geometry, D. SALAMON (ed)
194 Independent random variables and rearrangement invariant spaces, M. BRAVERMAN
195 Arithmetic of blowup algebras, WOLMER VASCONCELOS
196 Microlocal analysis for differential operators, A. GRIGIS & J. SJÖSTRAND
197 Two-dimensional homotopy and combinatorial group theory, C. HOG-ANGELONI et al
198 The algebraic characterization of geometric 4-manifolds, J.A. HILLMAN
199 Invariant potential theory in the unit ball of $\mathbf{C}^n$, MANFRED STOLL
200 The Grothendieck theory of dessins d'enfant, L. SCHNEPS (ed)
201 Singularities, JEAN-PAUL BRASSELET (ed)
202 The technique of pseudodifferential operators, H.O. CORDES
203 Hochschild cohomology of von Neumann algebras, A. SINCLAIR & R. SMITH
204 Combinatorial and geometric group theory, A.J. DUNCAN, N.D. GILBERT & J. HOWIE (eds)
205 Ergodic theory and its connections with harmonic analysis, K. PETERSEN & I. SALAMA (eds)
207 Groups of Lie type and their geometries, W.M. KANTOR & L. DI MARTINO (eds)
208 Vector bundles in algebraic geometry, N.J. HITCHIN, P. NEWSTEAD & W.M. OXBURY (eds)
209 Arithmetic of diagonal hypersurfaces over finite fields, F.Q. GOUVÊA & N. YUI
210 Hilbert C*-modules, E.C. LANCE
211 Groups 93 Galway / St Andrews I, C.M. CAMPBELL et al (eds)
212 Groups 93 Galway / St Andrews II, C.M. CAMPBELL et al (eds)
214 Generalised Euler-Jacobi inversion formula and asymptotics beyond all orders, V. KOWALENKO et al
215 Number theory 1992–93, S. DAVID (ed)
216 Stochastic partial differential equations, A. ETHERIDGE (ed)
217 Quadratic forms with applications to algebraic geometry and topology, A. PFISTER
218 Surveys in combinatorics, 1995, PETER ROWLINSON (ed)
220 Algebraic set theory, A. JOYAL & I. MOERDIJK
221 Harmonic approximation, S.J. GARDINER
222 Advances in linear logic, J.-Y. GIRARD, Y. LAFONT & L. REGNIER (eds)
223 Analytic semigroups and semilinear initial boundary value problems, KAZUAKI TAIRA
224 Computability, enumerability, unsolvability, S.B. COOPER, T.A. SLAMAN & S.S. WAINER (eds)
225 A mathematical introduction to string theory, S. ALBEVERIO, J. JOST, S. PAYCHA, S. SCARLATTI
226 Novikov conjectures, index theorems and rigidity I, S. FERRY, A. RANICKI & J. ROSENBERG (eds)
227 Novikov conjectures, index theorems and rigidity II, S. FERRY, A. RANICKI & J. ROSENBERG (eds)
228 Ergodic theory of $\mathbf{Z}^d$ actions, M. POLLICOTT & K. SCHMIDT (eds)
229 Ergodicity for infinite dimensional systems, G. DA PRATO & J. ZABCZYK
230 Prolegomena to a middlebrow arithmetic of curves of genus 2, J.W.S. CASSELS & E.V. FLYNN
231 Semigroup theory and its applications, K.H. HOFMANN & M.W. MISLOVE (eds)
232 The descriptive set theory of Polish group actions, H. BECKER & A.S. KECHRIS
233 Finite fields and applications, S. COHEN & H. NIEDERREITER (eds)
234 Introduction to subfactors, V. JONES & V.S. SUNDER
235 Number theory 1993–94, S. DAVID (ed)
236 The James forest, H. FETTER & B. GAMBOA DE BUEN
237 Sieve methods, exponential sums, and their applications in number theory, G.R.H. GREAVES et al
238 Representation theory and algebraic geometry, A. MARTSINKOVSKY & G. TODOROV (eds)
239 Clifford algebras and spinors, P. LOUNESTO
240 Stable groups, FRANK O. WAGNER
241 Surveys in combinatorics, 1997, R.A. BAILEY (ed)
242 Geometric Galois actions I, L. SCHNEPS & P. LOCHAK (eds)
243 Geometric Galois actions II, L. SCHNEPS & P. LOCHAK (eds)
244 Model theory of groups and automorphism groups, D. EVANS (ed)
245 Geometry, combinatorial designs and related structures, J.W.P. HIRSCHFELD et al
246 p-Automorphisms of finite p-groups, E.I. KHUKHRO
247 Analytic number theory, Y. MOTOHASHI (ed)
248 Tame topology and o-minimal structures, LOU VAN DEN DRIES
249 The atlas of finite groups: ten years on, ROBERT CURTIS & ROBERT WILSON (eds)
250 Characters and blocks of finite groups, G. NAVARRO
251 Gröbner bases and applications, B. BUCHBERGER & F. WINKLER (eds)
252 Geometry and cohomology in group theory, P. KROPHOLLER, G. NIBLO, R. STÖHR (eds)
253 The q-Schur algebra, S. DONKIN
254 Galois representations in arithmetic algebraic geometry, A.J. SCHOLL & R.L. TAYLOR (eds)
255 Symmetries and integrability of difference equations, P.A. CLARKSON & F.W. NIJHOFF (eds)
256 Aspects of Galois theory, HELMUT VÖLKLEIN et al
257 An introduction to noncommutative differential geometry and its physical applications 2ed, J. MADORE
258 Sets and proofs, S.B. COOPER & J. TRUSS (eds)
259 Models and computability, S.B. COOPER & J. TRUSS (eds)
260 Groups St Andrews 1997 in Bath, I, C.M. CAMPBELL et al
261 Groups St Andrews 1997 in Bath, II, C.M. CAMPBELL et al
263 Singularity theory, BILL BRUCE & DAVID MOND (eds)
264 New trends in algebraic geometry, K. HULEK, F. CATANESE, C. PETERS & M. REID (eds)

London Mathematical Society Lecture Note Series. 256

Aspects of Galois Theory

Edited by

Helmut Völklein (managing editor)
University of Florida

David Harbater
University of Pennsylvania

Peter Müller
University of Heidelberg

J.G. Thompson
University of Florida

CAMBRIDGE
UNIVERSITY PRESS

PUBLISHED BY THE PRESS SYNDICATE OF THE UNIVERSITY OF CAMBRIDGE
The Pitt Building, Trumpington Street, Cambridge, United Kingdom

CAMBRIDGE UNIVERSITY PRESS
The Edinburgh Building, Cambridge, CB2 2RU, UK   www.cup.cam.ac.uk
40 West 20th Street, New York, NY 10011-4211, USA   www.cup.org
10 Stamford Road, Oakleigh, Melbourne 3166, Australia

SCi
QA
214
.A87
1999

First published 1999

Printed in the United Kingdom at the University Press, Cambridge

A catalogue record for this book is available from the British Library

ISBN   0 521 63747 3 paperback

CONTENTS

Introduction   vii

Galois theory of semilinear transformations   1
Shreeram S. Abhyankar

Tools for the computation of families of coverings   38
Jean-Marc Couveignes

Some arithmetic properties of algebraic covers   66
Pierre Dèbes

Curves with infinite K-rational geometric fundamental group   85
Gerhard Frey, Ernst Kani & Helmut Völklein

Embedding problems and adding branch points   119
David Harbater

On beta and gamma functions associated with the
Grothendieck–Teichmüller groups   144
Yasutaka Ihara

Arithmetically exceptional functions and elliptic curves   180
Peter Müller

Tangential base points and Eisenstein power series   202
Hiroaki Nakamura

Braid-abelian tuples in $\mathrm{Sp}_n(K)$ 218
John Thompson & Helmut Völklein

Deformation of tame admissible covers of curves   239
Stefan Wewers

INTRODUCTION

This volume grew out of the 'UF GALOIS THEORY WEEK', a conference held at the University of Florida, Oct. 14–18, 1996. The conference was dedicated to the Inverse Galois Problem. The richness of this area stems from the fact that it attracts people from all kind of mathematical backgrounds, with methods ranging from explicit polynomial calculations and even numerical computer calculation to the structure and character theory of finite simple groups and up to the most abstract methods of algebraic/arithmetic geometry (like moduli stacks). In the original spirit of Galois, all these turn out to be just different aspects of the same matter.

Here is a brief description of the contents of this volume. Abhyankar continues his work on explicit classes of polynomials in characteristic $p > 0$ whose Galois groups comprise entire families of Lie type groups in characteristic p. In characteristic > 0, such polynomials have so far only been found for finitely many non-abelian simple groups except alternating groups. New methods for this, involving the determination of explicit equations for certain Hurwitz spaces, are developed in Couveignes' paper.

There is a way of realizing infinite series of Lie type groups as Galois groups over the rationals, based on Riemann's existence theorem and the notion of rigidity. It does not directly yield polynomials having these groups as Galois groups, however. New results in that direction are obtained in the paper of Thompson and Völklein, using a certain generalization of rigidity called the braid-abelian property and resulting in Galois realizations of the projective symplectic groups $\mathrm{PSp}(n, q)$ under certain restrictions on n and q.

The more abstract aspects come into play when considering the totality of Galois extensions — with certain properties — of a given field. This amounts to the study of certain (profinite) fundamental groups and absolute Galois groups. Harbater studies the fundamental group of an affine curve in positive characteristic. He determined the finite quotients of this fundamental group in his proof of Abhyankar's conjecture, but its full structure as profinite group remains mysterious. Ihara investigates the relationship between the absolute Galois group of the rationals and the so-called Grothendieck-Teichmüller group. The former is an arithmetic object, controlling the totality of Galois extensions of the rational field, whereas the latter is a geometric object (a certain subgroup of the automorphism group of the geometric fundamental group of the 3-punctured line). Ihara's paper provides evidence that these two groups may not be as closely related as was previously hoped for. Nakamura's article surveys related material.

Wewers' paper attempts to provide a bridge between the abstract and concrete approaches, by supplying a more elementary proof of Grothendieck's results on the relationship between the fundamental group of the punctured line in characteristic 0 and in positive characteristic. Dèbes studies general questions concerning the field of definition of a cover and models of the cover over such a field.

The articles of Frey/Kani/Völklein and of Müller apply methods developed for the Inverse Galois Problem to other areas. The former constructs infinite towers of unramified Galois curve covers defined over a fixed number field or finite field, with the additional property that there is a compatible system of rational points on these curves. In particular, this provides families of curves with 'many' rational points. Finally, Müller's paper investigates arithmetic properties of rational functions related to Hilbert's irreducibility theorem.

GALOIS THEORY OF SEMILINEAR TRANSFORMATIONS*

By

Shreeram S. Abhyankar

Mathematics Department, Purdue University, West Lafayette, IN 47907, USA;

e-mail: ram@cs.purdue.edu

Abstract. The general linear groups $GL(m, q)$ can be realized as Galois groups of certain vectorial (= q-additive) polynomials over rational function fields when the ground field contains $GF(q)$, where $m > 0$ is any integer and $q > 1$ is any power of any prime p. When calculated over the prime field as the ground field, these Galois groups get enlarged into the semilinear groups $\Gamma L(m, q)$. Similarly, for any integer $n > 0$, the Galois groups of the n-th iterates of these vectorials get enlarged from $GL(m, q, n)$ to $\Gamma L(m, q, n)$ where $GL(m, q, n)$ is the general linear group of the free module of rank m over the local ring $GF(q)[T]/T^n$ and $\Gamma L(m, q, n)$ is its semilinearization. Likewise, a corresponding enlargement to the semilinear symplectic groups $\Gamma Sp(2m, q)$ happens when dealing with suitable vectorials having the symplectic similitude groups $GSp(2m, q)$ as Galois groups. Much of this continues to hold when, instead of over rational function fields, the vectorials are considered over meromorphic function fields. A similar semilinear enlargement takes place when dealing with Galois groups between $SL(m, q)$ and $GL(m, q)$ or between $Sp(2m, q)$ and $GSp(2m, q)$. The calculation of these various Galois groups leads to a determination of the algebraic closures of the ground fields in the splitting fields of the corresponding vectorial polynomials.

Section 1: Introduction

Throughout this paper, let $k_p \subset K \subset \Omega$ be fields of characteristic $p > 0$ where Ω is an algebraic closure of K, let $q = p^u > 1$ be any power of p, let $m > 0$ be any integer, and to abbreviate frequently occurring expressions, for every integer $i \geq -1$, let us put

$$\langle i \rangle = 1 + q + q^2 + \cdots + q^i \text{ (convention: } \langle 0 \rangle = 1 \text{ and } \langle -1 \rangle = 0).$$

Moreover, for any nonconstant $\phi = \phi(Y) \in K[Y]$ we let

$$SF(\phi, K) = \text{the } \textbf{splitting field} \text{ of } \phi \text{ over } K \text{ in } \Omega$$

and

$$AC(k_p, \phi, K) = \text{the } \textbf{algebraic closure} \text{ of } k_p \text{ in } SF(\phi, K).$$

For various classes of separable ϕ, we shall determine the group $Gal(\phi, K)$ and the field $AC(k_p, \phi, K)$. Here K will mostly be a rational function field over k_p or a formal meromorphic series field over k_p. Also ϕ will mostly be a projective or subvectorial or vectorial polynomial over K.

---

*1991 Mathematical Subject Classification: 12F10, 14H30, 20D06, 20E22. This work was partly supported by NSF Grant DMS 91-01424 and NSA grant MDA 904-97-1-0010.

Recall that $f^*(Y)$ (resp: $\phi^*(Y)$ or $\phi^*(Y)$) in $K[Y]$ is said to be a **projective** (resp: **subvectorial** or **vectorial**) q-**polynomial** of q-**prodegree** (resp: q-**subdegree** or q-**degree**) m^* (where $m^* \geq 0$ is an integer) in Y with coefficients in K if it is of the form $f^*(Y) = \sum_{i=0}^{m^*} a_i^* Y^{\langle m^*-1-i \rangle}$ (resp: $\phi^*(Y) = \sum_{i=0}^{m^*} a_i^* Y^{q^{m^*-i}-1}$ or $\phi^*(Y) = \sum_{i=0}^{m^*} a_i^* Y^{q^{m^*-i}}$) with $a_i^* \in K$ for all i and $a_0^* \neq 0$. The phrase "of q-prodegree (resp: q-subdegree or q-degree) m^* in Y with coefficients in K" may be dropped or may be abbreviated to something like "in Y over K." Also the reference to q may be dropped. Note that $f^*(Y)$ (resp: $\phi^*(Y)$ or $\phi^*(Y)$) is **monic** $\Leftrightarrow a_0^* = 1$, and note that $f^*(Y)$ (resp: $\phi^*(Y)$ or $\phi^*(Y)$) is **separable** (i.e., its Y-discriminant is nonzero) $\Leftrightarrow a_{m^*}^* \neq 0$, and note that $\phi_Y^*(Y) = \phi_Y^*(0) = a_{m^*}^*$ where $\phi_Y^*(Y)$ is the Y-derivative of $\phi^*(Y)$. Also note that $f^*(Y) \to \phi^*(Y) = f^*(Y^{q-1})$ and $\phi^*(Y) \to \phi^*(Y) = Y\phi^*(Y)$ give bijections of projectives to subvectorials (= their **subvectorial associates**) to vectorials (= their **vectorial associates**).

To review what was said in Lemmas (2.4) and (2.5) of [A03] and Lemma (4.1.1) of [A08], for a moment let $f = f(Y)$ be a separable projective q-polynomial of q-prodegree m over K, let $\phi = \phi(Y) = f(Y^{q-1})$ and $\phi = \phi(Y) = Y\phi(Y)$, and let V be the set of all roots of ϕ in Ω, and note that then V is an m-dimensional $GF(q)$-vector-subspace of Ω; to see this, it suffices to observe that the cardinality of V is q^m and for all y, z in Ω and $\zeta \in GF(q)$ we have $\phi(y+z) = \phi(y) + \phi(z)$ and $\phi(\zeta z) = \zeta\phi(z)$. Let $\overline{V}$ be the set of all roots of f in Ω. Then $V \setminus \{0\}$ is the set of all roots of ϕ in Ω, and $y \mapsto y^{q-1}$ gives a surjective map $V \setminus \{0\} \to \overline{V}$ whose fibers are punctured 1-spaces, i.e., 1-spaces minus the zero vector. So we may identify $\overline{V}$ with the projective space associated with V. In particular, fixing $0 \neq y \in V$ and letting y' vary over all elements of V with $y'^{q-1} = y^{q-1}$ we see that $y'/y \in K(V)$ varies over all nonzero elements of $GF(q)$, and hence $GF(q) \subset K(V) = SF(\phi, K) = SF(\phi, K)$. It follows that any $g \in Gal(K(V), K)$ induces an automorphism g' of $GF(q)$, and for all $z \in V$ and $\zeta \in GF(q)$ we clearly have $g(\zeta z) = g'(\zeta)g(z)$; since g is clearly additive on V, we see that g induces on V a semilinear transformation, i.e., an element of $\Gamma L(V) = \Gamma L(m, q)$, and moreover this element belongs to $GL(V) = GL(m, q) \Leftrightarrow g'$ is identity. Thus in a natural manner $Gal(\phi, K) < \Gamma L(m, q)$. Clearly g' is identity for all $g \in Gal(K(V), K) \Leftrightarrow GF(q) \subset K$, and hence in the above identification $Gal(\phi, K) < GL(m, q) \Leftrightarrow GF(q) \subset K$. Thus we have the following:

Semilinearity Lemma (1.1). *Let $f = f(Y)$ be a separable projective q-polynomial of q-prodegree m in Y over K, let $\phi = \phi(Y) = f(Y^{q-1})$ and $\phi = \phi(Y) = Y\phi(Y)$, and let V be the set of all roots of ϕ in Ω. Then V is an m-dimensional $GF(q)$-vector-subspace of Ω with $GF(q) \subset K(V) = SF(\phi, K) = SF(\phi, K)$, and in a natural manner we may identify $Gal(\phi, K)$ with a subgroup of $\Gamma L(V) = \Gamma L(m, q)$; under this identification we have*

$Gal(\phi, K) < GL(m, q) \Leftrightarrow GF(q) \subset K$. *Likewise, we may identify* $Gal(f, K)$ *with a subgroup of* $P\Gamma L(m, q)$ *and then* $Gal(f, K)$ *becomes the image of* $Gal(\phi, K)$ *under the canonical epimorphism of* $\Gamma L(m, q)$ *onto* $P\Gamma L(m, q)$. *The Galois group* $Gal(\phi, K)$ *essentially equals the Galois group* $Gal(\phi, K)$ *except that the former acts on nonzero vectors while the latter acts on the entire vector space* V.

This lemma will be used tacitly. In particular, the said Galois groups will be regarded as subgroups of $\Gamma L(V) = \Gamma L(m, q)$ and its projectivization. In Section 2 we shall deal with vectorials whose Galois groups are between $SL(m, q)$ and $\Gamma L(m, q)$; this will be based on [A08]. In Section 3 we shall deal with iterates of some of the vectorials considered in Section 2; this will be based on [AS1]. In Section 4 we shall deal with vectorials whose Galois groups are between $Sp(2m, q)$ and $\Gamma Sp(2m, q)$; this will be based on [A04], [AL1] and [AL2]. For relevant general discussion about Galois Theory, see [A01], [A02] and [A07]. As a supplement to (1.1), in (2.5)(iii) of [A03] we proved the following:

Root Extraction Lemma (1.2). *Given any monic subvectorial* q-*poly-nomial* $\phi = \phi(Y)$ *of* q-*subdegree* m *in* Y *over* K, *there exists* $\Lambda \in SF(\phi, K)$ *such that* $\Lambda^{q-1} = (-1)^{(m-1)}\phi(0)$.

When $GF(q) \subset K$, the Galois groups of the vectorials over K to be considered in Section 2 will be between $SL(m, q)$ and $GL(m, q)$. Note that $SL(m, q) \vartriangleleft GL(m, q)$ with $GL(m, q)/SL(m, q) = Z_{q-1}$ and hence for every divisor d of $q - 1$ there is a unique group $GL^{(d)}(m, q)$ such that $SL(m, q) < GL^{(d)}(m, q) < GL(m, q)$ and $[GL(m, q) : GL^{(d)}(m, q)] = d$ where, as usual, $<$ and $\vartriangleleft$ denote subgroup and normal subgroup respectively, Z_{q-1} denotes a cyclic group of order $q - 1$, and : denotes index. Upon letting $PGL^{(d)}(m, q)$ to be the image of $GL^{(d)}(m, q)$ under the canonical epimorphism of $GL(m, q)$ onto $PGL(m, q)$ we see that $PGL^{(d)}(m, q)$ is the unique group between $PSL(m, q)$ and $PGL(m, q)$ such that $[PGL(m, q) : PGL^{(d)}(m, q)] = GCD(m, d)$.

Likewise $GL(m, q) \vartriangleleft \Gamma L(m, q)$ with $\Gamma L(m, q)/GL(m, q) = Z_u$ and hence for every divisor δ of u there is a unique group $\Gamma L_\delta(m, q)$ such that $GL(m, q) < \Gamma L_\delta(m, q) < \Gamma L(m, q)$ and $[\Gamma L_\delta(m, q) : GL(m, q)] = \delta$, where $P\Gamma L_\delta(m, q)$ is the image of $\Gamma L_\delta(m, q)$ under the canonical epimorphism of $\Gamma L(m, q)$ onto $P\Gamma L(m, q)$. Also we let $\Gamma SL_\delta(m, q)$ be the set of all subgroups I of $\Gamma L_\delta(m, q)$ such that $I \cap GL(m, q) = SL(m, q) \vartriangleleft I$ with $I/SL(m, q) = Z_\delta$, and we let $P\Gamma SL_\delta(m, q)$ be the set of images of the various members of $\Gamma SL_\delta(m, q)$ under the canonical epimorphism of $\Gamma L(m, q)$ onto $P\Gamma L(m, q)$; in Remark (4.4.1) of [A08] we have shown that $\Gamma SL_\delta(m, q)$ is a nonempty complete set of conjugate subgroups of $\Gamma L(m, q)$, and every I in $\Gamma SL_\delta(m, q)$ is a **split extension** of $SL(m, q)$ (i.e., some subgroup of I is mapped isomorphically onto $I/SL(m, q)$ by the residue class map of I onto $I/SL(m, q)$) such that

$\Gamma L_\delta(m, q)$ is generated by $\mathrm{GL}(m, q)$ and I. Finally we let $\Gamma L_\delta^{(d)}(m, q)$ be the set of all subgroups J of $\Gamma L_\delta(m, q)$ such that $J \cap \mathrm{GL}(m, q) = \mathrm{GL}^{(d)}(m, q) \lhd J$ with $J/\mathrm{GL}^{(d)}(m, q) = Z_\delta$ and $I < J$ for some I in $\Gamma \mathrm{SL}_\delta(m, q)$, and we let $P\Gamma L_\delta^{(d)}(m, q)$ be the set of images of the various members of $\Gamma L_\delta^{(d)}(m, q)$ under the canonical epimorphism of $\Gamma L(m, q)$ onto $P\Gamma L(m, q)$; in Remark (4.4.1) of [A08] we have shown that $\Gamma L_\delta^{(d)}(m, q)$ is a nonempty complete set of conjugate subgroups of $\Gamma L(m, q)$, and every J in $\Gamma L_\delta^{(d)}(m, q)$ is a split extension of $\mathrm{GL}^{(d)}(m, q)$ such that $\Gamma L_\delta(m, q)$ is generated by $\mathrm{GL}(m, q)$ and J; note that clearly $\Gamma L_\delta^{(q-1)}(m, q) = \Gamma \mathrm{SL}_\delta(m, q)$ and $\Gamma L_\delta^{(1)}(m, q) = \{\Gamma L_\delta(m, q)\}$.

To determine the Galois groups when $\mathrm{GF}(q)$ is not contained in K, we note that $\mathrm{SF}(Y^q - Y, K) = K(\mathrm{GF}(q))$ and we **let** $\delta(K)$ **be the unique divisor** of u such that

(1.3)

$$\mathrm{Gal}(Y^q - Y, K) = Z_{\delta(K)} \quad \text{i.e. equivalently} \quad [K(\mathrm{GF}(q)) : K] = \delta(K)$$

and we note that then (see Footnote 17 of [A08])

(1.4)
$$K \cap \mathrm{GF}(q) = \mathrm{GF}(p^{u/\delta(K)}).$$

Concerning $\delta(K)$, the following lemma is easily proved; see Propositions (4.2.3) to (4.2.5) of [A08].

Linear Enlargement Lemma (1.5). *For any separable projective q-polynomial $f = f(Y)$ of q-prodegree m in Y over K and its subvectorial associate $\phi = \phi(Y) = f(Y^{q-1})$ we have the following.*

(1.5.1) If $\mathrm{Gal}(\phi, K(\mathrm{GF}(q))) = SL(m, q)$, then $\mathrm{Gal}(\phi, K) \in \Gamma SL_{\delta(K)}(m, q)$ and $\mathrm{Gal}(f, K) \in P\Gamma SL_{\delta(K)}(m, q)$.

(1.5.2) If $\mathrm{Gal}(\phi, K(\mathrm{GF}(q))) = GL(m, q)$, then $\mathrm{Gal}(\phi, K) = \Gamma L_{\delta(K)}(m, q)$ and $\mathrm{Gal}(f, K) = P\Gamma L_{\delta(K)}(m, q)$.

(1.5.3) If $\mathrm{Gal}(\phi, K(\mathrm{GF}(q))) = GL^{(d)}(m, q)$ where d is a divisor of $q - 1$, and for some field K' between K and $\mathrm{SF}(\phi, K)$ we have $\delta(K') = \delta(K)$ and $\mathrm{Gal}(\phi, K'(\mathrm{GF}(q))) = SL(m, q)$, then $\mathrm{Gal}(\phi, K) \in \Gamma L_{\delta(K)}^{(d)}(m, q)$ and $\mathrm{Gal}(f, K) \in P\Gamma L_{\delta(K)}^{(d)}(m, q)$.

In determining $AC(k_p, \phi, K)$ we shall use the following obvious:

Algebraic Closure Lemma (1.6). *Just in this lemma let $k_p \subset K \subset \Omega$ be fields of any characteristic, which may or may not be zero, such that Ω is an algebraic closure of K. Let $\phi = \phi(Y)$ be a nonconstant separable polynomial in Y with coefficients in K, and let k^* be an algebraic field extension of k_p in $\mathrm{SF}(\phi, K)$ such that for every finite algebraic field extension k' of k^* in $\mathrm{SF}(\phi, K)$ we have $[K(k') : K(k^*)] = [k' : k^*]$ and $|\mathrm{Gal}(\phi, K(k'))| = |\mathrm{Gal}(\phi, K(k^*))|$. Then $AC(k_p, \phi, K) = k^*$.*

As a matter of terminology, we recall that a (noetherian) local ring S' is said to **dominate** a local ring S if S is a subring of S' and the **maximal**

ideal $M(S)$ of S is contained in the maximal ideal $M(S')$ of S', and we note that then the **residue field** $S/M(S)$ of S may be identified with a subfield of the residue field $S'/M(S')$ of S'; if under this identification, $S/M(S)$ coincides with $S'/M(S')$ then S' is said to be **residually rational** over S; thus in particular S' is residually rational over a subfield means that the subfield gets mapped isomorphically onto $S'/M(S')$ under the canonical epimorphism $S' \to S'/M(S')$.

It is a pleasure to thank Paul Loomis and Ganesh Sundaram for stimulating conversations concerning the material of this paper.

Section 2: Linear Groups

In this Section, to write down families of polynomials whose Galois groups are between $\mathrm{SL}(m, q)$ and $\Gamma\mathrm{L}(m, q)$, let $Y, X, T_1, T_2, \ldots$ be indeterminates over k_p. For every $e \geq 0$ let

$$K_e = k_p(X, T_1, \ldots, T_e)$$

and

K_e =the quotient field of an $(e + 1)$-dimensional regular local
domain R_e with $k_p \subset R_e$ and $M(R_e) = (X, T_1, \ldots, T_e)R_e$

and for every $e \geq 1$ and $0 \neq \tau \in k_p(T_1)$ let

$$K_{(e,\tau)} = k_p(X, \tau, T_2, \ldots, T_e).$$

We shall apply the considerations of Section 1 by taking $K = K_e$ or K_e or $K_{(e,\tau)}$ with suitable e and τ.

First, for $0 \leq e \leq m - 1$, consider the monic separable projective q-polynomial

$$f_e^{**} = f_e^{**}(Y) = Y^{(m-1)} + X + \sum_{i=1}^{e} T_i Y^{(i-1)}$$

of q-prodegree m in Y over K_e, and its subvectorial associate

$$\phi_e^{**} = \phi_e^{**}(Y) = f_e^{**}(Y^{q-1}) = Y^{q^m - 1} + X + \sum_{i=1}^{e} T_i Y^{q^i - 1}$$

and, for every divisor d of $q - 1$, let $f_e^{*(d)}$ and $\phi_e^{*(d)}$ be obtained by substituting $(-1)^{(m-1)} X^d$ for X in f_e^{**} and ϕ_e^{**} respectively, i.e., let

$$f_e^{*(d)} = f_e^{*(d)}(Y) = Y^{(m-1)} + (-1)^{(m-1)} X^d + \sum_{i=1}^{e} T_i Y^{(i-1)}$$

and

$$\phi_e^{*(d)} = \phi_e^{*(d)}(Y) = Y^{q^m-1} + (-1)^{\langle m-1\rangle}X^d + \sum_{i=1}^{e} T_i Y^{q^i-1}.$$

Next, for $1 \le e \le m-1$ and every $0 \ne \tau \in k_p(T_1)$ let $f_{(e,\tau)}^{**}$ and $\phi_{(e,\tau)}^{**}$ be obtained by substituting τ for T_1 in f_e^{**} and ϕ_e^{**} respectively, i.e., let

$$f_{(e,\tau)}^{**} = f_{(e,\tau)}^{**}(Y) = Y^{\langle m-1\rangle} + X + \tau Y + \sum_{i=2}^{e} T_i Y^{\langle i-1\rangle}$$

and

$$\phi_{(e,\tau)}^{**} = \phi_{(e,\tau)}^{**}(Y) = Y^{q^m-1} + X + \tau Y^{q-1} + \sum_{i=2}^{e} T_i Y^{q^i-1}$$

and, for every divisor d of $q-1$, let $f_{(e,\tau)}^{*(d)}$ and $\phi_{(e,\tau)}^{*(d)}$ be obtained by substituting $(-1)^{\langle m-1\rangle}X^d$ for X in $f_{(e,\tau)}^{**}$ and $\phi_{(e,\tau)}^{**}$ respectively, i.e., let

$$f_{(e,\tau)}^{*(d)} = f_{(e,\tau)}^{*(d)}(Y) = Y^{\langle m-1\rangle} + (-1)^{\langle m-1\rangle}X^d + \tau Y + \sum_{i=2}^{e} T_i Y^{\langle i-1\rangle}$$

and

$$\phi_{(e,\tau)}^{*(d)} = \phi_{(e,\tau)}^{*(d)}(Y) = Y^{q^m-1} + (-1)^{\langle m-1\rangle}X^d + \tau Y^{q-1} + \sum_{i=2}^{e} T_i Y^{q^i-1}.$$

Finally, for $1 \le e \le m-1$ and every $0 \ne \tau \in k_p(T_1)$ let $f_{(e,\tau)}^{*}$ and $\phi_{(e,\tau)}^{*}$ be obtained by substituting $((-1)^{\langle m-1\rangle}\tau^{q-1}, X)$ for (X, T_1) in f_e^{**} and ϕ_e^{**} respectively, i.e., let

$$f_{(e,\tau)}^{*} = f_{(e,\tau)}^{*}(Y) = Y^{\langle m-1\rangle} + (-1)^{\langle m-1\rangle}\tau^{q-1} + XY + \sum_{i=2}^{e} T_i Y^{\langle i-1\rangle}$$

and

$$\phi_{(e,\tau)}^{*} = \phi_{(e,\tau)}^{*}(Y) = Y^{q^m-1} + (-1)^{\langle m-1\rangle}\tau^{q-1} + XY^{q-1} + \sum_{i=2}^{e} T_i Y^{q^i-1}.$$

Concerning these polynomials, by MRT (= the Method of Ramification Theory) and MTR (= the Method of Throwing Away Roots), supplemented by Theorem I of [CaK] which we restate as Theorem (2.1*) below, in Theorems (2.3.1) to (2.3.5) of [A08] we respectively proved parts (2.1.1) to (2.1.5) of the following Theorem (2.1).

Theorem (2.1*) [Cameron-Kantor]. *If $m > 2$ and $H < GL(m, q)$ is such that its image under the canonical epimorphism of $GL(m, q)$ onto $PGL(m, q)$ is doubly transitive, then either $SL(m, q) < H$, or $(q, m) = (4, 2)$ with $A_7 \approx H < SL(4, 2) = GL(4, 2) \approx A_8$ (where $\approx$ denotes isomorphism, and A_7 and A_8 are the alternating groups on 7 and 8 letters respectively).*

Theorem (2.1). *For $1 \leq e \leq m - 1$ we have the following.*

(2.1.1) If $GF(q) \subset k_p$, then for every element $0 \neq \tau \in k_p(T_1)$ we have $Gal(\phi^*_{(e,\tau)}, K_{(e,\tau)}) = SL(m, q)$.

(2.1.2) If $GF(q) \subset k_p$, then for every element $0 \neq \tau \in k_p(T_1)$ we have $Gal(\phi^{**}_{(e,\tau)}, K_{(e,\tau)}) = GL(m, q)$.

(2.1.3) If $GF(q) \subset k_p$, then for every integer $\epsilon \geq e$ we have $Gal(\phi_e^{**}, K_\epsilon) = GL(m, q)$.

(2.1.4) If $GF(q) \subset k_p$, then for every element $0 \neq \tau \in k_p(T_1)$ and every divisor d of $q - 1$ we have $Gal(\phi^{*(d)}_{(e,\tau)}, K_{(e,\tau)}) = GL^{(d)}(m, q)$.

(2.1.5) If $GF(q) \subset k_p$, then for every integer $\epsilon \geq e$ and every divisor d of $q - 1$ we have $Gal(\phi_e^{*(d)}, K_\epsilon) = GL^{(d)}(m, q)$.

By using the Algebraic Closure Lemma (1.6), we shall now deduce the following consequences of the above Theorem.

Theorem (2.2). *For $1 \leq e \leq m - 1$ we have the following.*

(2.2.1) For every element $0 \neq \tau \in k_p(T_1)$ we have $AC(k_p, \phi^*_{(e,\tau)}, K_{(e,\tau)}) = k_p(GF(q))$.

(2.2.2) For every element $0 \neq \tau \in k_p(T_1)$ we have $AC(k_p, \phi^{**}_{(e,\tau)}, K_{(e,\tau)}) = k_p(GF(q))$.

(2.2.3) If $\epsilon \geq e$ is any integer such that R_ϵ is residually rational over k_p, then we have $AC(k_p, \phi_e^{**}, K_\epsilon) = k_p(GF(q))$.

(2.2.4) For every element $0 \neq \tau \in k_p(T_1)$ and every divisor d of $q - 1$, we have $AC(k_p, \phi^{*(d)}_{(e,\tau)}, K_{(e,\tau)}) = k_p(GF(q))$.

(2.2.5) If $\epsilon \geq e$ is any integer such that R_ϵ is residually rational over k_p, then for every divisor d of $q - 1$ we have $AC(k_p, \phi_e^{*(d)}, K_\epsilon) = k_p(GF(q))$.

To prove (2.2.1) or (2.2.2) or (2.2.4), let $1 \leq e \leq m-1$ and $0 \neq \tau \in k(T_1)$ be given, and respectively let $(\phi, G) = (\phi^*_{(e,\tau)}, SL(m, q))$ or $(\phi^{**}_{(e,\tau)}, GL(m, q))$ or $(\phi^{*(d)}_{(e,\tau)}, GL^{(d)}(m, q))$ where in the last case d is any divisor of $q - 1$. Upon letting $K = K_{(e,\tau)}$ and $k^* = k_p(GF(q))$, by (1.1) we see that $k^* \subset SF(\phi, K)$. Now we have $K(k^*) = k^*(X, \tau, T_2, \ldots, T_e)$ with $\tau \in k^*(T_1)$ and $GF(q) \subset k^*$, and given any finite algebraic field extension k' of k^* in $SF(\phi, K)$ we also have $K(k') = k'(X, \tau, T_2, \ldots, T_e)$ with $\tau \in k'(T_1)$ and $GF(q) \subset k'$, and hence respectively by (2.1.1) or (2.1.2) or (2.1.4) we see that $Gal(\phi, K(k')) = G = Gal(\phi, K(k^*))$. For any finite algebraic field extension k' of k^* in $SF(\phi, K)$ we clearly have $[K(k') : K(k^*)] = [k' : k^*]$. Therefore by (1.6) we conclude that $AC(k_p, \phi, K) = k^*$.

To prove (2.2.3) or (2.2.5), let $1 \leq r \leq m - 1$ and $\epsilon \geq e$ be given, and respectively let $(\phi, G) = (\phi_e^{**}, \mathrm{GL}(m,q))$ or $(\phi_e^{*(d)}, \mathrm{GL}^{(d)}(m,q))$ where in the second case d is any divisor of $q - 1$. Upon letting $K = K_\epsilon$ and $k^* = k_p(\mathrm{GF}(q))$, by (1.1) we see that $k^* \subset \mathrm{SF}(\phi, K)$. Moreover, upon letting R_ϵ^* to be the localization of the integral closure of R_ϵ in $K(k^*)$ at a maximal ideal in it we see that R_ϵ^* is an $(\epsilon + 1)$-dimensional regular local domain whose maximal ideal is generated by $(X, T_1, \ldots, T_\epsilon)$ and whose quotient field is $K(k^*)$, and we clearly have $\mathrm{GF}(q) \subset K(k^*)$, and given any finite algebraic field extension k' of k^* in $\mathrm{SF}(\phi, K)$, upon letting R_ϵ' to be the localization of the integral closure of R_ϵ^* in $K(k')$ at a maximal ideal in it we see that R_ϵ' is an $(\epsilon + 1)$-dimensional regular local domain whose maximal ideal is generated by $(X, T_1, \ldots, T_\epsilon)$ and whose quotient field is $K(k')$, and we clearly have $\mathrm{GF}(q) \subset K(k')$, and hence respectively by (2.1.3) or (2.1.5) we see that $\mathrm{Gal}(\phi, K(k')) = G = \mathrm{Gal}(\phi, K(k^*))$. Now, assuming R_ϵ to be residually rational over k_p, we see that R_ϵ^* is the integral closure of R_ϵ in $K(k^*)$, and R_ϵ^* is residually rational over k^*, and given any finite algebraic field extension k' of k^* in $\mathrm{SF}(\phi, K)$, we see that R_ϵ' is the integral closure of R_ϵ^* in $K(k')$, and R_ϵ' is residually rational over k', and also $[K(k') : K(k^*)] = [k' : k^*]$. Therefore again by (1.6) we conclude that $\mathrm{AC}(k_p, \phi, K) = k^*$.

In Theorems (4.3.1) to (4.3.5) of [A08] we deduced the following consequences of parts (2.1.1) to (2.1.5) of the above Theorem (2.1) together with the Linear Enlargement Lemma (1.5).

Theorem (2.3). *For $1 \leq e \leq m - 1$ we have the following.*

(2.3.1) For every element $0 \neq \tau \in k_p(T_1)$, upon letting $\delta = \delta(k_p)$, we have $\mathrm{Gal}(\phi_{(e,\tau)}^*, K_{(e,\tau)}) \in \Gamma SL_\delta(m,q)$ and $\mathrm{Gal}(f_{(e,\tau)}^*, K_{(e,\tau)}) \in P\Gamma SL_\delta(m,q)$.

(2.3.2) For every element $0 \neq \tau \in k_p(T_1)$, upon letting $\delta = \delta(k_p)$, we have $\mathrm{Gal}(\phi_{(e,\tau)}^{**}, K_{(e,\tau)}) = \Gamma L_\delta(m,q)$ and $\mathrm{Gal}(f_{(e,\tau)}^{**}, K_{(e,\tau)}) = P\Gamma L_\delta(m,q)$.

(2.3.3) For every integer $\epsilon \geq e$, upon letting $\delta = \delta(K_\epsilon)$, we have $\mathrm{Gal}(\phi_e^{**}, K_\epsilon) = \Gamma L_\delta(m,q)$ and $\mathrm{Gal}(f_e^{**}, K_\epsilon) = P\Gamma L_\delta(m,q)$. [Note that if either $R_\epsilon = k_p[[X, T_1, \ldots, T_\epsilon]]$ or $R_\epsilon = $ the localization of $k_p[X, T_1, \ldots, T_\epsilon]$ at the maximal ideal generated by $(X, T_1, \ldots, T_\epsilon)$ then R_ϵ is residually rational over k_p and we have $\delta(K_\epsilon) = \delta(k_p)$.]

(2.3.4) For every element $0 \neq \tau \in k_p(T_1)$ and every divisor d of $q - 1$, upon letting $\delta = \delta(k_p)$, we have $\mathrm{Gal}(\phi_{(e,\tau)}^{*(d)}, K_{(e,\tau)}) \in \Gamma L_\delta^{(d)}(m,q)$ and $\mathrm{Gal}(f_{(e,\tau)}^{*(d)}, K_{(e,\tau)}) \in P\Gamma L_\delta^{(d)}(m,q)$.

(2.3.5) For every integer $\epsilon \geq e$ and every divisor d of $q - 1$, upon letting $\delta = \delta(K_\epsilon)$, we have $\mathrm{Gal}(\phi_e^{*(d)}, K_\epsilon) \in \Gamma L_\delta^{(d)}(m,q)$ and $\mathrm{Gal}(f_e^{*(d)}, K_\epsilon) \in P\Gamma L_\delta^{(d)}(m,q)$. [Note that if either $R_\epsilon = k_p[[X, T_1, \ldots, T_\epsilon]]$ or $R_\epsilon = $ the localization of $k_p[X, T_1, \ldots, T_\epsilon]$ at the maximal ideal generated by $(X, T_1, \ldots, T_\epsilon)$ then R_ϵ is residually rational over k_p and we have $\delta(K_\epsilon) = \delta(k_p)$.]

Remark (2.4) [Local Surface Coverings].

(2.4.1). For $m > 1 = e$ we get the trinomials $f_1^{**} = Y^{(m-1)} + T_1 Y + X$ and $\phi_1^{**} = Y^{q^m} + T_1 Y^q + XY$, giving local coverings above a normal crossing of the branch locus in the local (X, T_1)-plane, dealt with in [A07] and [A08]; this is particularly significant with $R_2 = k_p[[X, T_1]]$; the above Theorems (2.2.3), (2.2.5), (2.3.3) and (2.3.5) give generalizations for the local $(\epsilon + 1)$-dimensional space; the following Theorems (2.4.3) and (2.4.5) are special cases of this. For $m > 1 = e$ and $\tau = 1$ we get the trinomials $f_{(1,1)}^* = Y^{(m-1)} + XY + (-1)^{(m-1)}$ and $\phi_{(1,1)}^* = Y^{q^m} + XY^q + (-1)^{(m-1)}Y$ giving unramified coverings of the affine line, and the trinomials $f_{(1,1)}^{**} = Y^{(m-1)} + Y + X$ and $\phi_{(1,1)}^{**} = Y^{q^m} + Y^q + XY$ giving unramified coverings of the once punctured affine line, dealt with in [A03] and [A08].

Remembering that now $m > 0$ is any integer, we conclude with the following consequences of the above theorems:

(2.4.2). *We have* $\mathrm{Gal}(\phi_{m-1}^{**}, K_{m-1}) = \Gamma L_\delta(m, q)$ *and* $\mathrm{Gal}(f_{m-1}^{**}, K_{m-1}) = P\Gamma L_\delta(m, q)$ *where* $\delta = \delta(k_p)$, *and we have* $AC(k_p, \phi_{m-1}^{**}, K_{m-1}) = k_p(GF(q))$.

(2.4.3). *We have* $\mathrm{Gal}(\phi_{m-1}^{**}, K_{m-1}) = \Gamma L_\delta(m, q)$ *and* $\mathrm{Gal}(f_{m-1}^{**}, K_{m-1}) = P\Gamma L_\delta(m, q)$ *where* $\delta = \delta(K_{m-1})$, *and moreover if* R_{m-1} *is residually rational over* k_p *then we have* $AC(k_p, \phi_{m-1}^{**}, K_{m-1}) = k_p(GF(q))$. *[Note that if either* $R_{m-1} = k_p[[X, T_1, \ldots, T_{m-1}]]$ *or* $R_{m-1} = $ *the localization of* $k_p[X, T_1, \ldots, T_{m-1}]$ *at the maximal ideal generated by* $(X, T_1, \ldots, T_{m-1})$ *then* R_{m-1} *is residually rational over* k_p *and we have* $\delta(K_{m-1}) = \delta(k_p).]$

(2.4.4). *We have* $\mathrm{Gal}(\phi_{m-1}^{*(d)}, K_{m-1}) \in \Gamma L_\delta^{(d)}(m, q)$ *and* $\mathrm{Gal}(f_{m-1}^{*(d)}, K_{m-1}) \in P\Gamma L_\delta^{(d)}(m, q)$ *where* d *is any divisor of* $q - 1$ *and* $\delta = \delta(k_p)$, *and we have* $AC(k_p, \phi_{m-1}^{*(d)}, K_{m-1}) = k_p(GF(q))$.

(2.4.5). *We have* $\mathrm{Gal}(\phi_{m-1}^{*(d)}, K_{m-1}) \in \Gamma L_\delta^{(d)}(m, q)$ *and* $\mathrm{Gal}(f_{m-1}^{*(d)}, K_{m-1}) \in P\Gamma L_\delta^{(d)}(m, q)$ *where* d *is any divisor of* $q - 1$ *and* $\delta = \delta(K_{m-1})$, *and moreover if* R_{m-1} *is residually rational over* k_p *then we have* $AC(k_p, \phi_{m-1}^{*(d)}, K_{m-1}) = k_p(GF(q))$. *[Note that if either* $R_{m-1} = k_p[[X, T_1, \ldots, T_{m-1}]]$ *or* $R_{m-1} = $ *the localization of* $k_p[X, T_1, \ldots, T_{m-1}]$ *at the maximal ideal generated by* $(X, T_1, \ldots, T_{m-1})$ *then* R_{m-1} *is residually rational over* k_p *and we have* $\delta(K_{m-1}) = \delta(k_p).]$

Namely, everything except the assertions about AC was noted as Theorems (4.4.2) to (4.4.5) of [A08]. For $m > 1$, the assertions about AC are special cases of Theorems (2.2.2) to (2.2.5) respectively. For $m = 1$, it is easy to see that if $GF(q) \subset k_p$ then $\mathrm{Gal}(\phi_0^{**}, K_0) = GL(1, q) = \mathrm{Gal}(\phi_0^{**}, K_0)$ and $\mathrm{Gal}(\phi_0^{*(d)}, K_0) = GL^{(d)}(1, q) = \mathrm{Gal}(\phi_0^{*(d)}, K_0)$ for every divisor d of $q - 1$, and from this the assertions about AC follow as in the proofs of Theorems (2.2.2) to (2.2.5).

Note (2.5) [From Local Surface Coverings to Affine Line Coverings]. As hinted in (2.4.1), the family of projective polynomials f_e^{**} was generalized from the $m > 1 = e$ case with $R_2 = k_p[[X, T_1]]$ when it is reduced to the trinomial $f_1^{**} = Y^{(m-1)} + T_1 Y + X$, giving a local covering above a normal crossing of the branch locus in the local (X, T_1)-plane, dealt with in [A07] and [A08]. Likewise, the families of projective polynomials $f_{(e,\tau)}^{**}$ and $f_{(e,\tau)}^{*}$ were generalized from the $m > 1 = e = \tau$ case when they are reduced to the trinomials $f_{(1,1)}^{*} = Y^{(m-1)} + XY + (-1)^{(m-1)}$ and $f_{(1,1)}^{**} = Y^{(m-1)} + Y + X$, giving unramified coverings of the affine line and the once punctured affine line respectively, dealt with in [A03] and [A08]. Out of this, the $m = 2$ and $q = p$ case of $f_{(1,1)}^{*}$, i.e., the trinomial $Y^{1+p} + XY + 1$, corresponds to the $t = 1$ case of the family of trinomials $Y^{p+t} + XY^t + 1$, where t is a positive integer prime to p, giving unramified coverings of the affine line, which was our starting point in [A01] and [A02].

Section 3: Iterated Linear Groups
In this Section, let

$$(3.1) \quad E = E(Y) = Y^{q^m} + \sum_{i=1}^{m} X_i Y^{q^{m-i}} \quad \text{with} \quad X_i \in K \text{ and } X_m \neq 0$$

be a monic separable vectorial q-polynomial of q-degree m in Y over K, where the elements $X_1, \dots, X_m$ need not be algebraically independent over k_p. When we want to assume that the elements $X_1, \dots, X_m$ are algebraically independent over k_p and $K = k_p(X_1, \dots, X_m)$, we may express this by saying that we are in the **generic** case. In the **general** (= not necessarily generic) case, let V be the set of all roots of E in Ω, and note that then V is an m-dimensional $\mathrm{GF}(q)$-vector-subspace of Ω. Let $X_{1,1}, \dots, X_{m,1}$ be a $\mathrm{GF}(q)$-basis of V. Then

$$(3.2) \quad Y^{q^m} + \sum_{i=1}^{m} X_i Y^{q^{m-i}} = \prod_{(\lambda_1, \dots, \lambda_m) \in \mathrm{GF}(q)^m} (Y - \lambda_1 X_{1,1} - \cdots - \lambda_m X_{m,1})$$

and hence

$$(3.3) \quad k_p[X_1, \dots, X_m] \subset k_p(\mathrm{GF}(q))[X_{1,1}, \dots, X_{m,1}]$$

and

$$(3.4) \quad \mathrm{SF}(E, K) = K(V) = K(\mathrm{GF}(q))(X_{1,1}, \dots, X_{m,1}).$$

As noted in (1.1), we also have

$$(3.5) \quad \mathrm{Gal}(E, K(\mathrm{GF}(q))) < \mathrm{GL}(V) = \mathrm{GL}(m, q)$$

and

(3.6) $\mathrm{Gal}(E, K) < \Gamma\mathrm{L}(V) = \Gamma\mathrm{L}(m, q).$

In the **generic** case, by (3.3) and (3.4) we see that the elements $X_{1,1}, \ldots,$ $X_{m,1}$ are algebraically independent over $k_p(\mathrm{GF}(q))$ and we have $\mathrm{SF}(E, K) = K(V) = k_p(\mathrm{GF}(q))(X_{1,1}, \ldots, X_{m,1})$, and hence $\mathrm{AC}(k_p, E, K) = k_p(\mathrm{GF}(q))$ and by (3.2) we see that every $\mathrm{GF}(q)$-linear automorphism of V corresponds to an element of $\mathrm{Gal}(E, K(\mathrm{GF}(q)))$, i.e., $\mathrm{GL}(m, q) < \mathrm{Gal}(E, K(\mathrm{GF}(q)))$, and therefore by (3.5) we get $\mathrm{Gal}(E, K(\mathrm{GF}(q))) = \mathrm{GL}(m, q)$ and obviously $\delta(k_p) = \delta(K)$ and hence, upon letting $\delta = \delta(k_p)$, in view of (1.5.2) we conclude that $\mathrm{Gal}(E, K) = \Gamma\mathrm{L}_\delta(m, q)$. Thus:

Theorem (3.7). *In the generic case, the elements* $X_{1,1}, \ldots, X_{m,1}$ *are algebraically independent over* k_p, *and we have* $\mathrm{SF}(E, K) = K(V) = k_p(GF(q))(X_{1,1}, \ldots, X_{m,1})$ *and* $AC(k_p, E, K) = k_p(GF(q))$ *and* $Gal(E, K(GF(q))) = GL(m, q)$, *and upon letting* $\delta = \delta(k_p)$ *we also have* $\delta = \delta(K)$ *and* $Gal(E, K) = \Gamma L_\delta(m, q)$.

Remark (3.8) [Polynomial and Power Series Invariants]. Thus, in the **generic** case, the splitting field of E over the m variable rational function field $K = k_p(X_1, \ldots, X_m)$ is the m variable rational function field $K^{[1]} = k^*(X_{1,1}, \ldots, X_{m,1})$ with $k^* = k_p(\mathrm{GF}(q))$, the groups $\mathrm{GL}(m, q)$ and $\Gamma\mathrm{L}_\delta(m, q)$ with $\delta = \delta(k_p)$ act on $K^{[1]}$ and, for these actions, the fixed fields (= the fields of invariants) are $K^* = k^*(X_1, \ldots, X_m)$ and K respectively. Since a polynomial ring is normal (= integrally closed in its quotient field), in view of (3.2) we also see that the integral closure of the m variable polynomial ring $k_p[X_1, \ldots, X_m]$ in the Galois extension $K^{[1]}$ of its quotient field K is the m variable polynomial ring $k^*[X_{1,1}, \ldots, X_{m,1}]$, the groups $\mathrm{GL}(m, q)$ and $\Gamma\mathrm{L}_\delta(m, q)$ act on the ring $k^*[X_{1,1}, \ldots, X_{m,1}]$ and, for these actions, the rings of invariants are the rings $k^*[X_1, \ldots, X_m]$ and $k_p[X_1, \ldots, X_m]$ respectively, where we recall that when a group G acts on a ring J, the ring of invariants is the subring $J^G = \{z \in J : \sigma(z) = z \text{ for all } \sigma \in G\}$. Likewise, upon letting R and $R^{[1]}$ to be the m-dimensional regular local domains obtained by localizing $k_p[X_1, \ldots, X_m]$ and $k^*[X_{1,1}, \ldots, X_{m,1}]$ at the maximal ideals generated by $X_1, \ldots, X_m$ and $X_{1,1}, \ldots, X_{m,1}$ respectively, and upon letting R^* to be the m-dimensional regular local domain obtained by localizing $k^*[X_1, \ldots, X_m]$ at the maximal ideal generated by $X_1, \ldots, X_m$, we see that $R^{[1]}$ is the integral closure of R in the Galois extension $K^{[1]}$ of its quotient field K, the local ring $R^{[1]}$ dominates the local ring R, the groups $\mathrm{GL}(m, q)$ and $\Gamma\mathrm{L}_\delta(m, q)$ act on the ring $R^{[1]}$ and, for these actions, the rings of invariants are the rings R^* and R respectively. Finally, upon letting R and $R^{[1]}$ to be the completions of R and $R^{[1]}$ respectively, and upon letting R^* to be the completion of R^*, we see that the m variable formal power series ring $R^{[1]} = k^*[[X_{1,1}, \ldots, X_{m,1}]]$ is the integral closure of the m variable formal power series ring $R = k_p[[X_1, \ldots, X_m]]$ in the m variable formal meromorphic series field $K^{[1]} = k^*((X_{1,1}, \ldots, X_{m,1}))$ which is the splitting field of E over the m variable formal meromorphic series field $K = k_p((X_1, \ldots, X_m))$ with the corresponding Galois group $\Gamma L_\delta(m, q)$, the groups $\mathrm{GL}(m, q)$ and $\Gamma L_\delta(m, q)$ act on the m variable formal power series ring $R^{[1]} = k[[X_{1,1}, \ldots, X_{m,1}]]$ and, for these actions, the rings of invariants are the m variable formal power series rings $R^* = k^*[[X_1, \ldots, X_m]]$ and $R = k_p[[X_1, \ldots, X_m]]$ respectively. Note that, continuing with the generic case, in the notation of Section 2 we have $E = \phi_{m-1}^{**}$ with $(X_1, \ldots, X_m) = (T_1, \ldots, T_{m-1}, X)$ and $K = K_{m-1}$, and so in Theorem (3.7) we have given a stand-alone proof of Theorem (2.4.2).

Remark (3.9) [Converse of Linearity]. As a partial converse of (1.1), we observe that if $E' = E'(Y)$ is any monic polynomial of degree q^m in Y with coefficients in Ω such that the set of all roots of E' in Ω is an m-dimensional $\mathrm{GF}(q)$-vector subspace V' of Ω, then E' must be a vectorial q-polynomial of q-degree m in Y over Ω. This follows from the identity (3.2) by substituting a $\mathrm{GF}(q)$-basis of V' for $X_{1,1}, \ldots, X_{m,1}$ in the RHS and noting that then E' must coincide with the LHS where $X_1, \ldots, X_m$ are suitable polynomial expressions in $X_{1,1}, \ldots, X_{m,1}$ with coefficients in $\mathrm{GF}(q)$.

Definition-Notation (3.10). Reverting to the **general** case, let $n > 0$ be an integer, and let $E^{[n]}$ be the n-th **iterate** of E, i.e., let $E^{[0]} = E^{[0]}(Y) = Y$, $E^{[1]} = E^{[1]}(Y) = E(Y)$, and $E^{[j]} = E^{[j]}(Y) = E(E^{[j-1]}(Y))$ for all $j > 1$. Let $V^{[0]} = \{0\} \subset V$, let $V^{[1]} = V$, and for every $j > 1$ let $V^{[j]}$ be the set of all roots of $E^{[j]}$ in Ω, and note that $E^{[j]}$ is a monic separable vectorial q-polynomial of q-degree mj in Y over K and hence $V^{[j]}$ is an (mj)-dimensional $\mathrm{GF}(q)$-vector-subspace of Ω. We get a $\mathrm{GF}(q)$-linear epimorphism $E : \Omega \to \Omega$ given by $z \mapsto E(z)$; for its j-th power $E^j : \Omega \to \Omega$, where $j \geq 0$ is any integer, we have $E^j(z) = E^{[j]}(z)$ for all $z \in \Omega$; likewise, for any integers $j \geq j' \geq 0$, we clearly have $V^{[j']} \subset V^{[j]}$ and $E^{j'}(V^{[j]}) = V^{[j-j']}$, and the restriction of $E^{j'}$ to $V^{[j]}$ gives a $\mathrm{GF}(q)$-linear epimorphism $V^{[j]} \to V^{[j-j']}$ with kernel $V^{[j']}$. Let $\overline{T}$ be the image of T under the canonical epimorphism $\mathrm{GF}(q)[T] \to \mathrm{GF}(q, n) = \mathrm{GF}(q)[T]/T^n$. For every $\gamma \in \mathrm{Aut}(\mathrm{GF}(q))$ and $\overline{r} = \sum_{i=0}^{n-1} r_i \overline{T}^i \in \mathrm{GF}(q, n)$ with $r_i \in \mathrm{GF}(q)$, let us put $\gamma(\overline{r}) = \sum_{i=0}^{n-1} \gamma(r_i)\overline{T}^i$; this gives a faithful action of $\mathrm{Aut}(\mathrm{GF}(q))$ on $\mathrm{GF}(q, n)$. For every $\overline{r} \in \mathrm{GF}(q, n)$ and $z \in \Omega$ we define $\overline{r}z \in \Omega$ by putting $\overline{r}z = \sum_{i=0}^{n-1} r_i E^i(z)$, and we note that this makes $V^{[n]}$ a $\mathrm{GF}(q, n)$-module, and then, for every $g \in \mathrm{Gal}(K(V^{[n]}), K)$ we have $g(\overline{r}z) = g'(\overline{r})g(z)$ where $g' \in \mathrm{Gal}(\mathrm{GF}(q), \mathrm{GF}(p^{u/\delta(K)}))$ is given by putting $g'(\zeta) = g(\zeta)$ for all $\zeta \in \mathrm{GF}(q)$, and hence in its action on $\mathrm{GF}(q, n)$ it is given by $g'(\overline{r}) = \sum_{i=0}^{n-1} g(r_i)\overline{T}^i \in \mathrm{GF}(q, n)$ for all $r \in \mathrm{GF}(q)[T]$. It follows that, in a natural manner, $\mathrm{Gal}(E^{[n]}, K)$ is a subgroup of $\Gamma\mathrm{L}_{\delta(K)}(V^{[n]})$ where, for any factor δ of u, by $\Gamma\mathrm{L}_{\delta}(V^{[n]})$ we denote the **group of all $\mathrm{GF}_\delta(q, n)$-semilinear automorphisms** of the module $V^{[n]}$, by which we mean all additive isomorphisms $\sigma : V^{[n]} \to V^{[n]}$ for which there exists $\sigma' \in \mathrm{Gal}(\mathrm{GF}(q), \mathrm{GF}(p^{u/\delta}))$ such that for all $\eta \in \mathrm{GF}(q, n)$ and $z \in V^{[n]}$ we have $\sigma(\eta z) = \sigma'(\eta)\sigma(z)$; note that now $\sigma \mapsto \sigma'$ gives an epimorphism $\Gamma\mathrm{L}_{\delta}(V^{[n]}) \to \mathrm{Gal}(\mathrm{GF}(q), \mathrm{GF}(p^{u/\delta}))$ whose kernel is the **group $\mathrm{GL}(V^{[n]})$ of all $\mathrm{GF}(q, n)$-linear automorphism** of $V^{[n]}$, by which we mean all additive isomorphisms $\sigma : V^{[n]} \to V^{[n]}$ such that for all $\eta \in \mathrm{GF}(q, n)$ and $z \in V^{[n]}$ we have $\sigma(\eta z) = \eta\sigma(z)$, i.e., $\mathrm{GL}(V^{[n]}) = \Gamma\mathrm{L}_1(V^{[n]})$. Observe that $\Gamma\mathrm{L}_{\delta}(V^{[n]})$ is a normal subgroup of the **group $\Gamma\mathrm{L}(V^{[n]})$ of all $\mathrm{GF}(q, n)$-semilinear automorphisms** of the module $V^{[n]}$, by which we mean all additive isomorphisms $\sigma : V^{[n]} \to V^{[n]}$ for which there exists $\sigma' \in \mathrm{Aut}(\mathrm{GF}(q))$ such that for all $\eta \in \mathrm{GF}(q, n)$ and $z \in V^{[n]}$ we have $\sigma(\eta z) = \sigma'(\eta)\sigma(z)$, i.e., $\Gamma\mathrm{L}(V^{[n]}) = \Gamma\mathrm{L}_u(V^{[n]})$; note that now $\sigma \mapsto \sigma'$ gives an epimorphism $\Gamma\mathrm{L}(V^{[n]}) \to \mathrm{Aut}(\mathrm{GF}(q)) = \mathrm{Gal}(\mathrm{GF}(q), \mathrm{GF}(p))$ whose kernel is the group $\mathrm{GL}(V^{[n]})$ and which is an extension of the above epimorphism $\Gamma\mathrm{L}_{\delta}(V^{[n]}) \to \mathrm{Gal}(\mathrm{GF}(q), \mathrm{GF}(p^{u/\delta}))$; also note that $\Gamma\mathrm{L}(V^{[n]})/\Gamma\mathrm{L}_{\delta}(V^{[n]}) = Z_{u/\delta}$ and $\Gamma\mathrm{L}_{\delta}(V^{[n]})/\mathrm{GL}(V^{[n]}) = Z_{\delta}$. Now identifying Galois groups of polynomials with the Galois groups of their splitting fields we see that $\mathrm{Gal}(E^{[n]}, K) \cap \mathrm{GL}(V^{[n]}) = \mathrm{Gal}(E^{[n]}, K(\mathrm{GF}(q)))$, and by the usual Galois correspondence we have $\mathrm{Gal}(E^{[n]}, K(\mathrm{GF}(q))) \lhd \mathrm{Gal}(E^{[n]}, K)$ with $\mathrm{Gal}(E^{[n]}, K)/\mathrm{Gal}(E^{[n]}, K(\mathrm{GF}(q))) = \mathrm{Gal}(K(\mathrm{GF}(q)), K) = Z_{\delta(K)}$. Thus

$$(3.11) \quad \begin{cases} \mathrm{Gal}(E^{[n]}, K) < \Gamma\mathrm{L}_{\delta(K)}(V^{[n]}) \lhd \Gamma\mathrm{L}(V^{[n]}) \\ \text{with } \Gamma\mathrm{L}(V^{[n]})/\Gamma\mathrm{L}_{\delta(K)}(V^{[n]}) = Z_{u/\delta(K)} \end{cases}$$

and

$$(3.12) \quad \begin{cases} \mathrm{Gal}(E^{[n]}, K) \cap \mathrm{GL}(V^{[n]}) = \mathrm{Gal}(E^{[n]}, K(\mathrm{GF}(q))) \lhd \mathrm{Gal}(E^{[n]}, K) \\ \text{with } \mathrm{Gal}(E^{[n]}, K)/\mathrm{Gal}(E^{[n]}, K(\mathrm{GF}(q))) = Z_{\delta(K)} \end{cases}$$

and

$$(3.13) \quad \begin{cases} \mathrm{GL}(V^{[n]}) \lhd \Gamma\mathrm{L}_{\delta(K)}(V^{[n]}) \\ \text{with } \Gamma\mathrm{L}_{\delta(K)}(V^{[n]})/\mathrm{GL}(V^{[n]}) = Z_{\delta(K)}. \end{cases}$$

Now by (3.12) we see that

$$(3.14) \quad \mathrm{Gal}(K(V^{[n]}), K) < \mathrm{GL}(V^{[n]}) \Leftrightarrow \mathrm{GF}(q) \subset K$$

and by (3.11) to (3.13) we see that

$$(3.15) \quad \begin{cases} \text{if } \mathrm{Gal}(E^{[n]}, K(\mathrm{GF}(q))) = \mathrm{GL}(V^{[n]}) \\ \text{then } \mathrm{Gal}(E^{[n]}, K) = \Gamma\mathrm{L}_{\delta(K)}(V^{[n]}). \end{cases}$$

Definition-Notation (3.16). Continuing with (3.10), in the **general** case, we can first take a $\mathrm{GF}(q)$-basis $X_{1,1}, \dots, X_{m,1}$ of V, and then iteratively we can take elements $X_{1,j}, \dots, X_{m,j}$ in Ω such that for all $j > 1$ we have $E(X_{1,j}) = X_{1,j-1}, \dots, E(X_{m,j}) = X_{m,j-1}$. It follows that $X_{1,n}, \dots, X_{m,n}$ is a free $\mathrm{GF}(q,n)$-basis of $V^{[n]}$, and hence $V^{[n]}$ is a free $\mathrm{GF}(q,n)$-module of rank m. Therefore we may identify $\mathrm{GL}(V^{[n]})$ with the **group $\mathrm{GL}(m,q,n)$ of all m by m invertible matrices over** $\mathrm{GF}(q,n)$. Likewise we may identify $\Gamma\mathrm{L}(V^{[n]})$ with the semidirect product $\Gamma\mathrm{L}(m,q,n) = \mathrm{GL}(m,q,n) \rtimes \mathrm{Aut}(\mathrm{GF}(q))$ where $\mathrm{Aut}(\mathrm{GF}(q))$ acts on $\mathrm{GL}(m,q,n)$ by componentwise application of the above explained action of $\mathrm{Aut}(\mathrm{GF}(q))$ on $\mathrm{GF}(q,n)$. For any factor δ of u, this identifies $\Gamma\mathrm{L}_{\delta}(V^{[n]})$ with the semidirect product $\Gamma\mathrm{L}_{\delta}(m,q,n) = \mathrm{GL}(m,q,n) \rtimes \mathrm{Gal}(\mathrm{GF}(q),\mathrm{GF}(p^{u/\delta}))$. Thus

$$(3.17) \qquad \mathrm{GL}(V^{[n]}) = \mathrm{GL}(m,q,n)$$

and

$$(3.18) \qquad \Gamma\mathrm{L}(V^{[n]}) = \Gamma\mathrm{L}(m,q,n) = \mathrm{GL}(m,q,n) \rtimes \mathrm{Aut}(\mathrm{GF}(q))$$

and

$$(3.19) \qquad \begin{aligned} \Gamma\mathrm{L}_{\delta(K)}(V^{[n]}) &= \Gamma\mathrm{L}_{\delta(K)}(m,q,n) \\ &= \mathrm{GL}(m,q,n) \rtimes \mathrm{Gal}(\mathrm{GF}(q),\mathrm{GF}(p^{u/\delta(K)})). \end{aligned}$$

Elements of $\mathrm{GL}(m,q,n)$ can be identified with expressions of the form $\xi_0 + \xi_1\overline{T} + \cdots + \xi_{n-1}\overline{T}^{n-1}$ with ξ_0 in $\mathrm{GL}(m,q)$ and $\xi_1, \dots, \xi_{n-1}$ in the set of all m by m matrices over $\mathrm{GF}(q)$, and hence

$$(3.20) \qquad |\mathrm{GL}(m,q,n)| = |\mathrm{GL}(m,q)|q^{m^2(n-1)}.$$

As an alternative proof of (3.20), we may note that $|\mathrm{GL}(m,q,n)|$ equals the cardinality of the set of all $\mathrm{GF}(q,n)$-bases of $\mathrm{GF}(q,n)^m$, and for any finite sequence $(y_1, \dots, y_\nu)$ in D^n where D is any local ring, upon letting $(\overline{y}_1, \dots, \overline{y}_\nu)$ to be the corresponding sequence in $\overline{D}^n$ where $\overline{D} = D/M(D)$, by Nakayama's Lemma we see that $(y_1, \dots, y_\nu)$ is a D-basis of D^n $\Leftrightarrow$ $(\overline{y}_1, \dots, \overline{y}_\nu)$ is a $\overline{D}$-basis of $\overline{D}^n$; now it suffices to observe that $|M(\mathrm{GF}(q,n))| = q^{(n-1)}$. Since $\mathrm{Aut}(\mathrm{GF}(q)) = Z_u$ and $\mathrm{Gal}(\mathrm{GF}(q),\mathrm{GF}(p^{u/\delta(K)})) = Z_{\delta(K)}$, we also have

$$(3.21) \qquad |\Gamma\mathrm{L}(m,q,n)| = |\mathrm{GL}(m,q,n)|u$$

and

$$(3.22) \qquad |\Gamma\mathrm{L}_{\delta(K)}(m,q,n)| = |\mathrm{GL}(m,q,n)|\delta(K).$$

In the proof of Proposition (5.3) of [A06], the following easy-to-prove fact was called:

Generalized Eisenstein Criterion (3.23). *Let B be a local domain dominated by an m-dimensional regular local domain B_1, let $(C_{1,1}, \ldots, C_{m,1})$ be a basis of $M(B_1)$, and let Δ be an algebraic closure of the quotient field L_1 of B_1. Let $F(Y) = Y^d + \sum_{l=1}^{d} A_l Y^{d-l}$ with $A_l \in M(B)$ for $1 \leq l < d$ and $A_d \in M(B)^2$ where $d > 0$ is any integer. Then $F(Y) - C_{1,1}$ is irreducible in $L_1[Y]$ and, upon taking $C' \in \Delta$ with $F(C') = C_{1,1}$ and upon letting $B' =$ the integral closure of B_1 in $L_1(C')$, we have that B' is an m-dimensional regular local domain with $M(B') = (C', C_{2,1}, \ldots, C_{m,1})B'$, and B' dominates B_1 and is residually rational over it.*

By double induction on i and j, or simple induction on $i + mj$, as an immediate consequence of (3.23) we get the following:

Corollary (3.24). *Let B be a local domain dominated by an m-dimensional regular local domain B_1, let $(C_{1,1}, \ldots, C_{m,1})$ be a basis of $M(B_1)$, and let Δ be an algebraic closure of the quotient field L_1 of B_1. For $1 \leq i \leq m$ and $1 < j \leq n$, let $F_{i,j}(Y) = Y^{d(i,j)} + \sum_{l=1}^{d(i,j)} A_{i,j,l} Y^{d(i,j)-l}$ with $A_{i,j,l} \in M(B)$ for $1 \leq l < d(i,j)$ and $A_{i,j,d(i,j)} \in M(B)^2$ where $d(i,j) > 0$ is any integer, and let $C_{i,j} \in \Delta$ be such that $F_{i,j}(C_{i,j}) = C_{i,j-1}$. For $1 < j \leq n$, let $L_j = L_1(C_{1,j}, \ldots, C_{m,j})$ and $B_j =$ the integral closure of B_1 in L_j. For $0 \leq i \leq m$ and $1 < j \leq n$, let $L_{i,j} = L_{j-1}(C_{1,j}, \ldots, C_{i,j})$ and $B_{i,j} =$ the integral closure of B_1 in $L_{i,j}$. [Note that now for $1 < j \leq n$ we obviously have $L_{0,j} = L_{j-1}$ and $B_{0,j} = B_{j-1}$ and $L_{m,j} = L_j = L_{j-1}(C_{1,j}, \ldots, C_{m,j})$ and $B_{m,j} = B_j =$ the integral closure of B_{j-1} in L_j, and for $1 \leq i \leq m$ and $1 < j \leq n$ we obviously have $L_{i,j} = L_{i-1,j}(C_{i,j})$ and $B_{i,j} =$ the integral closure of $B_{i-1,j}$ in $L_{i,j}$.] Then for $1 \leq i \leq m$ and $1 < j \leq n$, the polynomial $F_{i,j}(Y) - C_{i,j-1}$ is irreducible in $L_{i-1,j}[Y]$ (i.e., equivalently $[L_{i,j} : L_{i-1,j}] = d(i,j)$), and we have that $B_{i,j}$ is an m-dimensional regular local domain with $M(B_{i,j}) = (C_{1,j}, \ldots, C_{i,j}, C_{i+1,j-1}, \ldots, C_{m,j-1})B_{i,j}$, and $B_{i,j}$ dominates $B_{i-1,j}$ and is residually rational over it. (•) [Hence, in particular, for $1 < j \leq n$, the polynomials $F_{1,j}(Y) - C_{1,j-1}, \ldots, F_{m,j}(Y) - C_{m,j-1}$ are irreducible in $L_{j-1}[Y]$, the field degree $[L_j : L_{j-1}]$ equals the product $d(1,j) \ldots d(m,j)$, and we have that B_j is an m-dimensional regular local domain with $M(B_j) = (C_{1,j}, \ldots, C_{m,j})B_j$, and B_j dominates B_{j-1} and is residually rational over it.]*

To apply (3.24) to the **generic** case with notation as in (3.8), take $(B, B_1, L_1, \Delta, C_{1,1}, \ldots, C_{m,1}) = (R, R^{[1]}, K^{[1]}, \Omega, X_{1,1}, \ldots, X_{m,1})$, and for $1 \leq i \leq m$ and $1 < j \leq n$ take $F_{i,j} = E$ and $C_{i,j} = X_{i,j}$. For $1 < j \leq n$ let $K^{[j]} = L_j$ and $R^{[j]} = B_j$. Now $K^{[1]}(V^{[n]}) = K^{[n]}$ and $[K^{[n]} : K^{[1]}] = \prod_{1 < j \leq n}[K^{[j]} : K^{[j-1]}]$ and hence by the last bracketed part of (3.24) we see that $[K^{[1]}(V^{[n]}) : K^{[1]}] = q^{m^2(n-1)}$. Since $K \subset K^* \subset K^{[1]} \subset K^{[1]}(V^{[n]})$, and $K^{[1]}$ and $K^{[1]}(V^{[n]})$ are the respective splitting fields of E and $E^{[n]}$ over K, it follows that $|\text{Gal}(E^{[n]}, K^*)| = |\text{Gal}(E, K^*)|q^{m^2(n-1)}$ and $|\text{Gal}(E^{[n]}, K)| = |\text{Gal}(E, K)|q^{m^2(n-1)}$. Therefore by (3.7), (3.11), (3.12), (3.17), (3.19) and

(3.20) we get $\mathrm{Gal}(E^{[n]}, K^*) = \mathrm{GL}(m, q, n)$ and $\mathrm{Gal}(E^{[n]}, K) = \Gamma\mathrm{L}_\delta(m, q, n)$. Since $R^{[n]}$ is residually rational over $R^{[1]}$, by (3.7) we also see that $\mathrm{AC}(k_p, E^{[n]}, K) = k^*$. So we have proved the following:

Theorem (3.25). *In the generic case, the elements* $X_{1,1}, \ldots, X_{m,1}$ *are algebraically independent over* k_p, *and we have* $\mathrm{SF}(E^{[n]}, K) = K(V^{[n]}) = k_p(\mathrm{GF}(q))(X_{1,1}, \ldots, X_{m,1}, X_{1,n}, \ldots, X_{m,n})$ *and* $\mathrm{AC}(k_p, E^{[n]}, K) = k_p(\mathrm{GF}(q))$ *and* $\mathrm{Gal}(E^{[n]}, K(\mathrm{GF}(q))) = \mathrm{GL}(m, q, n)$, *and upon letting* $\delta = \delta(k_p)$ *we also have* $\delta = \delta(K)$ *and* $\mathrm{Gal}(E^{[n]}, K) = \Gamma\mathrm{L}_\delta(m, q, n)$.

Remark (3.26) [Iterated Polynomial and Power Series Invariants]. Thus, in the **generic** case, the splitting field of $E^{[n]}$ over the m variable rational function field $K = k_p(X_1, \ldots, X_m)$ is the field $K^{[n]} = k^*(X_{1,1}, \ldots, X_{m,1}, X_{1,n}, \ldots, X_{m,n})$ with $k^* = k_p(\mathrm{GF}(q))$, the groups $\mathrm{GL}(m, q, n)$ and $\Gamma\mathrm{L}_\delta(m, q, n)$ with $\delta = \delta(k_p)$ act on $K^{[n]}$ and, for these actions, the fixed fields are $K^* = k^*(X_1, \ldots, X_m)$ and K respectively. Upon letting R and R^* to be the m dimensional regular local domains obtained by localizing the polynomial rings $k_p[X_1, \ldots, X_m]$ and $k^*[X_1, \ldots, X_m]$ at the maximal ideals generated by $X_1, \ldots, X_m$ respectively, in view of (3.8) and (3.24) we see that the integral closure $R^{[n]}$ of R in $E^{[n]}$ is an m dimensional regular local domain with $M(R^{[n]}) = (X_{1,n}, \ldots, X_{m,n})R^{[n]}$, the local ring $R^{[n]}$ dominates the local ring R^* and is residually rational over it, the groups $\mathrm{GL}(m, q, n)$ and $\Gamma\mathrm{L}_\delta(m, q, n)$ act on the local ring $R^{[n]}$ and, for these actions, the rings of invariants are the local rings R^* and R respectively. Likewise, upon letting $\hat{R}$ and $\hat{R}^*$ to be the completions of R and R^* respectively, and upon letting $\hat{R}^{[n]}$ to be the completion of $R^{[n]}$, we see that the m variable formal power series ring $\hat{R}^{[n]} = k^*[[X_{1,n}, \ldots, X_{m,n}]]$ is the integral closure of the m variable formal power series ring $\hat{R} = k_p[[X_1, \ldots, X_m]]$ in the m variable formal meromorphic series field $\hat{K}^{[n]} = k^*((X_{1,n}, \ldots, X_{m,n}))$ which is the splitting field of $E^{[n]}$ over the m variable formal meromorphic series field $\hat{K} = k_p((X_1, \ldots, X_m))$ with the corresponding Galois group $\Gamma\mathrm{L}_\delta(m, q, n)$, the groups $\mathrm{GL}(m, q, n)$ and $\Gamma\mathrm{L}_\delta(m, q, n)$ act on the m variable formal power series ring $\hat{R}^{[n]} = k[[X_{1,n}, \ldots, X_{m,n}]]$ and, for these actions, the rings of invariants are the m variable formal power series rings $\hat{R}^* = k^*[[X_1, \ldots, X_m]]$ and $\hat{R} = k_p[[X_1, \ldots, X_m]]$ respectively.

Definition-Notation (3.27). Continuing with the **generic** case, for any divisor d of $q - 1$, let $E^{(d)}$ be obtained by substituting $(-1)^{(m-1)}X_m^d$ for X_m in E, i.e., let

$$E^{(d)} = E^{(d)}(Y) = Y^{q^m} + (-1)^{(m-1)}X_m^d Y + \sum_{i=1}^{m-1} X_i Y^{q^{m-1}}$$

and let $E^{(d)[n]}$ be the n-th **iterate** of $E^{(d)}$, i.e., let $E^{(d)[0]} = E^{(d)[0]}(Y) = Y$, $E^{(d)[1]} = E^{(d)[1]}(Y) = E^{(d)}(Y)$, and $E^{(d)[j]} = E^{(d)[j]}(Y) = E^{(d)}(E^{(d)[j-1]}(Y))$

for all $j > 1$. Now clearly $E^{(d)[n]}$ is a monic separable vectorial q-polynomial of q-degree mn in Y over K, and hence by the above displayed items (3.11), (3.12), (3.17) and (3.19) we see that $\text{Gal}(E^{(d)[n]}, K(\text{GF}(q))) < \text{GL}(m, q, n)$ and $\text{Gal}(E^{(d)[n]}, K) < \Gamma\text{L}_{\delta(K)}(m, q, n)$. To determine the exact values of these Galois groups, we first note that $\text{SL}(m, q)$ equals the quasi-p part $p(\text{GL}(m, q))$ (see Footnote 14 of [A08]). Moreover $\text{SL}(m, q) \lhd \text{GL}(m, q)$ with $\text{GL}(m, q))/\text{SL}(m, q) = Z_{q-1}$ and, by definition, $\text{GL}^{(d)}(m, q)$ is the unique group between $\text{SL}(m, q)$ and $\text{GL}(m, q)$ with $\text{GL}(m, q))/\text{GL}^{(d)}(m, q) = Z_d$. In the notation just before (3.20), by sending $\xi_0 + \xi_1 \overline{T} + \cdots + \xi_{n-1} \overline{T}^{n-1}$ to ξ_0 we get an epimorphism $\mu : \text{GL}(m, q, n) \to \text{GL}(m, q)$ whose kernel has order $q^{m^2(n-1)}$. Therefore, upon letting $\text{GL}^{(d)}(m, q, n) = \mu^{-1}(\text{GL}^{(d)}(m, q))$ we see that $\text{GL}^{(d)}(m, q, n)$ is the unique normal subgroup of $\text{GL}(m, q, n)$ for which we have $\text{GL}(m, q, n)/\text{GL}^{(d)}(m, q, n) = Z_d$, and upon letting $\text{SL}(m, q, n)) = \mu^{-1}(\text{SL}(m, q))$ we see that $p(\text{GL}(m, q, n)) = \text{SL}(m, q, n) = \text{GL}^{(q-1)}(m, q, n)$. Let us also note that $\text{GL}^{(d)}(m, q, n)$ can alternatively be characterized as the unique normal subgroup of $\text{GL}(m, q, n)$ for which we have $\text{SL}(m, q, n) \lhd \text{GL}^{(d)}(m, q, n)$ with $\text{GL}^{(d)}(m, q, n)/\text{SL}(m, q, n) = Z_{(q-1)/d}$. By (3.18) we have $\text{GL}(m, q, n) \lhd \Gamma\text{L}(m, q, n)$ with $\Gamma\text{L}(m, q, n)/\text{GL}(m, q, n) = Z_u$, hence for every divisor δ of u there is a unique group $\Gamma\text{L}_\delta(m, q, n)$ between $\text{GL}(m, q, n)$ and $\Gamma\text{L}(m, q, n)$ with $\Gamma\text{L}_\delta(m, q, n)/\text{GL}(m, q, n) = Z_\delta$. Let $\Gamma\text{SL}_\delta(m, q, n)$ be the set of all subgroups I of $\Gamma\text{L}_\delta(m, q, n)$ such that $I \cap \text{GL}(m, q, n) = \text{SL}(m, q, n) \lhd I$ with $I/\text{SL}(m, q, n) = Z_\delta$; in Remark (3.29) we shall show that $\Gamma\text{SL}_\delta(m, q, n)$ is a nonempty complete set of conjugate subgroups of $\Gamma\text{L}(m, q, n)$, and every I in $\Gamma\text{SL}_\delta(m, q, n)$ is a split extension of $\text{SL}(m, q, n)$ such that $\Gamma\text{L}_\delta(m, q, n)$ is generated by $\text{SL}(m, q, n)$ and I. Similarly, let $\Gamma\text{L}_\delta^{(d)}(m, q, n)$ be the set of all subgroups J of $\Gamma\text{L}_\delta(m, q, n)$ such that $J \cap \text{GL}(m, q, n) = \text{GL}^{(d)}(m, q, n) \lhd J$ with $J/\text{GL}^{(d)}(m, q, n) = Z_\delta$ and $I < J$ for some I in $\Gamma\text{SL}_\delta(m, q, n)$; in Remark (3.29) we shall show that $\Gamma\text{L}_\delta^{(d)}(m, q, n)$ is a nonempty complete set of conjugate subgroups of $\Gamma\text{L}(m, q, n)$, and every J in $\Gamma\text{L}_\delta^{(d)}(m, q, n)$ is a split extension of $\text{GL}^{(d)}(m, q, n)$ such that $\Gamma\text{L}_\delta(m, q, n)$ is generated by $\text{GL}(m, q, n)$ and J.

Note that clearly $\Gamma\text{L}_\delta^{(q-1)}(m, q, n) = \Gamma\text{SL}_\delta(m, q, n)$ and $\Gamma\text{L}_\delta^{(1)}(m, q, n) = \{\Gamma\text{L}_\delta(m, q, n)\}$. As a consequence of Theorem (3.25) we shall now deduce the following generalization of it:

Theorem (3.28). *In the generic case, for any divisor d of $q - 1$, we have $AC(k_p, E^{(d)[n]}, K) = k_p(\text{GF}(q))$ and $\text{Gal}(E^{(d)[n]}, K(\text{GF}(q)))$ $= \text{GL}^{(d)}(m, q, n)$, and upon letting $\delta = \delta(k_p)$ we also have $\delta = \delta(K)$ and $\text{Gal}(E^{(d)[n]}, K) \in \Gamma L_\delta^{(d)}(m, q, n)$.*

To deduce this, we first recall that by (1.1) we have $\text{GF}(q) \subset K(V^{[n]}) = \text{SF}(E^{[n]}, K)$, and moreover by (1.2) we can find $\Lambda \in K(V^{[n]})$ with $\Lambda^{q-1} = (-1)^{(m-1)} X_m$. Let $k^* = k_p(\text{GF}(q))$ and $K^* = K(\text{GF}(q))$. Now by (3.25) we have $AC(k_p, E^{[n]}, K) = k^*$ and $\text{Gal}(E^{[n]}, K^*) = \text{GL}(m, q, n)$. Let $\Lambda_d =$

$\Lambda^{(q-1)/d}$. Then clearly $AC(k_p, E^{[n]}, K(\Lambda_d)) = AC(k_p, E^{[n]}, K) = k^*$. Also we have the Galois extensions $K^* \subset K^*(\Lambda) \subset K^*(V^{[n]})$ and $K^* \subset K^*(\Lambda_d) \subset K^*(\Lambda)$ and therefore in view of the basic Galois correspondence we see that $\text{Gal}(E^{[n]}, K^*(\Lambda)) \triangleleft \text{Gal}(E^{[n]}, K^*(\Lambda_d)) \triangleleft \text{Gal}(E^{[n]}, K^*)$ and $\text{Gal}(E^{[n]}, K^*(\Lambda)) \triangleleft \text{Gal}(E^{[n]}, K^*)$ with $\text{Gal}(E^{[n]}, K^*)/\text{Gal}(E^{[n]}, K^*(\Lambda)) = \text{Gal}(K^*(\Lambda), K^*) = Z_{q-1}$ and $\text{Gal}(E^{[n]}, K^*)/\text{Gal}(E^{[n]}, K^*(\Lambda_d)) = \text{Gal}(K^*(\Lambda_d), K^*) = Z_d$. As indicated in (3.27), $SL(m, q, n)$ and $GL^{(d)}(m, q, n)$ are the only normal subgroups of $GL(m, q, n)$ of indices $q - 1$ and d respectively, and hence we get $SL(m, q, n) = \text{Gal}(E^{[n]}, K^*(\Lambda)) \triangleleft \text{Gal}(E^{[n]}, K^*(\Lambda_d)) = GL^{(d)}(m, q, n)$. We can find a $k_p(X_1, \ldots, X_{m-1})$-automorphism ν of Ω such that $\nu(\Lambda_d) = X_m$. Now $K = \nu(K(\Lambda_d)) \subset \nu(K^*(\Lambda_d)) = K^* \subset K'^* = \nu(K^*(\Lambda))$ where $K'^* = K'(\text{GF}(q))$ with $K' = K(\nu(\Lambda))$, and by applying ν to the coefficients of $E^{[n]}(Y)$ we get $E^{(d)[n]}(Y)$, and therefore $AC(k_p, E^{(d)[n]}, K) = k^*$ and $SL(m, q, n) = \text{Gal}(E^{(d)[n]}, K'^*) \triangleleft \text{Gal}(E^{(d)[n]}, K^*) = GL^{(d)}(m, q, n)$. Since $K \subset K'$, we clearly have $\text{Gal}(E^{[n]}, K') < \text{Gal}(E^{[n]}, K)$. Let $\delta = \delta(k_p)$. Then $\delta = \delta(K) = \delta(K')$ and hence by (3.11), (3.12), (3.17) and (3.19) we see that

$$\begin{cases} \text{Gal}(E^{(d)[n]}, K') \cap GL(m, q, n) = \text{Gal}(E^{(d)[n]}, K'^*) \triangleleft \text{Gal}(E^{(d)[n]}, K') \\ \text{with } \text{Gal}(E^{(d)[n]}, K')/\text{Gal}(E^{(d)[n]}, K'^*) = Z_\delta \\ \text{and } \text{Gal}(E^{(d)[n]}, K') < \Gamma L_\delta(m, q, n) \end{cases}$$

and

$$\begin{cases} \text{Gal}(E^{(d)[n]}, K) \cap GL(m, q, n) = \text{Gal}(E^{(d)[n]}, K^*) \triangleleft \text{Gal}(E^{(d)[n]}, K) \\ \text{with } \text{Gal}(E^{(d)[n]}, K)/\text{Gal}(E^{(d)[n]}, K^*) = Z_\delta \\ \text{and } \text{Gal}(E^{(d)[n]}, K) < \Gamma L_\delta(m, q, n) \end{cases}$$

and therefore we have $\text{Gal}(E^{(d)[n]}, K') \in \Gamma SL_\delta(m, q, n)$ and $\text{Gal}(E^{(d)[n]}, K) \in \Gamma L_\delta^{(d)}(m, q, n)$.

Remark (3.29) [Semilinearity]. To verify the properties of the sets $\Gamma SL_\delta(m, q, n)$ and $\Gamma L_\delta^{(d)}(m, q, n)$ asserted in (3.27) where d and δ are any divisors of $q - 1$ and u respectively, we first note that, as indicated in (3.18), the canonical epimorphism of $\Gamma L(m, q, n)$ onto $\text{Aut}(\text{GF}(q)) = Z_u$ with kernel $GL(m, q, n)$ splits; to see this, as on page 79 of [A02], we identify $\Gamma L(m, q, n)$ with $\{(g, \alpha) : g \in \text{Aut}(\text{GF}(q)) \text{ and } \alpha \in GL(m, q, n)\}$, and now upon letting $\Gamma(m, q, n) = \{(g, 1) : g \in \text{Aut}(\text{GF}(q))\}$ we see that $\Gamma(m, q, n)$ is a subgroup of $\Gamma L(m, q, n)$ which is mapped isomorphically onto $\text{Aut}(\text{GF}(q))$ by the canonical epimorphism of $\Gamma L(m, q, n)$ onto $\text{Aut}(\text{GF}(q))$. Let $\Gamma_\delta(m, q, n)$ be the unique subgroup of $\Gamma(m, q, n)$ of order δ. Then clearly $\Gamma L_\delta(m, q, n)$ is generated by $\Gamma_\delta(m, q, n)$ and $GL(m, q, n)$. Upon letting $I_\delta(m, q, n)$ be the subgroup of $\Gamma L_\delta(m, q, n)$ generated by $\Gamma_\delta(m, q, n)$ and

$SL(m, q, n)$ we see that $I_\delta(m, q, n) \in \Gamma SL_\delta(m, q, n)$. Likewise, upon letting $J_\delta^{(d)}(m, q, n)$ be the subgroup of $\Gamma L_\delta(m, q, n)$ generated by $\Gamma_\delta(m, q, n)$ and $GL^{(d)}(m, q, n)$ we see that $I_\delta(m, q, n) < J_\delta^{(d)}(m, q, n) \in \Gamma L_\delta^{(d)}(m, q, n)$. Let $\mu : GL(m, q, n) \to GL(m, q)$ be the epimorphism described in (3.27). Then $(g, \alpha) \mapsto (g, \det \mu(\alpha))$ gives an epimorphism σ of $\Gamma L(m, q, n)$ onto $\Gamma L(1, q)$ with kernel $SL(m, q, n)$, where we identify $\Gamma L(1, q)$ with $\{(g, \zeta) : g \in \mathrm{Aut}(GF(q))$ and $\zeta \in GF(q)^* = GL(1, q)\}$. Let $GL^{(d)}(1, q)$, $\Gamma L_\delta(1, q)$, $\Gamma SL_\delta(1, q)$ and $\Gamma L_\delta^{(d)}(1, q)$ be as in Section 1, let $\Gamma(1, q) = \{(g, 1) : g \in \mathrm{Aut}(GF(q))\} < \Gamma L(1, q)$, let $\Gamma_\delta(1, q)$ be the unique subgroup of $\Gamma(1, q)$ of order δ, let $I_\delta(1, q)$ be the subgroup of $\Gamma L_\delta(1, q)$ generated by $\Gamma_\delta(1, q)$ and $SL(1, q)$, and let $J_\delta(1, q)$ be the subgroup of $\Gamma L_\delta(1, q)$ generated by $\Gamma_\delta(1, q)$ and $GL^{(d)}(1, q)$. Now σ induces the usual bijection between the set of subgroups of $\Gamma L(m, q, n)$ containing $SL(m, q, n)$ and the set of all subgroups of $\Gamma L(1, q)$. In particular $\Gamma L_\delta^{(d)}(m, q, n) = \{J < \Gamma L(m, q, n) : \sigma(J) \in \Gamma L_\delta^{(d)}(1, q)\}$. Also we have $\sigma(\Gamma SL_\delta(m, q, n)) = \Gamma SL_\delta(1, q)$ and we have $\sigma(\Gamma L_\delta^{(d)}(m, q, n)) = \Gamma L_\delta^{(d)}(1, q)$. It is easy to see that $\Gamma SL_\delta(1, q)$ is the set of all conjugates of $I_\delta(1, q)$ in $\Gamma L(1, q)$, and $\Gamma L_\delta^{(d)}(1, q)$ is the set of all conjugates of $J_\delta^{(d)}(1, q)$ in $\Gamma L(1, q)$. ¿From this it follows that $\Gamma SL_\delta(m, q, n)$ is the set of all conjugates of $I_\delta(m, q, n)$ in $\Gamma L(m, q, n)$, and $\Gamma L_\delta^{(d)}(m, q, n)$ is the set of all conjugates of $J_\delta^{(d)}(m, q, n)$ in $\Gamma L(m, q, n)$. Clearly $I_\delta(m, q, n)$ is a split extension of $SL(m, q, n)$ such that $\Gamma L_\delta(m, q)$ is generated by $GL(m, q, n)$ and $I_\delta(m, q, n)$, and likewise $J_\delta^{(d)}(m, q, n)$ is a split extension of $GL^{(d)}(m, q, n)$ such that $\Gamma L_\delta(m, q, n)$ is generated by $GL(m, q, n)$ and $J_\delta^{(d)}(m, q, n)$.

Therefore $\Gamma SL_\delta(m, q, n)$ is a nonempty complete set of conjugate subgroups of $\Gamma L(m, q, n)$, and every I in $\Gamma SL_\delta(m, q, n)$ is a split extension of $SL(m, q, n)$ such that $\Gamma L_\delta(m, q, n)$ is generated by $GL(m, q, n)$ and I, and likewise $\Gamma L_\delta^{(d)}(m, q, n)$ is a nonempty complete set of conjugate subgroups of $\Gamma L(m, q, n)$, and every J in $\Gamma L_\delta^{(d)}(m, q, n)$ is a split extension of $GL^{(d)}(m, q, n)$ such that $\Gamma L_\delta(m, q, n)$ is generated by $GL(m, q)$ and J.

Remark (3.30) [Generalized Iteration]. Reverting to the **general** case, let us generalize the idea of the n-th iterate $E^{[n]}$ of E by introducing the concept of the **generalized r-th iterate** $E^{[r]}$ of E for any $r = r(T) = \sum r_i T^i \in \Omega[T]$ with $r_i \in \Omega$ (and $r_i = 0$ for all except a finite number of of i), by putting $E^{[r]} = E^{[r]}(Y) = \sum r_i E^{[i]}(Y)$. Note that, for the Y-derivative $E_Y^{[r]}(Y)$ of $E^{[r]}(Y)$ we clearly have $E_Y^{[r]}(Y) = E_Y^{[r]}(0) = r(X_m)$, and hence if $r(X_m) \neq 0$ then $E^{[r]}$ is a separable vectorial q-polynomial over Ω whose q-degree in Y equals m times the T-degree of r. Also note that the definition of $E^{[r]}$ remains valid for any vectorial E without assuming it to be monic or separable. Moreover, in such a general set-up, this makes the additive group of all vectorial q-polynomials $E = E(Y)$ in Y over Ω into a

$\Omega[T]$-**premodule** having all the properties of a module except additivity in E, i.e., for all $r' \in \Omega[T]$ we have $E^{[r+r']} = E^{[r]} + E^{[r']}$ and $E^{[rr']} = (E^{[r]})^{[r']}$, but in general for a vectorial q-polynomial E' over Ω we need not have $(E + E')^{[r]} = E^{[r]} + E'^{[r]}$. At any rate, $E^{[n]}$ of the previous notation corresponds to $E^{[T^n]}$ in the present notation. Reverting to the fixed monic separable vectorial E exhibited in (3.1), the said premodule structure makes Ω into a $\Omega[T]$-module when for every $r \in \Omega[T]$ and $z \in \Omega$ we define the "product" of r and z to be $E^{[r]}(z)$; in particular Ω becomes a $\mathrm{GF}(q)[T]$-module and as such we denote it by Ω_E. Now let us fix $s = s(T) \in \mathrm{GF}(p^{u/\delta(K)})[T]$ with $s(X_m) \neq 0$ and $\deg_T s = n$, and note that then $E^{[s]}$ is a separable vectorial q-polynomial of q-degree mn in Y over K. Let $V^{[s]}$ be the set of all roots of $E^{[s]}$ in Ω, and note that then $V^{[s]}$ is an (mn)-dimensional $\mathrm{GF}(q)$-vector-subspace of Ω. Let $\mathrm{GF}(q, s) = \mathrm{GF}(q)[T]/(s)$ where (s) is the ideal generated by s in $\mathrm{GF}(q)[T]$, and let $\omega : \mathrm{GF}(q)[T] \to \mathrm{GF}(q, s)$ be the canonical epimorphism. For every $\gamma \in \mathrm{Gal}(\mathrm{GF}(q), \mathrm{GF}(p^{u/\delta(K)}))$ and $r = \sum r_i T^i \in \Omega[T]$ with $r_i \in \mathrm{GF}(q)$, let us put $\gamma(\omega(r)) = \omega(\sum \gamma(r_i)T^i)$; this gives a faithful action of $\mathrm{Gal}(\mathrm{GF}(q), \mathrm{GF}(p^{u/\delta(K)}))$ on $\mathrm{GF}(q, s)$. Now $V^{[s]}$ is a submodule of Ω_E and as such it is annihilated by (s) and hence we may regard it as a $\mathrm{GF}(q, s)$-module; note that then, for every $r \in \Omega[T]$ and $z \in \Omega$, the "product" of $\omega(r)$ and z is given by $\omega(r)z = E^{[r]}(z) = \sum r_i E^{[i]}(z)$, and for every $g \in \mathrm{Gal}(K(V^{[s]}), K)$ we have $g(\omega(r)z) = \sum g(r_i)E^{[i]}(g(z)) = g'(\omega(r))g(z)$ where $g' \in \mathrm{Gal}(\mathrm{GF}(q), \mathrm{GF}(p^{u/\delta(K)}))$ is given by putting $g'(\zeta) = g(\zeta)$ for all $\zeta \in \mathrm{GF}(q)$, and hence in its action on $\mathrm{GF}(q, s)$ it is given by $g'(\omega(r)) = \omega(\sum g(r_i)T^i)$ for all $r \in \mathrm{GF}(q)[T]$. It follows that, in a natural manner, $\mathrm{Gal}(E^{[s]}, K)$ is a subgroup of $\Gamma \mathrm{L}_{\delta(K)}(V^{[s]})$ where, for any factor δ of $\delta(K)$, by $\Gamma \mathrm{L}_{\delta}(V^{[s]})$ we denote the **group of all $\mathrm{GF}_{\delta}(q, s)$-semilinear automorphisms** of the module $V^{[s]}$, by which we mean all additive isomorphisms $\sigma : V^{[s]} \to V^{[s]}$ for which there exists $\sigma' \in \mathrm{Gal}(\mathrm{GF}(q), \mathrm{GF}(p^{u/\delta}))$ such that for all $\eta \in \mathrm{GF}(q, s)$ and $z \in V^{[s]}$ we have $\sigma(\eta z) = \sigma'(\eta)\sigma(z)$; note that now $\sigma \to \sigma'$ gives an epimorphism $\Gamma \mathrm{L}_{\delta}(V^{[s]}) \to \mathrm{Gal}(\mathrm{GF}(q), \mathrm{GF}(p^{u/\delta}))$ whose kernel is the **group $\mathrm{GL}(V^{[s]})$ of all $\mathrm{GF}(q, s)$-linear automorphisms** of $V^{[s]}$, by which we mean all additive isomorphisms $\sigma : V^{[s]} \to V^{[s]}$ such that for all $\eta \in \mathrm{GF}(q, s)$ and $z \in V^{[s]}$ we have $\sigma(\eta z) = \eta \sigma(z)$, i.e., $\mathrm{GL}(V^{[s]}) = \Gamma \mathrm{L}_1(V^{[s]})$.

Also note that $\Gamma \mathrm{L}_{\delta}(V^{[s]})/\mathrm{GL}(V^{[s]}) = Z_{\delta}$ and $\Gamma \mathrm{L}_{\delta}(V^{[s]}) \lhd \Gamma \mathrm{L}_{\delta(K)}(V^{[s]})$ with $\Gamma \mathrm{L}_{\delta(K)}(V^{[s]})/\Gamma \mathrm{L}_{\delta}(V^{[s]}) = Z_{\delta(K)/\delta}$. Now identifying Galois groups of polynomials with the Galois groups of their splitting fields we see that

$$\mathrm{Gal}(E^{[s]}, K) \cap \mathrm{GL}(V^{[s]}) = \mathrm{Gal}(E^{[s]}, K(\mathrm{GF}(q)))$$

and by the usual Galois correspondence we have

$$\mathrm{Gal}(E^{[s]}, K(\mathrm{GF}(q))) \lhd \mathrm{Gal}(E^{[s]}, K)$$

with

$$\mathrm{Gal}(E^{[s]}, K)/\mathrm{Gal}(E^{[s]}, K(\mathrm{GF}(q))) = \mathrm{Gal}(K(\mathrm{GF}(q)), K) = Z_{\delta(K)}.$$

Thus

(3.31) $$\mathrm{Gal}(E^{[s]}, K) < \Gamma\mathrm{L}_{\delta(K)}(V^{[s]})$$

and

(3.32) $$\begin{cases} \mathrm{Gal}(E^{[s]}, K) \cap \mathrm{GL}(V^{[s]}) = \mathrm{Gal}(E^{[s]}, K(\mathrm{GF}(q))) \lhd \mathrm{Gal}(E^{[s]}, K) \\ \text{with } \mathrm{Gal}(E^{[s]}, K)/\mathrm{Gal}(E^{[s]}, K(\mathrm{GF}(q))) = Z_{\delta(K)} \end{cases}$$

and

(3.33) $$\begin{cases} \mathrm{GL}(V^{[s]}) \lhd \Gamma\mathrm{L}_{\delta(K)}(V^{[s]}) \\ \text{with } \Gamma\mathrm{L}_{\delta(K)}(V^{[s]})/\mathrm{GL}(V^{[s]}) = Z_{\delta(K)}. \end{cases}$$

Now by (3.32) we see that

(3.34) $$\mathrm{Gal}(E^{[s]}, K) < \mathrm{GL}(V^{[s]}) \Leftrightarrow \mathrm{GF}(q) \subset K$$

and by (3.31) to (3.33) we see that

(3.35) $$\begin{cases} \text{if } \mathrm{Gal}(E^{[s]}, K(\mathrm{GF}(q))) = \mathrm{GL}(V^{[s]}) \\ \text{then } \mathrm{Gal}(E^{[s]}, K) = \Gamma\mathrm{L}_{\delta(K)}(V^{[s]}). \end{cases}$$

Remark (3.36) [Moore Determinant]. In his path-breaking paper of 1896, E. H. Moore [Mor] showed that, in the **generic case and assuming** $\mathrm{GF}(q) \subset k_p$, the Galois group of the vectorial q-polynomial E of q-degree m is $\mathrm{GL}(m, q)$. Our proof of Theorem (3.7) was inspired by Moore's proof. To describe Moore's original proof, let us consider the determinant

$$Q_m(T_1, \ldots, T_{m+1}) = \det(T_{i+1}^{q^j})_{i,j=0,1,\ldots,m}$$

where i, j are respectively the row and column indices, and $T_1, \ldots, T_{m+1}$ are indeterminates over $\mathrm{GF}(q)$. For this determinant, which is now called the **Moore determinant**, Moore obtained the **Moore factorization**

$$Q_m(T_1, \ldots, T_{m+1}) = \prod_{l=0}^{m} P_l(T_1, \ldots, T_{l+1})$$

where $P_l(T_1, \ldots, T_{l+1})$ is the monic polynomial of degree q^l in T_{l+1} with coefficients in $\mathrm{GF}(q)[T_1, \ldots, T_l]$ given by

$$P_l(T_1, \ldots, T_{l+1}) = \prod_{(\lambda_1, \ldots, \lambda_l) \in \mathrm{GF}(q)^l} (T_{l+1} + \lambda_1 T_1 + \cdots + \lambda_l T_l).$$

Note that

$$P_0(T_1) = T_1 \quad \text{and} \quad P_1(T_1, T_2) = \prod_{\lambda_1 \in \mathrm{GF}(q)} (T_2 - \lambda_1 T_1)$$

and more generally

$$P_l(T_1, \ldots, T_{l+1}) = \prod_{(\lambda_1, \ldots, \lambda_l) \in \mathrm{GF}(q)^l} (T_{l+1} - \lambda_1 T_1 - \cdots - \lambda_l T_l).$$

To establish the said factorization, by elementary properties of determinants, replacing the m-th row $[T_{m+1}, T_{m+1}^q, \ldots, T_{m+1}^{q^m}]$ by the summation $\sum_{0 \le i \le m-1} (\lambda_{i+1}$ times the i-th row $[T_{i+1}, T_{i+1}^q, \ldots, T_{i+1}^{q^m}])$ we get zero. For every $(\lambda_1, \ldots, \lambda_m) \in \mathrm{GF}(q)^m$, for the said summation we have

$$\sum_{0 \le i \le m-1} \lambda_{i+1}[\ldots, T_{i+1}^{q^j}, \ldots] = [\ldots, (\sum_{0 \le i \le m-1} \lambda_{i+1} T_{i+1})^{q^j}, \ldots]$$

and hence $Q_m(T_1, \ldots, T_m, \lambda_1 T_1 + \cdots + \lambda_m T_m) = 0$. Therefore, as polynomials in T_{m+1}, the roots of Q_m and P_m coincide. Now by induction on m we get the desired factorization together with the fact that $Q_m(T_1, \ldots, T_{m+1})$ is a *nonzero* polynomial of degree q^m in T_{m+1} with coefficients in the polynomial ring $GF(q)[T_1, \ldots, T_m]$ and its leading coefficient is the *nonzero* polynomial $Q_{m-1}(T_1, \ldots, T_m)$. Moreover, $Q_m(T_1, \ldots, T_{m+1})$ is symmetric in $T_1, \ldots, T_{m+1}$ upto sign. For instance $Q_m(T_1, \ldots, T_{m+1})$ is a *nonzero* polynomial of degree q^m in T_1 with coefficients in $GF(q)[T_2, \ldots, T_{m+1}]$ and its leading coefficient is $(-1)^{m+1} Q_{m-1}(T_2, \ldots, T_{m+1})$. The inductive procedure used for proving the Moore factorization also shows that the value of the Moore determinant $Q_m(T_1, \ldots, T_{m+1})$ stays *nonzero* when for $T_1, \ldots, T_m$ we substitute any elements from an overfield of $GF(q)$ which are linearly independent over $GF(q)$. By the structure of the Moore determinant we see that

$$(*) \quad Q_m(T_1, \ldots, T_m, Y) = Q_{m-1}(T_1, \ldots, T_m) \left[Y^{q^m} + \sum_{i=1}^{m} H_i Y^{q^{m-i}} \right]$$

with $H_i = H_i(T_1, \ldots, T_m) \in \mathrm{GF}(q)[T_1, \ldots, T_m]$. By the Moore factorization we see that $T_1, \ldots, T_m$ are algebraic over $k_p(H_1, \ldots, H_m)$, and hence by a transcendence degree argument we conclude that the elements $H_1, \ldots, H_m$ are algebraically independent over k_p. Therefore, in view of the above equation $(*)$, we may identify $(X_1, \ldots, X_m)$ with $(H_1, \ldots, H_m)$ and $E(Y)$ with $Q_m(T_1, \ldots, T_m, Y)/Q_{m-1}(T_1, \ldots, T_m)$. Now, by the Moore factorization, the roots of $E(Y)$ constitute the $GF(q)$-vector-space V generated by $T_1, \ldots, T_m$, and hence $k_p(T_1, \ldots, T_m) = \mathrm{SF}(E, K)$ with $K = k_p(X_1, \ldots, X_m)$, and we may identify $(X_{1,1}, \ldots, X_{m,1})$ with $(T_1, \ldots, T_m)$ in view of (3.1) and (3.2). Clearly every g in $GL(V)$ gives rise to a k_p-automorphism of $k_p(T_1, \ldots, T_m)$ which leaves the symmetric functions $H_1 = X_1, \ldots, H_m = X_m$ unchanged and hence belong to $Gal(E, K)$. Therefore $Gal(E, K) = GL(V) = GL(m, q)$ which proves Moore's Theorem. As a consequence of the Moore factorization, we also get another proof of the partial converse of (1.1) noted in (3.9) to the effect that if $E' = E'(Y)$ is any monic polynomial of degree q^m in Y with coefficients in Ω such that the set of all roots of E' in Ω is an m-dimensional $GF(q)$-vector subspace V' of Ω then E' is a vectorial q-polynomial of q-degree m in Y over Ω: namely, upon letting $z_1, \ldots, z_m$ to be a basis of V', by the Moore factorization we see that the elements of V' are the roots of $Q_m = Q_m(z_1, \ldots, z_m, Y)$ regarded as a polynomial in Y over $GF(q)(z_1, \ldots, z_m)$; now the degree of Q_m in Y is q^m and its leading coefficient is $Q_{m-1}(z_1, \ldots, z_m) \neq 0$, and hence the given polynomial must be a constant multiple of Q_m; by (*) it is clear that Q_m is a vectorial q-polynomial of q-degree m in Y, and hence so is the given polynomial.

Note (3.37) [Question on Generalized Iteration]. In the **generic case and assuming** $GF(q) \subset k_p$, in Theorem (3.25) we generalized Moore's Theorem by showing that, for any integer $n > 0$, the Galois group of the n-th iterate $E^{[n]}$ of the vectorial q-polynomial E of q-degree m is the generalized general linear group $GL(m, q, n)$ consisting of all m by m matrices with invertible determinant over the local ring $GF(q)[T]/T^n$. For $m = 1$ this was proved by Carlitz [Car] in 1938 as part of his explicit class field theory over finite fields; the case of $m = 1$ was further enhanced by Drinfeld [Dri] in 1974; for a survey of the relevant work of Carlitz and Drinfeld see Goss [Gos] and Hayes [Hay]. Actually, referring to (3.30) and (3.34), Carlitz [Car] showed that if $m = 1$ and $s = s(T) \in GF(q)[T]$ with $s(X_m) \neq 0$ then $Gal(E^{[s]}, K) = GL(V^{[s]})$ with $E^{[s]}$ and $V^{[s]}$ as defined in (3.30), and **we may ask if this continues to hold also for** $m > 1$.

Section 4: Symplectic Groups

In this Section, to write down families of polynomials whose Galois groups are between $Sp(2m, q)$ and $\Gamma Sp(2m, q)$, let $Y, X, S, T_1, T_2, \ldots$ be indeterminates over k_p and, as in Section 2, for every $e \geq 0$ let

$$K_e = k_p(X, T_1, \ldots, T_e).$$

We shall apply the considerations of Section 1 by taking $K = K_e$ or $K_e(S)$ with suitable e.

First, for $0 \leq e \leq m - 1$, consider the monic separable projective q-polynomial

$$f_e^\sharp = f_e^\sharp(Y) = Y^{\langle 2m-1 \rangle} + S^{r(0)} + S^{r(m)} X Y^{\langle m-1 \rangle}$$

$$+ \sum_{i=1}^{e} \left(S^{r(m+i)} T_i^{q^i} Y^{\langle m-1+i \rangle} + S^{r(m-i)} T_i Y^{\langle m-1-i \rangle} \right)$$

of q-prodegree $2m$ in Y over $K_e(S)$ where $r = (r(0), \ldots, r(2m))$ is a sequence of nonnegative integers with

(4.1*)
$$r(2m) = 0$$

such that for some nonnegative integer t we have

(4.1**) $q^i r(m-i) = r(m+i) + t q^m \langle i - 1 \rangle$ for $0 \leq i \leq m$.

Next, for $0 \leq e \leq m - 1$, let

$$\phi_e^{\natural} = \phi_e^{\natural}(Y) = f_e^{\natural}(Y^{q-1}) = Y^{q^{2m}-1} + S^{r(0)} + S^{r(m)} X Y^{q^m - 1}$$
$$+ \sum_{i=1}^{e} \left(S^{r(m+i)} T_i^{q^i} Y^{q^{m+i}-1} + S^{r(m-i)} T_i Y^{q^{m-i}-1} \right)$$

and

$$\phi_e^{\natural} = \phi_e^{\natural}(Y) = Y \phi_e^{\natural}(Y) = Y^{q^{2m}} + S^{r(0)} Y + S^{r(m)} X Y^{q^m}$$
$$+ \sum_{i=1}^{e} \left(S^{r(m+i)} T_i^{q^i} Y^{q^{m+i}} + S^{r(m-i)} T_i Y^{q^{m-i}} \right)$$

be the subvectorial and vectorial associates of $f_e^{\natural}$ respectively, and for every divisor d of $q - 1$, let $f_e^{(d)}, \phi_e^{(d)}, \phi_e^{(d)}$ be obtained by substituting S^d for S in $f_e^{\natural}, \phi_e^{\natural}, \phi_e^{\natural}$ respectively, i.e., let

$$f_e^{(d)} = f_e^{(d)}(Y) = Y^{\langle 2m-1 \rangle} + S^{r(0)d} + S^{r(m)d} X Y^{\langle m-1 \rangle}$$
$$+ \sum_{i=1}^{e} \left(S^{r(m+i)d} T_i^{q^i} Y^{\langle m-1+i \rangle} + S^{r(m-i)d} T_i Y^{\langle m-1-i \rangle} \right)$$

and

$$\phi_e^{(d)} = \phi_e^{(d)}(Y) = Y^{q^{2m}-1} + S^{r(0)d} + S^{r(m)d} X Y^{q^m - 1}$$
$$+ \sum_{i=1}^{e} \left(S^{r(m+i)d} T_i^{q^i} Y^{q^{m+i}-1} + S^{r(m-i)d} T_i Y^{q^{m-i}-1} \right)$$

and

$$\phi_e^{(d)} = \phi_e^{(d)}(Y) = Y^{q^{2m}} + S^{r(0)d} Y + S^{r(m)d} X Y^{q^m}$$
$$+ \sum_{i=1}^{e} \left(S^{r(m+i)d} T_i^{q^i} Y^{q^{m+i}} + S^{r(m-i)d} T_i Y^{q^{m-i}} \right).$$

Finally, for $0 \leq e \leq m - 1$, let f_e, ϕ_e, ϕ_e be obtained by putting $S = 1$ in $f_e^\natural$, $\phi_e^\natural$, $\phi_e^\natural$ respectively, i.e., let

$$f_e = f_e(Y) = Y^{\langle 2m-1 \rangle} + 1 + XY^{\langle m-1 \rangle}$$
$$+ \sum_{i=1}^{e} \left(T_i^{q^i} Y^{\langle m-1+i \rangle} + T_i Y^{\langle m-1-i \rangle} \right)$$

and

$$\phi_e = \phi_e(Y) = Y^{q^{2m}-1} + 1 + XY^{q^m-1}$$
$$+ \sum_{i=1}^{e} \left(T_i^{q^i} Y^{q^{m+i}-1} + T_i Y^{q^{m-i}-1} \right)$$

and

$$\phi_e = \phi_e(Y) = Y^{q^{2m}} + Y + XY^{q^m}$$
$$+ \sum_{i=1}^{e} \left(T_i^{q^i} Y^{q^{m+i}} + T_i Y^{q^{m-i}} \right).$$

Note that

$$(4.1') \quad \begin{cases} \text{if } r(i) = \langle 2m - 1 \rangle - \langle i - 1 \rangle \text{ for } 0 \leq i \leq 2m \\ \text{then conditions (4.1*) and (4.1**) are satisfied with } t = q^m + 1 \end{cases}$$

and

$$(4.1'') \quad \begin{cases} \text{if } r(m + i) = 0 \text{ and } r(m - i) = q^{m-i}\langle i - 1 \rangle \text{ for } 0 \leq i \leq m \\ \text{then conditions (4.1*) and (4.1**) are satisfied with } t = 1. \end{cases}$$

Case $(4.1')$ arises when we homogenize f_e, i.e., when we put $f_e^\natural(Y) = S^{\langle 2m-1 \rangle} f_e(Y/S)$, and so we may call it the **homogeneous case**. By analogy, case $(4.1'')$ may be called the **twisted homogeneous case**.

Recall that a bivariate $\psi^*(Y, Z)$ in $K[Y, Z]$ is said to be a **bivectorial q-polynomial** of q-**bidegree** (m^*, m') (where $m^* \geq 0$ and $m' \geq 0$ are integers) in (Y, Z) over K if it is of the form

$$(4.2) \qquad \psi^*(Y, Z) = \sum_{i=0}^{m^*} \sum_{j=0}^{m'} a_{ij}^* Y^{q^i} Z^{q^j} \qquad \text{with } a_{ij}^* \in K \text{ for all } i, j$$

and $a_{m^* j'}^* \neq 0 \neq a_{i'm'}^*$ for some j', i'; note that then, for every overfield L of K with $\mathrm{GF}(q) \subset L$ and every $\mathrm{GF}(q)$-vector-subspace V^* of L, by sending every $(y, z) \in V^* \times V^*$ to $\psi^*(y, z)$ we get a map $\psi^* : V^* \times V^* \to L$ which is bilinear (over $\mathrm{GF}(q)$). Likewise, a bivariate $\psi^*(Y, Z)$ in $K[Y, Z]$ is said to be a **symplectic q-polynomial** of q-**symdegree** (m^*, n^*) (where $m^* \geq n^* \geq 0$ are integers) in (Y, Z) over K if it is of the form

$$(4.3) \quad \psi^*(Y, Z) = \sum_{i=0}^{m^*} \sum_{j=0}^{m^*} a_{ij}^* Y^{q^i} Z^{q^j} \quad \text{with} \begin{cases} a_{ji}^* = -a_{ij}^* \in K \text{ for all } i, j \\ \text{and } a_{i'i'}^* = 0 \text{ for all } i' \end{cases}$$

and $a_{m^* n^*}^* \neq 0 \neq a_{m^* n^*}^*$ and $a_{m^* j'}^* = 0 = a_{j' m^*}^*$ for all $j' \neq n^*$; note that then $\psi^*(Y, Z)$ is a bivectorial polynomial of q-bidegree (m^*, m^*) in (Y, Z) over K, and the above map $\psi^* : V^* \times V^* \to L$ is **alternating** and **antisymmetric**. [Recall that for vector spaces V and W over any field, a map $\theta : V \times V \to W$ is **alternating** means $\theta(x, x) = 0$ for all x in V, and **antisymmetric** means $\theta(z, y) = -\theta(y, z)$ for all y, z in V. Note that if θ is **biadditive** then for all y, z in V we have $\theta(y + z, y + z) = \theta(y, y) + \theta(y, z) + \theta(z, y) + \theta(z, z)$ and hence if θ is also alternating then $\theta(z, y) = -\theta(y, z)$ and therefore θ is antisymmetric.]

Recall that, given any vectorial q-polynomial

$$(4.4) \qquad \phi(Y) = \sum_{i=0}^{2m} a_i Y^{q^{2m-i}} \quad \text{with } a_i \in K \text{ and } a_0 \neq 0 \neq a_m$$

of q-degree $2m$ in Y over K, the **vectorial derivative** $\psi(Y, Z)$ of $\phi(Y)$ is defined by putting

$$(4.5) \qquad \psi(Y, Z) = Y^{q^m} \phi(Z) - Z^{q^m} \phi(Y)$$

and observe that then

$$(4.6) \qquad \begin{cases} \psi(Y, Z) \text{ is a symplectic } q\text{-polynomial} \\ \text{of } q\text{-symdegree } (2m, m) \text{ in } (Y, Z) \text{ over } K. \end{cases}$$

For a moment assume that $\phi(Y)$ is separable, and let V be the set of all roots of $\phi(Y)$ in Ω, and note that then V is a $2m$ dimensional GF(q)-vector-subspace of Ω. Let $\Gamma(Y, Z)$ be a symplectic q-polynomial of q-symdegree $(2m - 1, m - 1)$ in (Y, Z) over K. Then for every $0 \neq y \in V$, clearly there exists $z \in V$ with $\Gamma(y, z) \neq 0$. Now assume that $\psi(Y, Z) = \alpha \Gamma(Y, Z)^q - \Gamma(Y, Z)$ with $0 \neq \alpha \in K$. If $\alpha = 1$ then let $\beta = 1$ and if $\alpha \neq 1$ then let $\beta = \Gamma(y', z')^{-1}$ for some y', z' in V with $\Gamma(y', z') \neq 0$. Note that in both the cases we have $\beta \in K(V)$ with $\beta^{q-1} = \alpha$. For all y, z in V let $b(y, z) = \beta \Gamma(y, z)$. Then for all y, z in V we have $b(y, z)^q - b(y, z) = \beta[\alpha \Gamma(y, z)^q - \Gamma(y, z)] = \beta \psi(y, z) = 0$ and hence $b(y, z) \in$ GF(q). Therefore $b : V \times V \to$ GF(q) is **symplectic**. [Recall that for a vector space V over any field, a map b from $V \times V$ to the field is **symplectic** if it is bilinear, alternating, and **nondegenerate**, where nondegenerate means that for every $0 \neq y \in V$ we

have $b(y, z) \neq 0$ for some $z \in V$. For any such symplectic map: a **similitude**
of b is an element g of $\mathrm{GL}(V)$ for which there is a nonzero element g^* of
the field such that for all y, z in V we have $b(g(y), g(z)) = g^* b(y, z)$; an
isometry of b is a similitude g of b with $g^* = 1$; a **semisimilitude** of b
is an element g of $\Gamma\mathrm{L}(V)$ for which there is a nonzero element g^* of the
field and an automorphism g' of the field such that for all y, z in V we have
$b(g(y), g(z)) = g^* g'(b(y, z))$; in case the field is $\mathrm{GF}(q)$ and δ is a divisor of
u, the semisimilitude g is said to be of **type** δ if $g' \in \mathrm{Gal}(\mathrm{GF}(q), \mathrm{GF}(p^{u/\delta}))$;
for generalities about symplectic maps see the book [KLi], especially (2.1.7)
and 2.1.2 on pages 11 and 12 of that book.] Given any $g \in \mathrm{Gal}(\phi, K) =$
$\mathrm{Gal}(K(V), K)$, upon letting $g^* = \beta/g(\beta)$ we have $0 \neq g^* \in \mathrm{GF}(q)$ because
$0 \neq \beta^{q-1} = \alpha \in K$ (also if $\alpha = 1$ then obviously $g^* = 1$), and upon letting
$g'(\zeta) = g(\zeta)$ for every $\zeta \in \mathrm{GF}(q)$ we get $g' \in \mathrm{Gal}(\mathrm{GF}(q), \mathrm{GF}(p^{u/\delta(K)}))$
such that for all y, z in V we have $g(b(y, z)) = g'(b(y, z))$ and hence we
have $b(g(y), g(z)) = \beta\Gamma(g(y), g(z)) = g^* g(\beta)\Gamma(g(y), g(z)) = g^* g(\beta\Gamma(y, z)) =$
$g^* g(b(y, z)) = g^* g'(b(y, z))$. Thus we have proved the following refinement
of (4.7) of [AL2]:

Semisimilitude Lemma (4.7). *Let $f = f(Y)$ be a separable projective q-
polynomial of q-prodegree $2m$ in Y over K, let $\phi = \phi(Y) = f(Y^{q-1})$ and $\phi =$
$\phi(Y) = Y\phi(Y)$, let $\psi(Y, Z) \in K[Y, Z]$ be the vectorial derivative of $\psi(Y)$,
let V be the set of all roots of $\phi(Y)$ in an algebraic closure Ω of K, and note
that then V is a $2m$ dimensional $\mathrm{GF}(q)$-vector-subspace of Ω. Assume that
there exists a symplectic q-polynomial $\Gamma(Y, Z)$ of q-symdegree $(2m-1, m-1)$
in (Y, Z) over K such that (4.7^*) $\psi(Y, Z) = \alpha\Gamma(Y, Z)^q - \Gamma(Y, Z)$ with $0 \neq$
$\alpha \in K$. Then for every $0 \neq y \in V$ there exists $z \in L$ with $\Gamma(y, z) \neq 0$. If
$\alpha = 1$ then let $\beta = 1$ and if $\alpha \neq 1$ then let $\beta = \Gamma(y', z')^{-1}$ for some y', z'
in V with $\Gamma(y', z') \neq 0$. Note that in both cases we have $\beta \in K(V)$ with
$\beta^{q-1} = \alpha$. For all y, z in V let $b(y, z) = \beta\Gamma(y, z)$. Then $b : V \times V \to \mathrm{GF}(q)$
is symplectic, and in a natural manner we have $\mathrm{Gal}(\phi, K) < \Gamma Sp(2m, q) =$
the semisimilitude group of b, so that for each $g \in \mathrm{Gal}(\phi, K)$ there exists a
unique nonzero element g^* in $\mathrm{GF}(q)$ and a unique automorphism g' of $\mathrm{GF}(q)$
such that for all y, z in V we have $b(g(y), g(z)) = g^* g'(b(y, z))$. Moreover,
for every $g \in \mathrm{Gal}(\phi, K)$ we have $g' \in \mathrm{Gal}(\mathrm{GF}(q), \mathrm{GF}(p^{u/\delta(K)}))$, and if $\alpha = 1$
then for every $g \in \mathrm{Gal}(\phi, K)$ we also have $g^* = 1$. Thus, in particular, if
$\mathrm{GF}(q) \subset K$ then $\mathrm{Gal}(\phi, K) < GSp(2m, q) =$ the similitude group of b,
and if $\mathrm{GF}(q) \subset K$ and $\alpha = 1$ then $\mathrm{Gal}(\phi, K) < Sp(2m, q) =$ the isometry
group of b. Finally note that the Galois group $\mathrm{Gal}(\phi, K)$ essentially equals
the Galois group $\mathrm{Gal}(\phi, K)$ except that the former acts on nonzero vectors
while the latter acts on the entire vector space V.*

In the situation of (4.7), we clearly have:

$$(4.7.1) \quad \begin{cases} \mathrm{Gal}(\phi, K) < \Gamma\mathrm{Sp}_{\delta(K)}(2m, q) \\ \text{and } \mathrm{GSp}(2m, q) \lhd \Gamma\mathrm{Sp}_{\delta(K)}(2m, q) \lhd \Gamma\mathrm{Sp}(2m, q) \\ \text{with } \Gamma\mathrm{Sp}_{\delta(K)}(2m, q)/\mathrm{GSp}(2m, q) = Z_{\delta(K)} \end{cases}$$

where, for any factor δ of u, by $\Gamma\mathrm{Sp}_{\delta}(2m, q)$ we denote the group of all semisimilitudes of b of type δ. Note that, in the above notation, $g \mapsto (g', g^*)$ and $g \mapsto g'$ give canonical epimorphisms $\rho : \Gamma\mathrm{Sp}(2m, q) \rightarrow \Gamma\mathrm{L}(1, q)$ and $\kappa : \Gamma\mathrm{Sp}(2m, q) \rightarrow \mathrm{Aut}(\mathrm{GF}(q))$ with kernels $\mathrm{Sp}(2m, q)$ and $\mathrm{GSp}(2m, q)$ respectively, and the inverse image of $\mathrm{Gal}(\mathrm{GF}(q), \mathrm{GF}(p^{u/\delta}))$ under the second epimorphism κ is the group $\Gamma\mathrm{Sp}_{\delta}(2m, q)$; the group $\Gamma\mathrm{Sp}_{\delta}(2m, q)$ can also be characterized as the unique normal subgroup of $\Gamma\mathrm{Sp}(2m, q)$ such that $\mathrm{GSp}(2m, q) \lhd \Gamma\mathrm{Sp}_{\delta}(2m, q)$ with $\Gamma\mathrm{Sp}_{\delta}(2m, q)/\mathrm{GSp}(2m, q) = Z_{\delta}$; the image of $\Gamma\mathrm{Sp}_{\delta}(2m, q)$ under the canonical epimorphism $\Gamma\mathrm{Sp}(2m, q) \rightarrow \mathrm{P}\Gamma\mathrm{Sp}(2m, q)$ is denoted by $\mathrm{P}\Gamma\mathrm{Sp}_{\delta}(2m, q)$. By restricting κ to $\mathrm{GSp}(2m, q)$ we get the canonical epimorphism $\mathrm{GSp}(2m, q) \rightarrow \mathrm{GF}(q)^* = \mathrm{GF}(q) \setminus \{0\}$ with kernel $\mathrm{Sp}(2m, q) = p(\mathrm{GSp}(2m, q))$ (see the lines between (5.1) and (5.2) of [A04]), and hence, for any divisor d of $q - 1$, there is a unique normal subgroup $\mathrm{GSp}^{(d)}(2m, q)$ of $\mathrm{GSp}(2m, q)$ such that $\mathrm{Sp}(2m, q) \lhd \mathrm{GSp}^{(d)}(2m, q)$ with $\mathrm{GSp}(2m, q)/\mathrm{GSp}^{(d)}(2m, q) = Z_d$; the image $\mathrm{PGSp}^{(d)}(2m, q)$ of $\mathrm{GSp}^{(d)}(2m, q) = Z_d$ under the canonical epimorphism $\mathrm{GSp}(2m, q) \rightarrow \mathrm{PGSp}(2m, q)$ can be characterized as the unique group between $\mathrm{PSp}(2m, q) \rightarrow \mathrm{PGSp}(2m, q)$ such that $\mathrm{PGSp}^{(d)}(2m, q) = \mathrm{PSp}(2m, q)$ or $\mathrm{PGSp}(2m, q)$ according as d is even or odd; this follows from the fact that $\mathrm{PGSp}(2m, q)/\mathrm{PSp}(2m, q) = Z_2$ or Z_1 according as q is odd or even; note that if q is even then $\mathrm{PSp}(2m, q) = \mathrm{PGSp}(2m, q)$; see 2.1.2, 2.1.B, 2.1.C and 2.1.D of [KLi]. In the situation of (4.7), by the usual Galois correspondence we also see that:

$$(4.7.2) \quad \begin{cases} \mathrm{Gal}(\phi, K) \cap \mathrm{GSp}(2m, q) = \mathrm{Gal}(\phi, K(\mathrm{GF}(q))) \lhd \mathrm{Gal}(\phi, K) \\ \text{with } \mathrm{Gal}(\phi, K)/\mathrm{Gal}(\phi, K(\mathrm{GF}(q))) = Z_{\delta(K)}. \end{cases}$$

As an immediate consequence of (4.7.1) and (4.7.2) we see that, in the situation of (4.7):

$$(4.7.3) \quad \begin{cases} \text{if } \mathrm{Gal}(\phi, K(\mathrm{GF}(q))) = \mathrm{GSp}^{(d)}(2m, q) \\ \text{where } d \text{ is a divisor of } q - 1 \\ \text{then we have} \\ \mathrm{Gal}(\phi, K) \cap \mathrm{GSp}(2m, q) \\ = \mathrm{GSp}^{(d)}(2m, q) \lhd \mathrm{Gal}(\phi, K) < \Gamma\mathrm{Sp}_{\delta(K)}(2m, q) \\ \text{with } \mathrm{Gal}(\phi, K)/\mathrm{GSp}^{(d)}(2m, q) = Z_{\delta(K)}. \end{cases}$$

Since $\mathrm{GSp}^{(1)}(2m, q) = \mathrm{GSp}(2m, q)$ and $\Gamma\mathrm{Sp}_{\delta(K)}(2m, q)/\mathrm{GSp}(2m, q) = Z_{\delta(K)}$,

by taking $d = 1$ in (4.7.3) we see that, in the situation of (4.7):
(4.7.4)

$$\begin{cases} \text{if } \mathrm{Gal}(\phi, K(\mathrm{GF}(q))) = \mathrm{GSp}(2m, q) \\ \text{then we have} \\ \mathrm{Gal}(\phi, K) = \Gamma\mathrm{Sp}_{\delta(K)}(2m, q) \text{ and } \mathrm{Gal}(f, K) = \mathrm{P}\Gamma\mathrm{Sp}_{\delta(K)}(2m, q). \end{cases}$$

The above definitions of the various groups between $\mathrm{Sp}(2m, q)$ and $\Gamma\mathrm{Sp}(2m, q)$, or between $\mathrm{PSp}(2m, q)$ and $\mathrm{P}\Gamma\mathrm{Sp}(2m, q)$, as well as the definitions of the canonical epimorphisms $\rho : \Gamma\mathrm{Sp}(2m, q) \to \Gamma\mathrm{L}(1, q)$ and $\kappa : \Gamma\mathrm{Sp}(2m, q) \to \mathrm{Aut}(\mathrm{GF}(q))$ with kernels $\mathrm{Sp}(2m, q)$ and $\mathrm{GSp}(2m, q)$ respectively, apply for any symplectic map $b : V \times V \to \mathrm{GF}(q)$ where V is any $2m$ dimensional vector space over $\mathrm{GF}(q)$. Moreover, given any such symplectic map, for any divisor δ of u, let $\Gamma'\mathrm{Sp}_\delta(2m, q)$ be the set of all subgroups I of $\Gamma\mathrm{Sp}_\delta(2m, q)$ such that $I \cap \mathrm{GSp}(2m, q) = \mathrm{Sp}(2m, q) \triangleleft I$ with $I/\mathrm{Sp}(2m, q) = Z_\delta$, and let $\mathrm{P}\Gamma'\mathrm{Sp}_\delta(2m, q)$ be the set of images of the various members of $\Gamma'\mathrm{Sp}_\delta(2m, q)$ under the canonical epimorphism of $\Gamma\mathrm{Sp}(2m, q)$ onto $\mathrm{P}\Gamma\mathrm{Sp}(2m, q)$; in Remark (4.21) we shall show that $\Gamma'\mathrm{Sp}_\delta(2m, q)$ is a nonempty complete set of conjugate subgroups of $\Gamma\mathrm{Sp}(2m, q)$, and every I in $\Gamma'\mathrm{Sp}_\delta(2m, q)$ is a split extension of $\mathrm{Sp}(2m, q)$ such that $\Gamma\mathrm{Sp}_\delta(2m, q)$ is generated by $\mathrm{Sp}(2m, q)$ and I. Likewise, for any divisor d of $q-1$, let $\Gamma\mathrm{Sp}_\delta^{(d)}(2m, q)$ be the set of all subgroups J of $\Gamma\mathrm{Sp}_\delta(2m, q)$ such that $J \cap \mathrm{GSp}(2m, q) = \mathrm{GSp}^{(d)}(2m, q) \triangleleft J$ with $J/\mathrm{GSp}^{(d)}(2m, q) = Z_\delta$ and $I < J$ for some I in $\Gamma'\mathrm{Sp}_\delta(2m, q)$, and let $\mathrm{P}\Gamma\mathrm{Sp}_\delta^{(d)}(2m, q)$ be the set of images of the various members of $\Gamma\mathrm{Sp}_\delta^{(d)}(2m, q)$ under the canonical epimorphism of $\Gamma\mathrm{Sp}(2m, q)$ onto $\mathrm{P}\Gamma\mathrm{Sp}(2m, q)$; in Remark (4.21) we shall show that $\Gamma\mathrm{Sp}_\delta^{(d)}(2m, q)$ is a nonempty complete set of conjugate subgroups of $\Gamma\mathrm{Sp}(2m, q)$, and every J in $\Gamma\mathrm{Sp}_\delta^{(d)}(2m, q)$ is a split extension of $\mathrm{GSp}^{(d)}(2m, q)$ such that $\Gamma\mathrm{Sp}_\delta(2m, q)$ is generated by $\mathrm{GSp}(2m, q)$ and J; note that clearly $\Gamma\mathrm{Sp}_\delta^{(q-1)}(2m, q) = \Gamma'\mathrm{Sp}_\delta(2m, q)$ and $\Gamma\mathrm{Sp}_\delta^{(1)}(2m, q) = \{\Gamma\mathrm{Sp}_\delta(2m, q)\}$.

Since $\mathrm{GSp}^{(q-1)}(2m, q) = \mathrm{Sp}(2m, q)$, by taking $d = q - 1$ in (4.7.3) we see that, in the situation of (4.7):

$$\text{(4.7.5)} \quad \begin{cases} \text{if } \mathrm{Gal}(\phi, K(\mathrm{GF}(q))) = \mathrm{Sp}(2m, q) \\ \text{then we have} \\ \mathrm{Gal}(\phi, K) \in \Gamma'\mathrm{Sp}_{\delta(K)}(2m, q) \\ \text{and } \mathrm{Gal}(f, K) \in \mathrm{P}\Gamma'\mathrm{Sp}_{\delta(K)}(2m, q). \end{cases}$$

Finally, because of the usual Galois correspondence, by (4.7.4) and (4.7.5) we see that, in the situation of (4.7):

$$(4.7.6) \quad \begin{cases} \text{if } \mathrm{Gal}(\phi, K(\mathrm{GF}(q))) = \mathrm{GSp}^{(d)}(2m,q) \\ \text{where } d \text{ is a factor of } q-1 \text{ and} \\ \text{for some field } K' \text{ between } K \text{ and } \mathrm{SF}(\phi,K) \text{ we have} \\ \delta(K') = \delta(K) \text{ and } \mathrm{Gal}(\phi, K'(\mathrm{GF}(q))) = \mathrm{Sp}(2m,q) \\ \text{then we have} \\ \mathrm{Gal}(\phi, K) \in \mathrm{\Gamma Sp}^{(d)}_{\delta(K)}(2m,q) \\ \text{and } \mathrm{Gal}(f, K) \in \mathrm{P\Gamma Sp}^{(d)}_{\delta(K)}(2m,q). \end{cases}$$

As an example of a vectorial derivative, for $0 \le e \le m-1$, in (3.8) of [AL1] we considered the bivariate polynomial

$$(4.8) \qquad \psi_e^\sharp(Y,Z) = Y^{q^m}\phi_e^\sharp(Z) - Z^{q^m}\phi_e^\sharp(Y)$$

and, by applying the Mantra of [A05], in (3.9) of [AL1] we obtained the factorization

$$(4.9) \qquad \psi_e^\sharp(Y,Z) = S^{-tq^m}\Gamma_e^\sharp(Y,Z)^q - \Gamma_e^\sharp(Y,Z)$$

where, as noted in (3.10) of [AL1], we have

$$(4.10) \quad \begin{aligned} &\Gamma_e^\sharp(Y,Z) \\ &= \sum_{i=1}^{e}\sum_{j=0}^{i-1}\left(Z^{q^{m+j}}Y^{q^{m-i+j}} - Y^{q^{m+j}}Z^{q^{m-i+j}}\right)S^{b(i,j)+tq^{m-1}}T_i^{q^j} \\ &\quad + \sum_{j=0}^{m-1}\left(Z^{q^{m+j}}Y^{q^j} - Y^{q^{m+j}}Z^{q^j}\right)S^{tq^j\{m-j-1\}}. \end{aligned}$$

Again, as noted in (3.11) of [AL1],

$$(4.11) \quad \begin{cases} \text{for } 0 \le e \le m-1 \text{ we have that} \\ \Gamma_e^\sharp \text{ is a polynomial of degree } q^{2m-1} \text{ in } Z \\ \text{with coefficients in } \mathrm{GF}(p)[Y,S,T_1,\dots,T_e] \text{ and in it} \\ \text{the coefficient of the highest } Z\text{-degree term is } (YS^t)^{q^{m-1}}. \end{cases}$$

Finally, as noted in (3.12) of [AL1] and (4.12) of [AL2], as a consequence of (4.8), (4.9) and (4.11) we have the following:

Root Extraction Theorem (4.12). *For $0 \le e \le m-1$, there exists a nonzero root y_e of $\phi_e^\sharp(Y)$ in any splitting field $L_e^\sharp$ of $\phi_e^\sharp(Y)$ over $K_e(S)$, and given any such y_e there exists a root z_e of $\phi_e^\sharp(Y)$ in $L_e^\sharp$ such that*

$\Gamma_e^\sharp(y_e, z_e) \neq 0$, *and for every such* z_e *we have* $\Gamma_e^\sharp(y_e, z_e)^{q-1} = S^{tq^m}$ *(note that for any* $y_e \in L_e^\sharp$ *and* $z_e \in L_e^\sharp$ *we obviously have* $\Gamma_e^\sharp(y_e, z_e) \in L_e^\sharp$*). More-over, for every divisor* d *of* $(q-1)/GCD(t, q-1)$*, there exist integers* σ, τ *with* $\sigma t q^m + \tau(q-1) = (q-1)/d$*, and if* $GF(q) \subset k_p$ *then, given any such roots* y_e, z_e *and any such integers* σ, τ*, there exists* $\lambda \in GF(q) \subset k_p$ *such that for* $\Lambda_e = \lambda \Gamma_e^\sharp(y_e, z_e)^\sigma S^\tau$ *we have* $\Lambda_e \in L_e^\sharp$ *with* $\Lambda_e^d = S$*, and in a natural manner we have* $Gal(\phi_e, K_e) < Gal(\phi_e^{(d)}, K_e(S)) \lhd Gal(\phi_e^\sharp, K_e(S))$ $Gal(\phi_e^\sharp, K_e(S))/Gal(\phi_e^{(d)}, K_e(S)) = Z_d$.

As another example of a vectorial derivative, for $0 \leq e \leq m - 1$, we consider the bivectorial polynomial

(4.13) $$\psi_e(Y, Z) = Y^{q^m}\phi_e(Z) - Z^{q^m}\phi_e(Y)$$

and we note that, upon letting $\Gamma_e(Y, Z)$ be obtained by putting $S = 1$ in $\Gamma_e^\sharp(Y, Z)$, by (4.9) and (4.10) we get the factorization

(4.14) $$\psi_e(Y, Z) = \Gamma_e(Y, Z)^q - \Gamma_e(Y, Z)$$

where

(4.15)
$$\Gamma_e(Y, Z) = \sum_{i=1}^{e}\sum_{j=0}^{i-1}\left(Z^{q^{m+j}}Y^{q^{m-i+j}} - Y^{q^{m+j}}Z^{q^{m-i+j}}\right)T_i^{q^j}$$
$$+ \sum_{j=0}^{m-1}\left(Z^{q^{m+j}}Y^{q^j} - Y^{q^{m+j}}Z^{q^j}\right).$$

Also we note that, by (4.11),

(4.16) $$\begin{cases} \text{for } 0 \leq e \leq m - 1 \text{ we have that} \\ \Gamma_e \text{ is a polynomial of degree } q^{2m-1} \text{ in } Z \\ \text{with coefficients in } GF(p)[Y, T_1, \ldots, T_e] \text{ and in it} \\ \text{the coefficient of the highest } Z\text{-degree term is } Y^{q^{m-1}}. \end{cases}$$

Now, as hinted above, for $0 \leq e \leq m - 1$, by (4.8) and (4.13) we see that the bivariate polynomials $\psi_e^\sharp(Y, Z)$ and $\psi_e(Y, Z)$ are the vectorial derivatives of the monic separable vectorial q-polynomials $\phi_e^\sharp(Y)$ and $\phi_e(Y)$ of q-degree $2m$ in Y over $K_e(S)$ and K_e respectively, and by (4.10) and (4.15) we see that $\Gamma_e^\sharp(Y, Z)$ and $\Gamma_e(Y, Z)$ are symplectic q-polynomials of q-symdegree $(2m - 1, m - 1)$ in (Y, Z) over $K_e(S)$ and K_e respectively, and by (4.9) and (4.14) we see that condition (4.7*) is satisfied with $\alpha = S^{-tq^m}$ and $\alpha = 1$ respectively. Therefore by (4.7) we get the following restatement of (4.17) of [AL2]:

Theorem (4.17). *If $GF(q) \subset k_p$ then, for $0 \le e \le m-1$, in a natural manner we have $Gal(\phi_e^\sharp, K_e(S)) < GSp(2m, q)$ and $Gal(\phi_e, K_e) < Sp(2m, q)$.*

By various factorization arguments given in [A04], [AL1] and [AL2], in (3.6) of [AL2] we concluded with the following:

Supplemented Symplectic Rank Theorem (4.18). *For $1 \le e \le m-1$, the groups $Gal(f_e^\sharp, K_e(S))$ and $Gal(f_e, K_e)$ are transitive permutation groups of Rank 3 with subdegrees 1, $q\langle 2m - 3\rangle$ and q^{2m-1}, i.e., each of them is a transitive permutation group whose one-point stabilizer has orbits of sizes 1, $q\langle 2m - 3\rangle$ and q^{2m-1}. Moreover, if $m = 2 = q$ then $|Gal(f_1^\sharp, K_1(S))| \equiv 0$ (mod 6!) and $|Gal(f_1, K_1)| \equiv 0$ (mod 6!).*

Recall that an **antiflag** in a finite dimensional vector space is a pair of a vector and a one-codimensional subspace not containing it. A group of linear transformations is said to be **antiflag transitive** if it transitively permutes all antiflags. Concerning such groups, in Theorem II of [CaK], or in greater detail in Proposition (6.2) of [CaK], the following result is proved:

Theorem (4.19*) [Cameron-Kantor]. *Let $G < GSp(2m, q)$ be antiflag transitive and, upon letting Θ_{2m} be the natural epimorphism of $GL(2m, q)$ onto $PGL(2m, q)$, assume that $\Theta_{2m}(G)$ is a primitive Rank 3 permutation group, and also assume that in case of $m = 2 = q$ the group $\Theta_{2m}(G)$ is not isomorphic to the alternating group A_6. Then $Sp(2m, q) < G$. [Note that if $G < GSp(2m, q)$ is such that $\Theta_{2m}(G)$ is a transitive permutation group of Rank 3 then G is automatically antiflag transitive; the redundancy in our hypothesis is there because we are only stating a part of Theorem II of [CaK]; also note that we shall use this theorem in conjunction with (4.18), and then the primitivity assumption can be dispensed with, because a transitive permutation group of Rank 3 with subdegrees 1, $q\langle 2m - 3\rangle$ and q^{2m-1} is automatically primitive.]*

As asserted in (5.2) and (5.3) of [AL2], as a consequence of (4.7), (4.12), (4.17), (4.18) and (4.19*), we have deduced the following Theorem.

Theorem (4.19). *For $1 \le e \le m - 1$ we have the following.*

(4.19.1) If $GF(q) \subset k_p$ then in a natural manner we have $Gal(\phi_e, K_e) = Sp(2m, q)$ and $Gal(f_e, K_e) = PSp(2m, q)$.

(4.19.2) If $GF(q) \subset k_p$ and $GCD(t, q - 1) = 1$ then in a natural manner we have $Gal(\phi_e^\sharp, K_e(S)) = GSp(2m, q)$ and $Gal(f_e^\sharp, K_e(S)) = PGSp(2m, q)$.

(4.19.3) If $GF(q) \subset k_p$ and $GCD(t, q-1) = 1$ then, for every divisor d of $q - 1$, in a natural manner we have $Gal(\phi_e^{(d)}, K_e(S)) = GSp^{(d)}(2m, q)$ and $Gal(f_e^{(d)}, K_e(S)) = PGSp^{(d)}(2m, q)$.

For a moment, given any integer e with $1 \le e \le m - 1$, either let $\phi = \phi_e$ and $f = f_e$ and $K = K_e$, or assuming $GCD(t, q - 1) = 1$ let $\phi = \phi_e^\sharp$ and $f = f_e^\sharp$ and $K = K_e(S)$, or assuming $GCD(t, q - 1) = 1$ and given any divisor d of $q - 1$ let $\phi = \phi_e^{(d)}$ and $f = f_e^{(d)}$ and $K = K_e(S)$. Now upon letting

$k^* = k_p(\mathrm{GF}(q))$, by (1.1) we know that $k^* \subset \mathrm{SF}(\phi, K)$, and for any finite algebraic field extension k' of k^* in $\mathrm{SF}(\phi, K)$ we obviously have $[K(k') : K(k^*)] = [k' : k^*]$, and upon letting $\delta = \delta(k_p)$, we obviously have $\delta = \delta(K)$. In case of $\phi = \phi_e$, by (4.19.1) we have $\mathrm{Gal}(\phi, K(k')) = \mathrm{Sp}(2m, q) = \mathrm{Gal}(\phi, K(k^*))$ and $\mathrm{Gal}(f, K(k')) = \mathrm{PSp}(2m, q) = \mathrm{Gal}(f, K(k^*))$, and hence by (1.6) we get $\mathrm{AC}(k_p, \phi, K) = k^*$, and by (4.7.5) we get $\mathrm{Gal}(\phi, K) \in \Gamma'\mathrm{Sp}_\delta(2m, q)$ and $\mathrm{Gal}(f, K) \in \mathrm{P}\Gamma'\mathrm{Sp}_\delta(2m, q)$. Similarly, in case of $\phi = \phi_e^\natural$, by (4.19.2) we have $\mathrm{Gal}(\phi, K(k')) = \mathrm{GSp}(2m, q) = \mathrm{Gal}(\phi, K(k^*))$ and $\mathrm{Gal}(f, K(k')) = \mathrm{PGSp}(2m, q) = \mathrm{Gal}(f, K(k^*))$, and hence by (1.6) we get $\mathrm{AC}(k_p, \phi, K) = k^*$, and by (4.7.4) we get $\mathrm{Gal}(\phi, K) \in \Gamma\mathrm{Sp}_\delta(2m, q)$ and $\mathrm{Gal}(f, K) \in \mathrm{P}\Gamma\mathrm{Sp}_\delta(2m, q)$. Finally, in case of $\phi = \phi_e^{(d)}$, by (4.19.3) we have $\mathrm{Gal}(\phi, K(k')) = \mathrm{GSp}^{(d)}(2m, q) = \mathrm{Gal}(\phi, K(k^*))$ and $\mathrm{Gal}(f, K(k')) = \mathrm{PGSp}^{(d)}(2m, q) = \mathrm{Gal}(f, K(k^*))$, and hence by (1.6) we get $\mathrm{AC}(k_p, \phi, K) = k^*$, and by (4.12) we can find $\Lambda \in \mathrm{SF}(\phi, K)$ with $\Lambda^{(q-1)/d} = S$; now upon letting $K' = K(\Lambda)$ we obviously have $\delta(K') = \delta(K)$, and by the usual Galois correspondence we see that $\mathrm{Gal}(\phi, K'(k^*)) \lhd \mathrm{Gal}(\phi, K(k^*))$ with $\mathrm{Gal}(\phi, K(k^*))/\mathrm{Gal}(\phi, K'(k^*)) = \mathrm{Gal}(K'(k^*), K(k^*)) = Z_{(q-1)/d}$; since $p(\mathrm{GSp}^{(d)}(2m, q)) = \mathrm{Sp}(2m, q)$ with $\mathrm{GSp}^{(d)}(2m, q))/\mathrm{Sp}(2m, q) = Z_{(q-1)/d}$, we must have $\mathrm{Gal}(\phi, K'(k^*)) = \mathrm{Sp}(2m, q)$; hence $\mathrm{Gal}(\phi, K) \in \Gamma\mathrm{Sp}_\delta^{(d)}(2m, q)$ by (4.7.6) and therefore $\mathrm{Gal}(f, K) \in \Gamma\mathrm{PSp}_\delta^{(d)}(2m, q)$. Thus we have proved the following Theorem.

Theorem (4.20). *For $1 \le e \le m - 1$ we have the following.*

(4.20.1) We have $AC(k_p, \phi_e, K_e) = k_p(GF(q))$ and in a natural manner we have $Gal(\phi_e, K_e(GF(q))) = Sp(2m, q)$ and $Gal(f_e, K_e(GF(q))) = PSp(2m, q)$, and upon letting $\delta = \delta(k_p)$ we have $\delta = \delta(K_e)$ and in a natural manner we have $Gal(\phi_e, K_e) \in \Gamma'Sp_\delta(2m, q)$ and $Gal(f_e, K_e) \in P\Gamma'Sp_\delta(2m, q)$.

(4.20.2) If $GCD(t, q-1) = 1$ then we have $AC(k_p, \phi_e^\natural, K_e(S)) = k_p(GF(q))$ and in a natural manner we have $Gal(\phi_e^\natural, K_e(GF(q))(S)) = GSp(2m, q)$ and $Gal(f_e^\natural, K_e(GF(q))(S)) = PGSp(2m, q)$, and upon letting $\delta = \delta(k_p)$ we have $\delta = \delta(K_e(S))$ and in a natural manner we have $Gal(\phi_e^\natural, K_e) = \Gamma Sp_\delta(2m, q)$ and $Gal(f_e^\natural, K_e) = P\Gamma Sp_\delta(2m, q)$.

(4.20.3) If $GCD(t, q - 1) = 1$ then, for every divisor d of $q - 1$, we have $AC(k_p, \phi_e^{(d)}, K_e(S)) = k_p(GF(q))$ and in a natural manner we have $Gal(\phi_e^{(d)}, K_e(GF(q))(S)) = GSp^{(d)}(2m, q)$ and $Gal(f_e^{(d)}, K_e(GF(q))(S)) = PGSp^{(d)}(2m, q)$, and upon letting $\delta = \delta(k_p)$ we have $\delta = \delta(K_e(S))$ and in a natural manner we have $Gal(\phi_e^{(d)}, K_e) \in \Gamma Sp_\delta^{(d)}(2m, q)$ and $Gal(f_e^{(d)}, K_e) \in \Gamma PSp_\delta^{(d)}(2m, q)$.

Remark (4.21) [Symplectic Semilinearity]. Given any symplectic map $b : V \times V \to \mathrm{GF}(q)$ where V is any $2m$ dimensional vector space over $\mathrm{GF}(q)$, to verify the properties of the sets $\Gamma'\mathrm{Sp}_\delta(2m, q)$ and $\Gamma\mathrm{Sp}_\delta^{(d)}(2m, q)$ asserted between (4.7.4) and (4.7.5) where d and δ are any divisors of $q - 1$ and u respectively, we first note that the canonical epimorphism $\kappa : \Gamma\mathrm{Sp}(2m, q) \to \mathrm{Aut}(\mathrm{GF}(q))$ with kernel $\mathrm{GSp}(2m, q)$ splits; to see this, we take a symplectic basis (see page 24 of [KLi]) of V, i.e., a basis $(x_1, \ldots, x_m, y_1, \ldots, y_m)$ of V such that $b(x_i, x_j) = b(y_i, y_j) = 0$ for all i, j, and $b(x_i, y_j) = 1$ or 0 according as $i = j$ or $i \neq j$, and as on page 79 of [A02] we identify $\Gamma\mathrm{Sp}(2m, q)$ with $\{(g, \alpha) : g \in \mathrm{Aut}(\mathrm{GF}(q)) \text{ and } \alpha \in \mathrm{GSp}(2m, q)\}$, and we note that then κ is given by $(g, \alpha) \mapsto g$, and hence upon letting $\Gamma'(2m, q) = \{(g, 1) : g \in \mathrm{Aut}(\mathrm{GF}(q))\}$ we see that $\Gamma'(2m, q)$ is a subgroup of $\Gamma\mathrm{Sp}(2m, q)$ which is mapped isomorphically onto $\mathrm{Aut}(\mathrm{GF}(q))$ by κ. Let $\Gamma'_\delta(2m, q)$ be the unique subgroup of $\Gamma'(2m, q)$ of order δ. Then clearly $\Gamma\mathrm{Sp}_\delta(2m, q)$ is generated by $\Gamma'_\delta(2m, q)$ and $\mathrm{GSp}(2m, q)$. Upon letting $I'_\delta(2m, q)$ be the subgroup of $\Gamma\mathrm{Sp}_\delta(2m, q)$ generated by $\Gamma'_\delta(2m, q)$ and $\mathrm{Sp}(2m, q)$ we see that $I'_\delta(2m, q) \in \Gamma'\mathrm{Sp}_\delta(2m, q)$. Likewise, upon letting $J'^{(d)}_\delta(2m, q)$ be the subgroup of $\Gamma\mathrm{Sp}_\delta(2m, q)$ generated by $\Gamma'_\delta(2m, q)$ and $\mathrm{GSp}^{(d)}(2m, q)$ we see that $I'_\delta(2m, q) < J'^{(d)}_\delta(2m, q) \in \Gamma\mathrm{Sp}^{(d)}_\delta(2m, q)$. Obviously κ factors through the canonical epimorphism $\rho : \Gamma\mathrm{Sp}(2m, q) \to \Gamma\mathrm{L}(1, q)$ with kernel $\mathrm{Sp}(2m, q)$.

Let $\mathrm{GL}^{(d)}(1, q), \Gamma\mathrm{L}_\delta(1, q), \Gamma\mathrm{SL}_\delta(1, q), \Gamma\mathrm{L}^{(d)}_\delta(1, q), \Gamma(1, q), \Gamma_\delta(1, q), I_\delta(1, q)$ and $J_\delta(1, q)$ be as in (3.29). Now ρ induces the usual bijection between the set of subgroups of $\Gamma\mathrm{Sp}(2m, q)$ containing $\mathrm{Sp}(2m, q)$ and the set of all subgroups of $\Gamma\mathrm{L}(1, q)$. In particular we have $\Gamma\mathrm{Sp}^{(d)}_\delta(2m, q) = \{J < \Gamma\mathrm{Sp}(2m, q) : \rho(J) \in \Gamma\mathrm{L}^{(d)}_\delta(1, q)\}$. Also we have $\rho(\Gamma'\mathrm{Sp}_\delta(2m, q)) = \Gamma\mathrm{L}_\delta(1, q)$ and we have $\rho(\Gamma\mathrm{Sp}^{(d)}_\delta(2m, q)) = \Gamma\mathrm{L}^{(d)}_\delta(1, q)$. Now it is easy to see that $\Gamma\mathrm{SL}_\delta(1, q)$ is the set of all conjugates of $I_\delta(1, q)$ in $\Gamma\mathrm{L}(1, q)$, and $\Gamma\mathrm{L}^{(d)}_\delta(1, q)$ is the set of all conjugates of $J^{(d)}_\delta(1, q)$ in $\Gamma\mathrm{L}(1, q)$. ¿From this it follows that $\Gamma'\mathrm{Sp}_\delta(2m, q)$ is the set of all conjugates of $I'_\delta(2m, q)$ in $\Gamma\mathrm{Sp}(2m, q)$, and $\Gamma\mathrm{Sp}^{(d)}_\delta(2m, q)$ is the set of all conjugates of $J'^{(d)}_\delta(2m, q)$ in $\Gamma\mathrm{Sp}(2m, q)$. Clearly $I'_\delta(2m, q)$ is a split extension of $\mathrm{Sp}(2m, q)$ such that $\Gamma\mathrm{Sp}_\delta(2m, q)$ is generated by $\mathrm{GSp}(2m, q)$ and $I'_\delta(2m, q)$, and likewise $J'^{(d)}_\delta(2m, q)$ is a split extension of $\mathrm{GSp}^{(d)}(2m, q)$ such that $\Gamma\mathrm{Sp}_\delta(2m, q)$ is generated by $\mathrm{GSp}(2m, q)$ and $J'^{(d)}_\delta(2m, q)$.

Therefore $\Gamma'Sp_\delta(2m, q)$ is a nonempty complete set of conjugate subgroups of $\Gamma Sp(2m, q)$, and every I in $\Gamma'Sp_\delta(2m, q)$ is a split extension of $Sp(2m, q)$ such that $\Gamma Sp_\delta(2m, q)$ is generated by $GSp(2m, q)$ and I, and in the same manner $\Gamma Sp^{(d)}_\delta(2m, q)$ is a nonempty complete set of conjugate subgroups of $\Gamma Sp(2m, q)$, and every J in $\Gamma Sp^{(d)}_\delta(2m, q)$ is a split extension of $GSp^{(d)}(2m, q)$ such that $\Gamma Sp_\delta(2m, q)$ is generated by $GSp(2m, q)$ and J.

Remark (4.22) [Symplectic Similitude]. Referring to (4.20.1) and assuming $\mathrm{GF}(q) \subset k_p$, for $1 = e \leq m - 1$ we have the vectorial quintinomial $\phi_1 = Y^{q^{2m}} + T_1^q Y^{q^{m+1}} + XY^{q^m} + T_1 Y^{q^{m-1}} + Y$ which, in [A04], was our starting point of getting an unramified covering of the affine line over $k_p(T_1)$ having $\mathrm{Sp}(2m,q)$ as Galois group. Referring to (4.20.2) and assuming $\mathrm{GF}(q) \subset k_p$, in [AL1] and [AL2] this was modified to the vectorial quintinomial

$$\phi_1^{\parallel} = Y^{q^{2m}} + S^{r(m+1)}T_1^q Y^{q^{m+1}} + S^{r(m)}XY^{q^m} + S^{r(m-1)}T_1 Y^{q^{m-1}} + S^{r(0)}Y$$

with $\mathrm{GCD}(t, q-1) = 1$ giving an unramified covering of the once punctured affine line over $k_p(X, T_1)$ having $\mathrm{GSp}(2m,q)$ as Galois group. Referring to (4.20.1) but without assuming $\mathrm{GF}(q) \subset k_p$, in the excluded case of $e = 0$ we get the trinomial $\phi_0 = Y^{q^{2m}} + XY^{q^m} + Y$ which can be obtained by substituting $(2, q^m)$ for (m, q) in the trinomial ϕ_0^* mentioned in (2.4.1) and hence by (2.2.1) and (2.3.1) we see that: $\mathrm{AC}(k_p, \phi_0, K_0) = k_p(\mathrm{GF}(q))$ and $\mathrm{Gal}(\phi_0, K_0(\mathrm{GF}(q))) = \mathrm{SL}(2,q) = \mathrm{Sp}(2,q)$, and upon letting $\delta = \delta(k_p)$ we have $\delta = \delta(K_0)$ and $\mathrm{Gal}(\phi_0, K_0) \in \Gamma\mathrm{SL}(2,q) = \Gamma\mathrm{Sp}(2,q)$.

Note (4.23) [Division Points]. It may be noted that most of our equations are universal in the sense that the only coefficients involved are $0, 1, -1$. Hence it may be possible to "lift" some of them to characteristic zero without changing their Galois groups. In particular, in this manner, there should result nice explicit equations over the rational numbers having $\mathrm{GSp}(2m,q)$ as Galois group. Some of these may very well match (coincide) with some of the coverings which Serre [Se2] has recently obtained by adjoining the coordinates of division points of abelian varieties. Indeed, this unpublished work of Serre [Se2] was the motivation which led us to deform our original Sp equations into GSp equations. The origins of this work of Serre may be traced back to the 1870 book of Jordan where Jordan generalized Jacobi's elliptic function calculations to hyperelliptic function calculations. This was the birth of symplectic groups and explains why Jordan called them linear abelian groups in honor of Abel, who generalized hyperelliptic functions to abelian functions which are inverses of algebraic integrals. In his 1901 book [Dic] Dickson continued to follow Jordan's terminology. It was almost seventy years later that Weyl [Wey] renamed them symplectic groups. Jacobi's elliptic function calculations were brought to fruition by Serre in his 1972 paper [Se1]; see Theorem 19.1 on page 366 of Silverman's book [Sil] for a statement of Serre's result; this book of Silverman also outlines a procedure for obtaining equations of fields attached to division points of elliptic curves. Likewise, Jordan's hyperelliptic function calculations were greatly refined in Serre's 1985 unpublished work [Se2] which is the higher dimensional extension of his 1972 paper [Se1]. Now Drinfeld's work [Dri] on "Drinfeld Modules" seems to have been inspired by Serre's work [Se1] on division points of elliptic curves, eventually extended by him [Se2] to division points of abelian varieties. In turn, our description of the module $E^{[s]}$ in (3.30) is based on the ideas of Drinfeld Modules. For a discussion of Drinfeld Modules and their relationship with division points of elliptic curves and abelian varieties see Goss [Gos]. Very briefly, the roots of the separable vectorial q-polynomial E of q-degree $2m$ exhibited in (3.1) form a $2m$ dimensional $\mathrm{GF}(q)$-vector-space on which the Galois group of E acts. The said Galois group also acts on the roots of $E^{[s]}$ discussed in (3.30) which are the analogues of "s-division points of E." Indeed, we have used the letter E to remind ourselves of elliptic curves in case of $m = 1$ and more generally of $2m$ dimensional abelian varieties. Turning the table around, starting with vectorial polynomials, we may visualize an m dimensional abelian variety as a characteristic zero house to be built around a $2m$ dimensional $\mathrm{GF}(p)$-vector space. This will match with the fact that the p-division points of an m dimensional abelian variety over a field of characteristic different from p do form such a vector space.

REFERENCES

[A01]   S. S. Abhyankar, *Coverings of algebraic curves*, American Journal of Mathematics **79** (1957), 825-856.

[A02]   S. S. Abhyankar, *Galois theory on the line in nonzero characteristic*, Bulletin of the American Mathematical Society **27** (1992), 68-133.

[A03]   S. S. Abhyankar, *Nice equations for nice groups*, Israel Journal of Mathematics **88** (1994), 1-24.

[A04]   S. S. Abhyankar, *More nice equations for nice groups*, Proceedings of the American Mathematical Society **124** (1996), 2977-2991.

[A05]   S. S. Abhyankar, *Factorizations over finite fields*, Finite Fields and Applications, London Mathematical Society, Lecture Note Series **233** (1996), 1-21.

[A06]   S. S. Abhyankar, *Projective polynomials*, Proceedings of the American Mathematical Society **125** (1997), 1643-1650.

[A07]   S. S. Abhyankar, *Local fundamental groups of algebraic varieties*, Proceedings of the American Mathematical Society **125** (1997), 1635-1641.

[A08]   S. S. Abhyankar, *Semilinear transformations*, Proceedings of the American Mathematical Society, (To Appear).

[AL1]   S. S. Abhyankar and P. A. Loomis, *Once more nice equations for nice groups*, Proceedings of the American Mathematical Society **126** (1998), 1885-1896.

[AL2]   S. S. Abhyankar and P. A. Loomis, *Twice more nice equations for nice groups*, (To Appear).

[AS1]   S. S. Abhyankar and G. S. Sundaram, *Galois theory of Moore-Carlitz-Drinfeld modules*, C. R. Acad. Sci. Paris **325** (1997), 349-353.

[CaK]   P. J. Cameron and W. M. Kantor, *2-Transitive and antiflag transitive collineation groups of finite projective spaces*, Journal of Algebra **60** (1979), 384-422.

[Car]   L. Carlitz, *A class of polynomials*, Transactions of the American Mathematical Society **43** (1938), 167-182.

[Dri]   V. G. Drinfeld, *Elliptic Modules*, Math. Sbornik **94** (1974), 594-627.

[Dic]   L. E. Dickson, *Linear Groups*, Teubner, 1901.

[Gos]   D. Goss, *Basic Structures of Function Field Arithmetic*, Springer-Verlag, 1996.

[Hay]   D. Hayes, *A brief introduction to Drinfeld modules*, The Arithmetic of Function Fields (eds. D. Goss et al) (1992), 1-32.

[Jor]   C. Jordan, *Traité des Substitutions et des Équationes Algébriques*, Gauthier-Villars, Paris, 1870.

[KLi]   P. Kleidman and M. W. Liebeck, *The Subgroup Structure of the Finite Classical Groups*, Cambridge University Press, Cambridge, 1990.

[Mor]   E. H. Moore, *A two-fold generalization of Fermat's theorem*, Bulletin of the American Mathematical Society **2** (1896), 189-199.

[Se1]   J.-P. Serre, *Propriétés galoisiennes des points d'ordre fini des courbes elliptiques*, Inventiones Mathematicae **15** (1972), 259-331.

[Se2]   J.-P. Serre, *Résumé des cours et travaux*, Annuaire du Collège de France **85-86** (1985).

[Sil]   J. H. Silverman, *The Arithmetic of Elliptic Curves*, Springer Verlag, Berlin, 1986.

[Wey]   H. Weyl, *The Classical Groups*, Princeton University Press, Princeton, 1939.

Tools for the computation of families of coverings

Jean-Marc Couveignes[*]

March 5, 1999

Abstract

We list several techniques for efficient computation of families of coverings and we illutrate them on an example.

1 Introduction

The computation of algebraic models for coverings of the line is interesting both for theoretical reasons (e.g. the inverse Galois problem) and computational ones (as a test example for computer algebra tools). In many cases one reduces to solving a zero dimensional algebraic system (see [22, 2, 21] for many examples). This can be achieved using Buchberger algorithm. For many reasons, however, one would like to avoid using such an expensive algorithm from the point of view of complexity. In particular, the algebraic system one can associate to a covering does not provide a very sharp characterization and usually admits many solutions having nothing to do with the initial problem. Indeed, such a system may easily encode multiplicities but certainly not such discrete invariants as the monodromy group. On the other hand, famous work by Atkin and Swinnerton-Dyer achieves quite nontrivial computations using methods from numerical analysis [1] and similar methods were applied succesfully in different contexts. Some time ago Ralph Dentzer asked about how to compute an algebraic model for a covering of the sphere ramified above four points with monodromy group M_{24} given in [20]. This computation was achieved by Granboulan in [14] with a lot of numerical methods. At that time I collected several tricks and constructions in order to help with this computation, but this was not published because of the length of the result itself and also because the computational challenge appeared

---

[*]Délégation Générale pour l'Armement and Algorithmique Arithmétique Expérimentale, Université Bordeaux I

Key words: coverings, configuration spaces, stable curves, iterative methods

to be the most important. It seems that this information, however, may be of some use to other people performing similar computations. Also I have developed it a little bit further and I give in this work an illustration of it on a simple though non trivial example. It gave me an opportunity to consider these computations from a more conceptual point of view. In particular I realized the importance of explicit patching à la Harbater [15], and I tried to detail the algorithmic aspects of it for genus zero coverings. In this special case one can use the explicit description of the moduli spaces [12] to deal efficiently with *not necessarily Galois* coverings. The reason why patching is efficient is that it allows the computation of formal fibers without computing the extension of the base. The latter may really be huge in non-rigid cases (degree 144 in [14]). On the contrary, the extension of the basis is derived from the model for the fiber.

The paper is organized as follows. In section 2 we present down to earth techniques for computing an algebraic model for a covering. These techniques were known to Fricke. We illustrate them on simple examples. In section 3 we define the main family of coverings that will serve us as an example and start studying its combinatorial properties. The main point there is Hurwitz braid action. In section 4 and 5 we show how to compute an algebraic model for our family of coverings from the consideration of degenerate ones. Once a model has been computed for a degenerate cover, we first compute an analytic deformation of it which consists of a one parameter family of coverings, the parameter taking its values in a real interval. This is the purpose of section 4 and does not require more than linear algebra computations. From this analytic family we derive an algebraic one in section 5. This again reduces to linear algebra. In section 6 we give a more general and more conceptual description of our method. It uses the explicit description of the compactification of moduli spaces of curves given in [12] for the case of genus zero curves. We explain in particular how to compute in advance the degree of the coefficients that appear in the algebraic model we are looking for. Geometrically, these degrees are expressed in terms of the "thickness" of intersections in some formal curves.

It should be clear that the example we present is a toy that we chose for its simplicity and for such a small covering there exist simpler, faster methods.

The methods presented here apply to any genus zero covering of the sphere minus r points. Computations will be more difficult for higher genera, however, (except small values) because of the lack of a sufficiently explicit description for the corresponding moduli spaces.

We hope the reader will be convinced that the rich recent theory together with old computational methods make the computation of coverings much easier than it appears provided one does not rely too much on Buchberger's algorithm, as useful as it is.

I thank Helmut Völklein for his careful reading of a first version of this work. I thank Louis Granboulan and Lily Khadjavi for several useful corrections and comments.

2   Two coverings ramified over three points

In this section we shall compute an algebraic model for two coverings of the sphere minus three points. This will be useful in the next sections. We take this opportunity to recall how people have been efficiently computing simple coverings since the last century.

Let $\mathbb{P}_1(\mathbb{C})$ be the projective line over the field of complex numbers and let R_1, R_2, R_3 be three distinct points on it. By a coordinate on $\mathbb{P}_1(\mathbb{C})$ we mean a generator of its function field. For P a point and z a coordinate we denote by $z(P)$ the value of z at P. There is a unique coordinate z such that $z(R_1) = 0$, $z(R_2) = 1$, $z(R_3) = \infty$. Let β be the point with z-coordinate equal to $z(\beta) = i + 1/2$ and let Σ_1, Σ_2 and Σ_3 be the three loops represented on figure 1 in the plane with coordinate z.

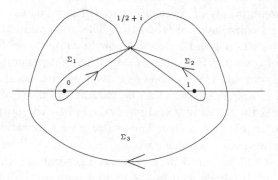

Figure 1: $\pi_1(\mathbb{P}_1 - \{R_1, R_2, R_3\}, b)$

Let ρ_1, ρ_2, ρ_3 be the three permutations below

$$
\begin{aligned}
\rho_1 &= [1,2,3,4,5,6,7], \\
\rho_2 &= [1,2], \\
\rho_3 &= (\rho_2 \rho_1)^{-1}.
\end{aligned}
$$

Consider the covering of $\mathbb{P}_1(\mathbb{C}) - \{R_1, R_2, R_3\}$ with monodromy (ρ_1, ρ_2, ρ_3) in the basis $(\Sigma_1, \Sigma_2, \Sigma_3)$.

The Riemann-Hurwitz formula shows that this is a genus zero covering that is a map $f : \mathbb{P}_1 \to \mathbb{P}_1$ unramified outside $\{R_1, R_2, R_3\}$. This map f can be represented as a rational fraction F provided we pick a coordinate on the left and a coordinate on the right. Let S_1 be the unique point above R_1 and S_2 the unique point with multiplicity 1 above R_3 and S_3 the unique point with multiplicity 6 above R_3. Let x be the coordinate which takes the values 0, 1 and ∞ at S_1, S_2, S_3 respectively.

We now can represent the covering f as a rational fraction $F(x) = z$ which satisfies

$$F(x) = K_1 \frac{x^7}{x-1}$$

with K_1 a constant and $F(x) - 1$ has a double zero i.e. $K_1 x^7 - (x-1)$ has a double zero. We therefore solve the system $\{K_1 x^7 - x + 1 = 0, 7 K_1 x^6 - 1 = 0\}$ and trivialy find the unique solution $x = 7/6$ and $K_1 = 6^6/7^7$. We thus get

$$F(x) = \frac{6^6}{7^7} \frac{x^7}{x-1} \tag{1}$$

and

$$F(x) - 1 = \frac{6^6}{7^7} \frac{(x - \frac{7}{6})^2 (x^5 + \frac{7}{3}x^4 + \frac{49}{12}x^3 + \frac{343}{54}x^2 + \frac{12005}{1296}x + \frac{16807}{1296})}{x - 1}. \tag{2}$$

We observe that the coefficients in the expressions above are rational. This could have been deduced *a priori* using a rigidity criterion.

The reader who is familiar with the theory of Grothendieck's *dessins d'enfant* as presented in [22] may like to see the *dessin* corresponding to f. It is the preimage of $[0, 1]$ by F. In figure 2, bullets correspond to points above 0 and arrows to points above 1.

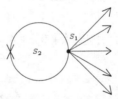

Figure 2: Dessin

We now consider a slightly more difficult example. Let τ_1, τ_2, τ_3 be the three permutations below

$$\begin{aligned}
\tau_1 &= [1], [2,3,4], [5], [6,7], \\
\tau_2 &= [1,2,5,6], \\
\tau_3 &= (\tau_2 \tau_1)^{-1}.
\end{aligned}$$

They define a genus zero covering $g : \mathbb{P}_1 \to \mathbb{P}_1$ which is unramified outside $\{R_1, R_2, R_3\}$ with monodromy (τ_1, τ_2, τ_3) in the basis $(\Sigma_1, \Sigma_2, \Sigma_3)$. We call T_1 the unique point with multiplicity 3 above R_1 and T_2 the unique point with multiplicity 4 above R_2 and T_3 the unique point above R_3. We take y to be the unique coordinate on $\mathbb{P}_1$ which takes values 1, 0 and ∞ at T_1, T_2 and T_3 respectively. We also call T_4 the unique point with multiplicity 2 above R_1.

The covering g may be represented as a polynomial function $G(y) = z$ which satisfies

$$G(y) = K_2(y-1)^3(y - y(T_4))^2 C(y) \text{ and } G(y) - 1 = K_2 y^4 A(y)$$

where $C(y)$ is a degree 2 monic polynomial and $A(y)$ is a degree 3 monic polynomial and K_2 is a constant. We set $C(y) = y^2 - ay + b$ and $A(y) = y^3 - cy^2 + dy - e$ and $y(T_4) = f$ and we write the identity

$$K_2(y-1)^3(y-f)^2(y^2 - ay + b) - 1 = K_2 y^4(y^3 - cy^2 + dy - e). \quad (3)$$

We differentiate the identity above with respect to the variable y and find

$$(y-f)(y-1)^2(7y^3+(-6a-4-5f)y^2+(4fa+2f+3a+5b)y-fa-2b-3fb)=y^3(7y^3-6cy^2+5dy-4e)$$

and since $f \neq 0$ we deduce that y^3 divides $7y^3 + (-6a - 4 - 5f)y^2 + (4fa + 2f + 3a + 5b)y - fa - 2b - 3fb$ and thus that

$$-6a - 4 - 5f = 0 \text{ and } 4fa + 2f + 3a + 5b = 0 \text{ and } fa + 2b + 3fb = 0.$$

We solve the system above and find that f is one of the three solutions of

$$10f^3 + 12f^2 + 9f + 4 = 0 \quad (4)$$

and

$$C(y) = y^2 - (-\frac{5}{6}f - \frac{2}{3})y + \frac{2}{3}f^2 + \frac{19}{30}f + \frac{2}{5}$$

and

$$A(y) = y^3 - (\frac{7}{6}f + \frac{7}{3})y^2 + (\frac{14}{5}f + \frac{7}{5})y - \frac{7}{4}f \quad (5)$$

and

$$K_2 = \frac{60}{7f - 4}. \quad (6)$$

Simple combinatorial considerations (see [5] or [6, Theorem 2]) show that the field of moduli of our covering is real. There is a single real solution to equation 4. We therefore take f to be this real solution. This finishes the computation of a model for the covering g. We draw the corresponding dessin on figure 3.

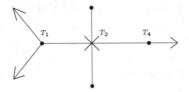

Figure 3: Dessin

3   Topological description

In this section we describe a simple family of coverings branched over 4 points and we start studying it from the point of view of combinatorics. We start with an integer $d = 7$ and four partitions of d, namely $\mathcal{P}_1 = \{1, 1, 2, 3\}$, $\mathcal{P}_2 = \{4, 1, 1, 1\}$, $\mathcal{P}_3 = \{2, 1, 1, 1, 1, 1\}$, and $\mathcal{P}_4 = \{6, 1\}$. Associated to these data we consider the set of isomorphism classes of connected coverings of the sphere ramified over four ordered points, of degree d and with ramification data given by the four above partitions in this order. The (not *a priori* connected) topological configuration space for such coverings is called a Hurwitz space. Its construction and finer ones are given in [16, 11, 10, 9, 23]. A nice introduction to these questions is [24, Chapter 10].

Our goal in this section is to obtain topological information on the Hurwitz space associated to the data above through a simple combinatorial study.

We say that two permutation vectors on d letters $(\zeta_1, \zeta_2, \zeta_3, \zeta_4)$ and $(\nu_1, \nu_2, \nu_3, \nu_4)$ are conjugate if there is a $u \in \mathcal{S}_d$ such that $\nu_i = {}^u\zeta_i$ for $i \in \{1, 2, 3, 4\}$.

We call $\tilde{\mathcal{P}}_i$ the conjugacy class in $\mathcal{S}_d$ associated to the partition $\mathcal{P}_i$. We first collect all vectors of permutations on d letters $(\zeta_1, \zeta_2, \zeta_3, \zeta_4)$ up to conjugacy, such that the following conditions hold

1. $\zeta_i \in \tilde{\mathcal{P}}_i$ for $i \in \{1, 2, 3, 4\}$

2. $\zeta_4 \zeta_3 \zeta_2 \zeta_1 = 1$

3. the ζ_i generate a transitive subgroup G of $\mathcal{S}_d$.

Being transitive of prime degree, G is primitive. Since it contains a transposition it must be the full symmetric group on 7 letters [25, Theorem 13.3].

There exist various formulae for counting vectors of permutations ([3, 4, 17, 13] among many others). Unfortunately we also need to exhibit all these vectors and the known methods do not provide an elegant feature for this task.

In general we may just do an exhaustive search with a computer. For the example under study, however, we can find all solutions by hand. A useful though trivial tool is the following "thickening" lemma. We first give a few natural definitions.

Definition 1 *Let d be a positive integer and S a permutation in S_d. Let $\mathcal{I}$ be a non empty subset of $\{1, 2, 3, ..., d\}$. We define a permutation $S_{|\mathcal{I}}$ in the following manner. For any $x \in \mathcal{I}$ we call K the smallest positive integer k such that $S^k(x)$ is in $\mathcal{I}$ and we set $S_{|\mathcal{I}}(x) = S^K(x)$. We call $[x, S(x), ..., S^{K-1}(x)]_{|\mathcal{I}}$ the tail of x. If x is not in $\mathcal{I}$ we set $S_{|\mathcal{I}}(x) = x$. We call $S_{|\mathcal{I}}$ the restriction of S to $\mathcal{I}$.*

Lemma 1 *Let ζ and S be two permutations in S_d and let $\mathcal{I} = Supp(\zeta)$ be the support of ζ and $S_{|\mathcal{I}}$ the restriction of S to it. Then the restriction $(\zeta S)_{|\mathcal{I}}$ of the product ζS is equal to the product $\zeta S_{|\mathcal{I}}$. This means that the product ζS is obtained from the product $\zeta S_{|\mathcal{I}}$ by replacing every x not fixed by ζ by its tail. Similarly $(S\zeta)_{|\mathcal{I}} = S_{|\mathcal{I}}\zeta$.*

This is useful when multiplying a fixed permutation by a permutation whose cycle lengths depend on a few parameters. For example

Corollary 1 *Let m, n, p be three positive integers. Then the product $[1, m + n + 1, m + 1] * [1, 2, ..., m + n][m + n + 1, ..., m + n + p]$ is $[1, 2, ..., m][m + 1, ..., m + n, m + n + 1, ..., m + n + p]$.*

Note that the restriction of a product is not the product of restrictions. However lemma 1 is of some theoretical interest. Indeed, consider the set $\mathcal{A}$ of infinite sequences of permutations $(\sigma_n)_{n \in \mathbb{N}, n > 0}$ such that $\sigma_n \in S_n$ and the restriction of σ_{n+1} to $\{1, 2, ..., n\}$ is σ_n. This is the inverse limit of the *sets* S_n with respect to the restriction maps. This $\mathcal{A}$ is not a group by the above remark. However, lemma 1 implies that the group S_∞ of permutations of the positive integers with bounded support acts on $\mathcal{A}$ by action on coordinates. Indeed for any permutation τ of degree m and any $\sigma = (\sigma_n)_n$ in $\mathcal{A}$ we define $\tau.\sigma$ as follows. For any $n \geq m$ set $\mu_n = \tau\sigma_n$ and for any $n < m$ take μ_n to be the restriction of μ_m to $\{1, 2, ..., n\}$. Then $\tau.\sigma = (\mu_n)_n$ is in $\mathcal{A}$. One can also define a right action. These actions clearly have no fixed points.

Elements of $\mathcal{A}$ admit a more geometric description. Consider pairs of the form $(\mathcal{U}, \iota)$ where $\mathcal{U}$ is a finite or enumerable disjoint union of oriented circles and ι is an injection of the set of positive integers into $\mathcal{U}$. Two such pairs $(\mathcal{U}_1, \iota_1)$ and $(\mathcal{U}_2, \iota_2)$ are said to be equivalent if and only if there is an orientation preserving homeomorphism h from $\mathcal{U}_1$ to $\mathcal{U}_2$ such that $\iota_2 = h \circ \iota_1$. We call such an equivalence class a propermutation. The left and right actions of S_∞ on propermutations can be seen as cutting and glueing circles.

As elementary as they are, these considerations allow mental computation with permutations.

We proceed as in [7] and find that there are exactly 48 vectors satisfying conditions 1, 2, 3 above (up to conjugacy). We give these vectors in the following definition in which a residue class modulo a positive integer N is identified with its smallest *positive* element.

Definition 2 *For any* $k \bmod 7$ *a residue class modulo* 7 *we denote by* a_k *the vector* $(\zeta_1, \zeta_2, \zeta_3, \zeta_4)$ *with*

$$
\begin{aligned}
\zeta_1 &= [1],[2],[3,4],[5,6,7], \\
\zeta_2 &= [1,2,3,5], \\
\zeta_3 &= [k \bmod 7, k+1 \bmod 7], \\
\zeta_4 &= (\zeta_3\zeta_2\zeta_1)^{-1}.
\end{aligned}
$$

We say that the 7 *such vectors form the A family.*

For any $k \bmod 7$ *a residue class modulo* 7 *we denote by* b_k *the vector* $(\zeta_1, \zeta_2, \zeta_3, \zeta_4)$ *with*

$$
\begin{aligned}
\zeta_1 &= [1],[2],[3,4,5],[6,7], \\
\zeta_2 &= [1,2,3,6], \\
\zeta_3 &= [k \bmod 7, k+1 \bmod 7], \\
\zeta_4 &= (\zeta_3\zeta_2\zeta_1)^{-1}.
\end{aligned}
$$

We say that the 7 *such vectors form the B family.*

For any $k \bmod 7$ *a residue class modulo* 7 *we denote by* c_k *the vector* $(\zeta_1, \zeta_2, \zeta_3, \zeta_4)$ *with*

$$
\begin{aligned}
\zeta_1 &= [1],[2,3,4],[5],[6,7], \\
\zeta_2 &= [1,2,5,6], \\
\zeta_3 &= [k \bmod 7, k+1 \bmod 7], \\
\zeta_4 &= (\zeta_3\zeta_2\zeta_1)^{-1}.
\end{aligned}
$$

We say that the 7 *such vectors form the C family.*

For any $k \bmod 3$ *a residue class modulo* 3 *we denote by* d_k *the vector* $(\zeta_1, \zeta_2, \zeta_3, \zeta_4)$ *with*

$$
\begin{aligned}
\zeta_1 &= [1],[2],[3,4],[5,6,7], \\
\zeta_2 &= [1,2,3,4], \\
\zeta_3 &= [k \bmod 3, 5], \\
\zeta_4 &= (\zeta_3\zeta_2\zeta_1)^{-1}.
\end{aligned}
$$

We say that the 3 *such vectors form the D family.*

For any $k \bmod 3$ *a residue class modulo* 3 *we denote by* e_k *the vector* $(\zeta_1, \zeta_2, \zeta_3, \zeta_4)$ *with*

$$
\begin{aligned}
\zeta_1 &= [1],[2],[3,4,5],[6,7], \\
\zeta_2 &= [1,2,3,5], \\
\zeta_3 &= [k \bmod 3, 6], \\
\zeta_4 &= (\zeta_3\zeta_2\zeta_1)^{-1}.
\end{aligned}
$$

We say that the 3 *such vectors form the E family.*

For any $k \bmod 5$ *a residue class modulo* 5 *we denote by* f_k *the vector* $(\zeta_1, \zeta_2, \zeta_3, \zeta_4)$ *with*

$$
\begin{aligned}
\zeta_1 &= [1],[2,3],[4,5,6],[7], \\
\zeta_2 &= [1,2,4,6], \\
\zeta_3 &= [k \bmod 5, 7], \\
\zeta_4 &= (\zeta_3\zeta_2\zeta_1)^{-1}.
\end{aligned}
$$

We say that the 5 *such vectors form the F family.*

For any k mod 5 *a residue class modulo* 5 *we denote by* g_k *the vector* $(\zeta_1, \zeta_2, \zeta_3, \zeta_4)$ *with*

$$
\begin{array}{rcl}
\zeta_1 & = & [1,2],[3],[4,5,6],[7], \\
\zeta_2 & = & [1,3,4,6], \\
\zeta_3 & = & [k \bmod 5, 7], \\
\zeta_4 & = & (\zeta_3\zeta_2\zeta_1)^{-1}.
\end{array}
$$

We say that the 5 *such vectors form the* G *family.*

For any k mod 5 *a residue class modulo* 5 *we denote by* h_k *the vector* $(\zeta_1, \zeta_2, \zeta_3, \zeta_4)$ *with*

$$
\begin{array}{rcl}
\zeta_1 & = & [1],[2,3,4],[5,6],[7], \\
\zeta_2 & = & [1,2,5,6], \\
\zeta_3 & = & [k \bmod 5, 7], \\
\zeta_4 & = & (\zeta_3\zeta_2\zeta_1)^{-1}.
\end{array}
$$

We say that the 5 *such vectors form the* H *family.*

For any k mod 5 *a residue class modulo* 5 *we denote by* i_k *the vector* $(\zeta_1, \zeta_2, \zeta_3, \zeta_4)$ *with*

$$
\begin{array}{rcl}
\zeta_1 & = & [1,2,3],[4],[5,6],[7], \\
\zeta_2 & = & [1,4,5,6], \\
\zeta_3 & = & [k \bmod 5, 7], \\
\zeta_4 & = & (\zeta_3\zeta_2\zeta_1)^{-1}.
\end{array}
$$

We say that the 5 *such vectors form the* I *family.*

We now compute the action of braids on these 48 vectors. To this end we consider the configuration space $X_{0,4} = \mathbb{P}_1^4 - \Delta$ of spheres minus four pairwise distinct points. A point $Q = (Q_1, Q_2, Q_3, Q_4)$ in $X_{0,4}$ corresponds to the sphere $\mathbb{P}_1 - (Q_1, Q_2, Q_3, Q_4)$. If Z is a coordinate on $\mathbb{P}_1$ we denote by $Z(Q)$ the vector $(Z(Q_1), Z(Q_2), Z(Q_3), Z(Q_4))$ and we call it the Z-coordinate of Q. We pick such a coordinate Z and choose as a base point for $X_{0,4}$ the point $P = (P_1, P_2, P_3, P_4)$ with Z-coordinate $Z(P) = (0, 1, 2, \infty)$. We also choose a base point b on the corresponding sphere $\mathbb{P}_1 - (P_1, P_2, P_3, P_4)$. We take for b the whole upper half plane in the Z-coordinate. This makes sense because the upper half plane is a contractible set. We also pick generators $(\Gamma_1, \Gamma_2, \Gamma_3, \Gamma_4)$ for $\pi_1(\mathbb{P}_1 - \{P_1, P_2, P_3, P_4\}, b)$ as on figure 4.

The fundamental group $\pi_1(X_{0,4}, P)$ is generated by braids $t_{1,2}$, $t_{2,3}$, $t_{3,4}$ defined in the classical way. For example $t_{1,2}$ is represented by the map $t_{1,2}(u)$ with Z-coordinates $Z(t_{1,2}(u)) = (1/2 - 1/2e^{2i\pi u}, 1/2 + 1/2e^{2i\pi u}, 2, \infty)$ for $u \in [0, 1]$. The action on monodromy vectors in these basis is then given by $t_{1,2}((\zeta_1, \zeta_2, \zeta_3, \zeta_4)) = (^{\zeta_2\zeta_1}\zeta_1, ^{\zeta_2\zeta_1}\zeta_2, \zeta_3, \zeta_4)$. See [16]. Straightforward calculation then gives the following fact.

Fact 1 *For any* k *a residue modulo* 7

$$
\begin{array}{rcl}
t_{1,2}(a_k) & = & a_{k-1} \\
t_{1,2}(b_k) & = & b_{k-1} \\
t_{1,2}(c_k) & = & c_{k-1}.
\end{array}
$$

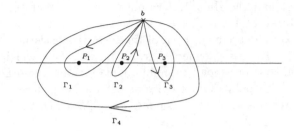

Figure 4: $\pi_1(\mathbb{P}_1 - \{P_1, P_2, P_3, P_4\}, b)$ in the Z-coordinate

For any k a residue modulo 3

$$t_{1,2}(d_k) = d_{k-1}.$$

For any k a residue modulo 4

$$t_{1,2}(e_k) = e_{k-1}.$$

For any k a residue modulo 5

$$t_{1,2}(f_k) = f_{k-1}$$
$$t_{1,2}(g_k) = g_{k-1}$$
$$t_{1,2}(h_k) = h_{k-1}$$
$$t_{1,2}(i_k) = i_{k-1}.$$

The action of $t_{2,3}$ on our 48 vectors is given below (trivial cycles are omitted)

$$[B_1,B_2,C_5][C_1,A_1,A_2][C_7,I_1,D_2,H_5,C_6][A_3,A_4,D_1,I_4,I_5]$$

$$[A_7,F_1,F_2,E_3,A_5][B_7,H_1,H_2,D_3,B_6][B_3,B_5,E_1,G_3,G_4][C_2,C_4,G_1,E_2,F_4].$$

The above fact gives us the combinatorial description of the configuration space $\mathcal{M}$ parametrizing our family. Since the coverings we consider have no automorphisms we even have a covering of universal curves:

$$\begin{array}{ccc} \mathcal{M} & \xleftarrow{\Gamma} & \mathcal{T} \\ \Lambda \downarrow & & \downarrow \Phi \\ X_{0,4} & \longleftarrow & X_{0,5} \end{array}$$

Following [8] we embed the moduli space of spheres minus four points $M_{0,4} = \mathbb{P}_1 - \{R_1, R_2, R_3\}$ in the configuration space $X_{0,4}$ as the subvariety of points (R_1, R_2, Q, R_3) for $Q \in \mathbb{P}_1 - \{R_1, R_2, R_3\}$. The restriction of Λ to the curve $M_{0,4}$ is a covering of curves also called $\Lambda : \mathcal{H} \to \mathbb{P}_1 - \{R_1, R_2, R_3\}$. The curve $\mathcal{H}$ is the moduli space of our family of coverings and is often called a Hurwitz space. This Hurwitz space is mapped by Λ onto the moduli space $M_{0,4}$ of spheres minus four points.

From fact 1 and the Hurwitz formula we deduce that the curve $\mathcal{H}$ has genus zero. Recall that z is the coordinate such that $z(R_1) = 0$, $z(R_2) = 1$ and $z(R_3) = \infty$. The preimage by Λ of the segment consisting of points with z-coordinates in $[1, \infty]$ is a connected graph on the sphere which we represent below. The bullets correspond to points above ∞ and the other vertices to points over 1. The points above 0 are associated with faces. To any point above ∞ corresponds one of the nine families A, B, C, D, E, F, G, H, and I.

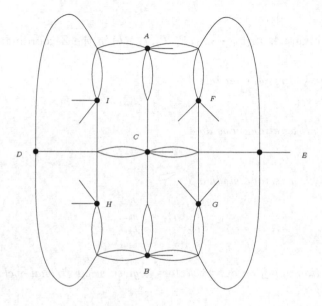

Figure 5: The Hurwitz space $\mathcal{H}$

4   Patching

In this section we compute an analytic model for the family of coverings presented in section 3.

Let again $\mathbb{P}_1(\mathbb{C})$ be the projective line over the field of complex numbers and let U_1, U_2, U_3, U_4 be four distinct points on it. We call Z the unique coordinate such that $Z(U_1) = 0$, $Z(U_2) = 1$, $Z(U_4) = \infty$. We also set $\lambda = Z(U_3)$ and $\chi = 1/\lambda$. For convenience we introduce another coordinate $W = Z/\lambda$ such that $W(U_1) = 0$, $W(U_2) = \chi$, $W(U_3) = 1$, and $W(U_4) = \infty$.

Assume first that λ is a real greater than 1 and let us choose as a base point b the whole upper half plane in the Z-coordinates (which is also the upper half plane in the W-coordinate since $Z = \lambda W$ and λ is real positive). We also pick generators $(\Theta_1, \Theta_2, \Theta_3, \Theta_4)$ for $\pi_1(\mathbb{P}_1(\mathbb{C}) - \{U_1, U_2, U_3, U_4\}, b)$ as represented on figure 6. Note that these data depend continuously on λ. For $\lambda = 2$ we find ourselves in the situation of figure 4.

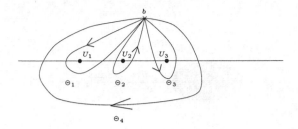

Figure 6: $\pi_1(\mathbb{P}_1(\mathbb{C}) - \{U_1, U_2, U_3, U_4\}, b)$ in the Z-coordinate

We define $\mu = K_3\chi^{1/7}$ where K_3 is a constant that will be chosen most conveniently latter and $\chi^{1/7}$ is the unique real seventh root of χ. The reason for introducing this μ is that we are going to study the Hurwitz space $\mathcal{H}$ locally at the point C represented on figure 5. This point corresponds to the C family and it is mapped onto $R_3 \in \mathbb{P}_1$ by the Hurwitz map Λ. The ramification index of Λ at C is 7. Therefore μ is a local parameter at C on the Hurwitz space and we expect all coordinates arising in the algebraic model we are looking for to be Laurent series in μ.

For any $\mu \in]0, 1[$ we call $\phi_\mu : \mathbb{P}_1 \to \mathbb{P}_1$ the covering with monodromy $c_5 = (\zeta_1, \zeta_2, \zeta_3, \zeta_4)$

$$
\begin{aligned}
\zeta_1 &= [1], [2, 3, 4], [5], [6, 7], \\
\zeta_2 &= [1, 2, 5, 6], \\
\zeta_3 &= [1, 5], \\
\zeta_4 &= (\zeta_3\zeta_2\zeta_1)^{-1}.
\end{aligned}
$$

in the basis of $\pi_1(\mathbb{P}_1 - \{U_1, U_2, U_3, U_4\}, b)$ given in figure 6. Again, one can show that this covering is defined over the field of real numbers (see [5, 6]). We call V_1 the unique point above U_1 with multiplicity 2 and V_2 the unique point above U_1 with multiplicity 3. We call V_3 the unique point above U_2 with multiplicity 4 and V_4 the unique point above U_3 with multiplicity 2 and V_5 the unique point above U_4 with multiplicity 1 and V_6 the unique point above U_4 with multiplicity 6.

We now try to understand what happens when μ tends to zero.

If we look at things from the point of view of W-coordinates we see that $W(U_2) = \chi$ tends to $0 = W(U_1)$ while $W(U_3) = 1$ and $W(U_4) = \infty$. In the limit we get a sphere minus three points $U_1 = U_2$, U_3 and U_4 and a fundamental group $\pi_1(\mathbb{P}_1(\mathbb{C}) - \{U_1 = U_2, U_3, U_4\}, b)$ generated by $\Gamma_{1,2} = \Gamma_1\Gamma_2, \Gamma_3, \Gamma_4$ as on figure 7. Indeed, when U_1 and U_2 coalesce, turning around the resulting point is just turning around U_1 then U_2. Topologically, this is equivalent to punching a big grey hole containing U_1 and U_2 or equivalently removing the segment $[U_1, U_2]$. Turning around this big hole is equivalent to turning around U_1 and then around U_2.

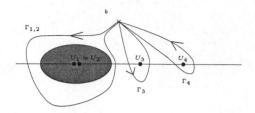

Figure 7: $\pi_1(\mathbb{P}_1 - \{U_1 = U_2, U_3, U_4)\}, b)$

We conclude that when μ tends to zero the covering ϕ_μ tends to a covering of the sphere minus three points with monodromy $(\zeta_2\zeta_1, \zeta_3, \zeta_4)$ in the basis $(\Gamma_{1,2}, \Gamma_3, \Gamma_4)$ of $\pi_1(\mathbb{P}_1 - \{U_1 = U_2, U_3, U_4\}, b)$ shown on figure 7. Coming back to section 2 we see that $\zeta_2\zeta_1 = \rho_1$ and $\zeta_3 = \rho_2$ and $\zeta_4 = \rho_3$. Therefore ϕ_μ tends to f when μ tends to zero.

In order to take advantage of this we shall write down a model for ϕ_μ. As usual, we must chose coordinates on each side. As for the right-hand side we shall of course consider W-coordinates. We note that when μ tends to zero W tends to the coordinate z of section 2 since it takes values 0, 1 and ∞ at the three ramification points of the limit covering.

On the left-hand side we must pin three points to 0, 1 and ∞. We must be careful to choosing three points that do remain distinct when μ tends to zero. For example, points above U_1 and U_2 may well coalesce since U_1 and U_2 coalesce. On the other hand, a point above U_1 and a point above U_3 will not coalesce. Two points above U_3 will not coalesce either. This corresponds *mutatis mutandis* to the notion of *admissible* families of points introduced in Définition 3 of [7]. The key mathematical idea underneath is to be found in sections 2 and 3 of [11].

In our situation we see that $\{V_3, V_5, V_6\}$ form an admissible family because V_5 and V_6 map to U_4 which does not coalesce to any other point. Indeed when μ tends to zero then V_3 tends to the point S_1 of section 2 because V_3 is above U_2 and U_2 tends to R_1 and S_1 is the unique point above R_1. Similarly V_5 tends to S_2 because V_5 is above U_4 and U_4 tends to R_3 and the multiplicity of V_5 is equal to 1 and is not affected as μ tends to zero because U_4 is simple (i.e. does not coalesce) in the W-coordinate, see [11]. Also V_6 tends to S_3. We call X the coordinate which takes values 0, 1 and ∞ at V_3, V_5 and V_6 respectively. Then X tends to the coordinate x of section 2.

The covering ϕ_μ is now represented by a rational fraction $\Phi_\mu(X) = W$ such that

$$\Phi_\mu(X) = \frac{K_4(X^2 - r_1 X + r_2)(X - X(V_1))^2(X - X(V_2))^3}{X - 1} \tag{7}$$

We shall denote by ν_μ the valuation associated to μ. We know that $\nu_\mu(\chi) = 7$ and from the above

$$\nu_\mu(r_2) + 2\nu_\mu(X(V_1)) + 3\nu_\mu(X(V_2)) = 7. \tag{11}$$

On the other hand since V_3 and all the points above U_1 coalesce we know that $\nu_\mu(r_2)$, $\nu_\mu(X(V_1))$ and $\nu_\mu(X(V_2))$ are positive integers. We deduce that $\nu_\mu(r_2) = 2$ and $\nu_\mu(X(V_1)) = \nu_\mu(X(V_2)) = 1$. We write

$$X(V_2) = \mu(v_{2,0} + v_{2,1}\mu + v_{2,2}\mu^2 + ...)$$

In order to complete the picture we now look at the situation from the point of view of the coordinate Z. We see that $Z(U_3) = \lambda$ tends to $\infty = Z(U_4)$ when μ tends to 0 while $Z(U_1) = 0$ and $Z(U_2) = 1$.

At the end we get a sphere minus three points U_1, U_2 and $U_3 = U_4$ and a fundamental group $\pi_1(\mathbb{P}_1(\mathbb{C}) - \{U_1, U_2, U_3 = U_4\}, b)$ generated by $\Gamma_1, \Gamma_2, \Gamma_{3,4} = \Gamma_3\Gamma_4$ as on figure 8. Indeed, when U_3 and U_4 coalesce, turning around the resulting point is ·just turning around U_3 then U_4. Topologically, this is equivalent to punching a big grey hole containing U_3 and U_4 or equivalently removing the segment $[U_3, U_4]$. Turning around this big hole is equivalent to turning around U_3 and then around U_4.

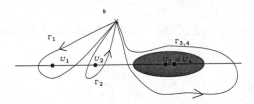

Figure 8: $\pi_1(\mathbb{P}_1 - \{U_1, U_2, U_3 = U_4)\}, b)$

We conclude that when μ tends to zero the covering ϕ_μ tends to a covering of the sphere minus three points with monodromy $(\zeta_1, \zeta_2, \zeta_4\zeta_3)$ in the basis $(\Gamma_1, \Gamma_2, \Gamma_{3,4})$ of $\pi_1(\mathbb{P}_1 - \{U_1, U_2, U_3 = U_4\}, b)$ shown on figure 8. Coming back to section 2 we see that $\zeta_1 = \tau_1$ and $\zeta_2 = \tau_2$ and $\zeta_4\zeta_3 = \tau_3$. Therefore ϕ_μ tends to g when μ tends to zero.

In order to take advantage of this we shall write down a model for ϕ_μ with adapted coordinates on each side. As for the right-hand side we shall of course consider Z-coordinates. We note that when μ tends to zero Z tends to the coordinate z of section 2 since it takes values 0, 1 and ∞ at the three ramification points of the limit covering.

On the left-hand side we must pin down again three points that do not pairwise coalesce, for example $\{V_3, V_2, V_6\}$. We pick a coordinate Y that takes values 0, 1 and ∞ at V_3, V_2 and V_6 respectively. This implies that X/Y is constant equal to $X(V_2)/Y(V_2)$. We set $\kappa = \frac{X(V_2)}{\mu} = v_{2,0} + O(\mu)$ and we have

$$X = \mu \kappa Y \tag{12}$$

Note also that Y tends to the coordinate y of section 2.

The covering ϕ_μ can be now represented by a rational fraction $\Psi_\mu(Y) = Z$ which is related to $\Phi_\mu(X) = W$ by $Z = W\lambda$ and $X = \mu\kappa Y$. We replace X by $\mu\kappa Y$ in equations 7, 8 and divide out by χ and find

$$\Psi_\mu(Y) = \frac{K_3^7 K_4 \kappa^7 (Y^2 - \frac{r_1}{\mu\kappa}Y + \frac{r_2}{\mu^2\kappa^2})(Y - \frac{X(V_1)}{\mu\kappa})^2 (Y-1)^3}{\mu\kappa Y - 1} \tag{13}$$

and

$$\Psi_\mu(Y) - 1 = \frac{K_3^7 K_4 \kappa^7 Y^4 (Y^3 - \frac{s_1}{\mu\kappa}Y^2 + \frac{s_2}{\mu^2\kappa^2}Y - \frac{s_3}{\mu^3\kappa^3})}{\mu\kappa Y - 1} \tag{14}$$

As μ tends to zero Ψ_μ tends to the rational fraction G given in section 2. Comparing the leading coefficients in 13 and 3, and using 6 we find

$$K_3^7 \frac{6^6}{7^7} v_{2,0}^7 = \frac{60}{7f - 4}.$$

We now take advantage of the freedom in chosing K_3 assuming K_3 is the unique real root of

$$K_3^7 = \frac{60.7^7}{6^6(7f - 4)}$$

so that

$$v_{2,0}^7 = 1.$$

Since our covering is real we deduce that $v_{2,0} = 1$ so

$$X(V_2) = \mu(1 + O(\mu)).$$

Comparing 13, 14 and 5 we find

and

$$\Phi_\mu(X) - \chi = \frac{K_4 X^4 (X^3 - s_1 X^2 + s_2 X - s_3)}{X - 1} \tag{8}$$

and

$$\Phi_\mu - 1 = \frac{K_4 (X - X(V_4))^2 (X^5 - u_1 X^4 + u_2 X^3 - u_3 X^2 + u_4 X - u_5)}{X - 1} \tag{9}$$

Where K_4, r_1, r_2, s_1, s_2, s_3, u_1,...,u_5 depend on μ.

As μ tends to zero the covering ϕ_μ tends to f while the coordinates X and W tend to x and y. Therefore the rational fraction Φ_μ tends to F. From the comparison of formulae 1 and 2 on the one hand and 7, 8, 9 on the other hand we deduce the following

$$
\begin{aligned}
r_1 &= O(\mu) \\
r_2 &= O(\mu) \\
X(V_1) &= O(\mu) \\
X(V_2) &= O(\mu) \\
s_1 &= O(\mu) \\
s_2 &= O(\mu) \\
s_3 &= O(\mu) \\
X(V_4) &= \frac{7}{6} + O(\mu) \\
u_1 &= -\frac{7}{3} + O(\mu) \\
u_2 &= \frac{49}{12} + O(\mu) \\
u_3 &= -\frac{343}{54} + O(\mu) \\
u_4 &= \frac{12005}{1296} + O(\mu) \\
u_5 &= -\frac{16807}{1296} + O(\mu) \\
K_4 &= \frac{6^6}{7^7} + O(\mu)
\end{aligned}
$$

Further if we set $X = 0$ in equations 7 and 8 we find

$$\chi = K_4 r_2 X(V_1)^2 X(V_2)^3. \tag{10}$$

$$r_1 = \mu(-\frac{2}{3} - \frac{5}{6}f + O(\mu))$$

$$r_2 = \mu^2(\frac{2}{5} + \frac{19}{30}f + \frac{2}{3}f^2 + O(\mu))$$

$$X(V_1) = \mu(f + O(\mu))$$

$$X(V_4) = \mu(1 + O(\mu))$$

$$s_1 = \mu(\frac{7}{3} + \frac{7}{6}f + O(\mu))$$

$$s_2 = \mu^2(\frac{7}{5} + \frac{14}{5}f + O(\mu))$$

$$s_3 = \mu^3(\frac{7}{4}f + O(\mu))$$

This time we have first order approximations for all the coefficients arising in our algebraic model. Computing higher order approximations now reduces to linear algebra by Hensel's Lemma. We just plug formal developements into equations 7, 8, 9 and develop in the variable μ. We get a non singular linear system of equations in the next order terms etc.

For example we find

$$X(V_2) = \mu - (8/21 + 5f/42)\mu^2 +$$
$$+(101/2205 + 529f/4410 - 5f^2/252)\mu^3 + \ldots$$
$$X(V_1) = f\mu - (5f/21 + 11f^2/42)\mu^2 +$$
$$+(29/4410 + 229f/17640 + 811f^2/4410)\mu^3 + \ldots$$

5   Looking for algebraic dependencies

In this section we shall derive an algebraic model from the analytic one obtained in section 4.

5.1   General procedure

In section 4 we obtained an analytic model for the covering ϕ_μ with coefficients in the complete field $\mathbb{Q}(f)((\mu))$. We know that this model is actually defined over the algebraic closure of $\mathbb{Q}(\lambda)$ in $\mathbb{Q}(f)((\mu))$. We shall now look for algebraic dependencies between the various coefficients arising in the expression for ϕ_μ. This will give us a model for the Hurwitz space $\mathcal{H}$ defined in section 3.

We pick two functions ($X(V_1)$ and $X(V_2)$ for example) and look for algebraic dependencies between them. Let us call $\mathbb{C}(\mathcal{H})$ the field of functions on the curve $\mathcal{H}$ (our Hurwitz space). This is a genus zero function field over $\mathbb{C}$.

To ease notations we shall set $v_1 = X(V_1)$ and $v_2 = X(V_2)$. Assume the degree of the extension $\mathbb{C}(\mathcal{H})/\mathbb{C}(v_1)$ is d_1 and the degree of $\mathbb{C}(\mathcal{H})/\mathbb{C}(v_2)$ is d_2. Then there exists a polynomial $E(X_1, X_2)$ in two variables, with coefficients in $\mathbb{C}$ and degree in X (resp. in Y) equal to d_2 (resp. d_1) such that $E(v_1, v_2) = 0$. If we know the expansions of v_1 and v_2 with enough accuracy (i.e. more than $(d_1 + 1)(d_2 + 1)$) finding such a relation is just a matter of linear algebra. We try successive increasing values for d_1 and d_2. For $d_1 = d_2 = 3$ we find

$$
\begin{aligned}
E(v_1, v_2) &= -36v_2v_1^3 + 24v_2^3 + 60v_1^3 + 54v_2^2v_1 + 72v_1^2v_2 - 18v_2^3v_1 - 36v_2^2v_1^2 \\
&= 0. \tag{15}
\end{aligned}
$$

We shall see in section 6 that the degrees d_1 and d_2 can be computed *a priori* from the monodromy ζ. We therefore need not compute infinitly many terms in our expansions. With bounded accuracy we can obtain enough dependencies to determine all the coefficients in our equation. This will provide us with a proof that the equation holds since we *a priori* know that such an equation does exist. Considerations in section 6 will enable us to choose functions like v_1 and v_2 that make the degrees d_1 and d_2 minimal.

The curve given by equation 15 is expected to be of genus 0. We therefore look for a parametrization using an algorithm due to Noether, Poincaré and Vessiot— see [18] for a complete algorithmic survey on this question, including problems of rationality. In our case, of course, finding a parameter is particularly trivial since 15 has a unique triple point.

We find that $T = v_2/v_1$ is a parameter and

$$
v_1 = \frac{10 + 12T + 9T^2 + 4T^3}{3T(2 + 2T + T^2)} \quad \text{and } v_2 = Tv_1.
$$

Let d_0 be the degree of the field extension $\mathbb{C}(\mathcal{H})/\mathbb{C}(v_1, v_2)$. We may reasonably expect this d_0 to be small (the irreducibility of equation 15 implies $d_0 = 1$). Let d_3 be the degree of $\mathbb{C}(\mathcal{H})/\mathbb{C}(r_1)$. We now look for algebraic dependencies between T and r_1 with degree d_0 in r_1 and d_3 in T. For $d_0 = 1$ and $d_3 = 6$ we find

$$
r_1 = -\frac{2(T + 4)(4T^3 + 9T^2 + 12T + 10)(T + 1)^2}{3(T^4 + 6T^3 + 21T^2 + 16T + 6)}.
$$

Similarly we find

$$
r_2 = \frac{(4T^3 + 9T^2 + 12T + 10)^2}{3(T^2 + 2T + 2)(T^4 + 6T^3 + 21T^2 + 16T + 6)}
$$

and this is enough for our purpose since $\mathbb{C}(\mathcal{H}) = \mathbb{C}(v_1, v_2) = \mathbb{C}(T)$.

Indeed we set $\gamma_T(X) = (X^2 - r1X + r2)(X - v_1)^2(X - v_2)^3/(X - 1)$ and factor its derivative with respect to X. This derivative has roots v_1, v_2 with multiplicity 2, 0 with multiplicity 3 and $v_4 = X(V_4)$. We deduce an expression for v_4 :

$$v_4 = \frac{2(T^6 + 6T^5 + 21T^4 + 56T^3 + 51T^2 + 30T + 10)}{3T(T^4 + 6T^3 + 21T^2 + 16T + 6)}.$$

We have $K_4 = 1/\gamma_T(v_4)$, $\chi = \gamma_T(0)/\gamma_T(v_4)$, and $\Phi_T(X) = \gamma_T(X)/\gamma_T(v_4)$. Note that we now write Φ_T rather than Φ_μ since T is the right algebraic parameter.

The singular value χ is given as a rational fraction in T which is unramified outside $\{0, 1, \infty\}$ i.e. a Belyi function. The associated dessin is the one on figure 5 and the monodromy is the one described in section 3. Note in particular that the factorisation of χ and $\chi - 1$ fits with fact 1. Indeed $\chi = \frac{\chi_0}{\chi_\infty}$ and $\chi - 1 = \frac{\chi_1}{\chi_\infty}$ with

$$\chi_0 = -(T^4 + 6T^3 + 21T^2 + 16T + 6)^5 T^4 (4T^3 + 9T^2 + 12T + 10)^7$$

and

$$\chi_\infty = (T-1)^5(T^4 + 8T^3 + 36T^2 + 40T + 20)^4(T+1)^2(2T^2 + T + 2)^2(2T^4 + 8T^3 + 15T^2 + 8T + 2)^3$$
$$(T^3 + 3T^2 + 6T + 10)(2T^6 + 12T^5 + 51T^4 + 94T^3 + 111T^2 + 60T + 20)$$

and

$$\chi_1 = -16(-80 - 288T - 48T^2 + 2432T^3 + 7896T^4 + 13776T^5 + 15656T^6 + 12432T^7 + 6711T^8 + 2222T^9$$
$$+ 477T^{10} + 60T^{11} + 4T^{12})(2 + 2T + T^2)^3(T^6 + 6T^5 + 21T^4 + 56T^3 + 51T^2 + 30T + 10)^5.$$

and

$$K_4 = -729 \frac{T^6(T^4 + 6T^3 + 21T^2 + 16T + 6)^6(T^2 + 2T + 2)^6}{\chi_\infty}.$$

We can now replace T by values in $\mathbb{Q}$ and find coverings defined over $\mathbb{Q}$ in our family. The key point here is that our family is defined over $\mathbb{Q}$. More precisely the Hurwitz space is irreducible and defined over $\mathbb{Q}$ and further has many rational points. To check our computations, we just make sure that formulae 7, 8 and 9 hold.

5.2   Using numerical approximations

In the previous paragraph we computed an algebraic model for our family of coverings from an analytic one using linear algebra computations over the field $\mathbb{Q}(f)$. Indeed we were dealing with series in $\mathbb{Q}(f)((\mu))$ with bounded accuracy. The computations, however, will be greatly accelerated if we work in $\mathbb{C}((\mu))$

instead of $\mathbb{Q}(f)((\mu))$ approximating complex numbers to some fixed accuracy. This accuracy depends now on the height of the equations we expect to find. We have no nice upper bound for this height. We just try. We compute an estimate for f and write down expansions for v_1 and v_2 with approximate coefficients in $\mathbb{C}$. For the linear algebra part (looking for dependencies) we no longer use Gauss algorithm but least squares. This gives a vector $\mathcal{V}$ that minimizes the L_2-norm of $M\mathcal{V}-b$ for a given matrix M and vector b. Of course least squares always give a solution and we need a criterion for this solution to be relevant, depending on the accuracy. We may give a quantified criterion but there is a very simple and efficient qualitative one: something interesting is happening if and only if a small perturbation in the data (i.e. close to zero according to current accuracy) induces a big gap in the minimal norm (i.e. change of magnitude). We thus obtain approximations for the coefficients in the linear equation we are looking for. Since these coefficients are expected to belong to $\mathbb{Q}(f)$ we then look for integer linear relations between 1, θ, θ^2 and any such coefficient c. This is achieved using the famous LLL algorithm [19]. Indeed such a relation corresponds to a small vector in the orthogonal lattice to the vector $(N, \lfloor N\theta \rfloor, \lfloor N\theta^2 \rfloor, \lfloor Nc \rfloor)$ where N is a large integer (close to the inverse of the accuracy). And LLL is designed for finding such small vectors. Another possibility is to perform all the computations using successively all the conjugates of f in $\mathbb{C}$ and then form the symmetric functions of the results. We then obtain approximations of *rational* numbers and may find their exact values thanks to continued fraction algorithm. This is the strategy adopted in [1].

There is an important variant to the method described above. Once we have computed Φ_μ as a rational fraction with coefficients in $\mathbb{C}((\mu))$ with a small accuracy (in the μ-adic topology) we may stop there the computation of expansions and replace μ by a small complex number μ_0. This will give us an approximation according to the ordinary absolute value in $\mathbb{C}$ of the rational fraction Φ_{μ_0} branched at $\{0, 1, \infty, \lambda_0\}$. We then plug this approximation into equations 7, 8 and 9 (where λ is fixed to the value λ_0) and apply an iterative numerical method like Newton's method to get an arbitrarily accurate approximation. We may then replace λ_0 by a very close value λ_1 in equations 7 and 8. The rational fraction Φ_{μ_0} is then a close approximation of a rational fraction Φ_{μ_1} branched at $\{0, 1, \infty, \lambda_1\}$ and the latter one is found by applying Newton's method to equations 7 and 8 with λ fixed to λ_1 and initial value Φ_{μ_0} for Φ_{μ_1}. We can move slowly this way in our Hurwitz space and list vectors of values of the various functions r_1, r_2, v_1,... at many points in it. If Q_1, Q_2, ..., Q_k are these points we represent any such function by an element in the algebra $\mathbb{C}^k$ and we can look for algebraic relations as before. We have replaced Taylor expansion at one point by interpolation at many points. We may of course mix the two approaches. We have to be careful that our points Q_i should be well distributed on the Hurwitz space. Otherwise we are going to loose much accuracy since too points that are close to each other lead to almost the same equations. We use least squares as before with the same criterion for testing the relevance of the result. This approach was used succesfully in [14].

6   Stable curves

In this section we shall give elements for the generalization of the method presented in section 4. We shall also explain how one can compute the degrees of coefficients in some algebraic model for a family of coverings by mere consideration of its monodromy. This will be useful when looking for algebraic depencies since it will tell us what is the degree of the relations we are looking for. In particular we shall be able to select functions that satisfy algebraic relations with smallest possible degree.

The key point in section 4 was that the degeneracy of a sphere minus four points could be seen from two different points of views (namely according to Z or W coordinates). There is a standard way to reconcile these two points of views. It is connected with the compactification of the moduli space of genus zero r-pointed curves. This compactification is very explicitly described in [12] in terms of trees of projective lines. In this section we shall assume that the reader has some familiarity with this work. We just recall that the authors construct a fine moduli space for r-pointed curves from the consideration of all possible cross-ratios between four marked points on the sphere. One defines the cross-ratio $[U_1, U_2, U_3, U_4]$ as $(z_3 - z_1)(z_4 - z_2)/(z_3 - z_2)/(z_4 - z_1)$ where z_1, z_2, z_3, z_4 are the value of any coordinate z at U_1, U_2, U_3, U_4. Note that $[0, \infty, x, 1] = x$. The relations between these cross-ratios are of two types. Those coming from the action of the symmetric group $\mathcal{S}_4$ on the cross-ratio $[U_1, U_2, U_3, U_4]$ by permutation of the indices plus a relation involving five points

$$[U_2, U_5, U_3, U_4].[U_1, U_2, U_3, U_4] = [U_1, U_5, U_3, U_4].$$

This last relation is connected with the algebraic group law on $\mathbb{P}_1 - \{U_3, U_4\}$. If we consider the projective variant of these equations (introducing numerators and denominators for all cross-ratios) we obtain a projective variety $\bar{M}_{0,r}$ which contains the moduli space $M_{0,r}$ as an open subset.

There also exists a universal curve $\bar{M}_{0,r+1} \to \bar{M}_{0,r}$ corresponding to "forgetting the last point".

The degenerate curves can be described as follows. Assume that we have a projective line minus four points U_1, U_2, U_3, U_4 and let U_1 and U_2 coalesce. The object we get at the end is obtained in the following way : replace U_1 and U_2 by another line crossing the first one and put U_1 and U_2 on it.

Now let $A = \mathbb{C}[[\mu]]$ be the local ring of Laurent series in the parameter μ and $Q = \mathbb{C}((\mu))$ its quotient ring. Let $\phi_\mu : \mathcal{C}_\mu \to \mathbb{P}_1 - \{U_1, U_2, U_3, U_4\}$ be a covering defined over Q (i.e. a family of coverings parametrized by the

Figure 9: A 4-pointed tree of two projective lines

local parameter μ). We assume that the cross-ratio $[U_1, U_4, U_2, U_3]$ is equal to μ^e for some positive integer e. Let W be the coordinate that takes values 0, μ^e, 1, ∞ at U_1, U_2, U_3, U_4 respectively. Let $Z = W/\mu^e$ be the coordinate that takes values 0, 1, μ^{-e}, ∞ at U_1, U_2, U_3, U_4 respectively. The associated field extension to ϕ_μ is $Q(W) \subset Q(\mathcal{C}_\mu)$. By restricting μ to values in $[0,1]$ we may define the monodromy $\zeta = (\zeta_1, \zeta_2, \zeta_3, \zeta_4)$ of ϕ_μ as in section 4. We assume $(\zeta_2\zeta_1)^e = 1$. We can always reduce to this case after base change (i.e. replacing μ by $\mu^{\frac{1}{o}}$ for some integer o). We now consider the ring $\mathcal{R}_0$ generated by W and $T = 1/Z$ over A. This ring $\mathcal{R}_0$ is equal to $\mathbb{C}[[\mu]][W, T]/(WT - \mu^e)$. Let also $\mathcal{R}_1 = A[W]$ and $\mathcal{R}_2 = A[T]$. Then $Spec(\mathcal{R}_1)$ and $Spec(\mathcal{R}_2)$ glue together along $Spec(\mathcal{R}_0)$ to form a fibered surface $\mathcal{S}$ over $Spec(A)$ the generic fiber of which is a smooth curve of genus zero while the special fiber $\mathcal{S}_s$ is made of two genus zero curves $\mathcal{V}$ and $\mathcal{W}$ crossing at the point O with coordinates $W = T = \mu = 0$. The local ring at O is $\mathbb{C}[[\mu, W, T]]/(WT - \mu^e)$. The Zarisky closures of the points U_1, U_2, U_3 and U_4 define horizontal divisors on $\mathcal{S}$. We represent the situation in figure 10.

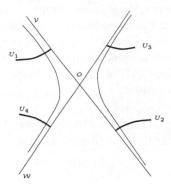

Figure 10: The fibered surface $\mathcal{S}$

We may now take the normal closure of $\mathcal{S}$ in the field extension $Q(W) \subset Q(\mathcal{C}_\mu)$ and obtain a fibered surface $\mathcal{T}$ over $Spec(A)$ and a map $\varphi : \mathcal{T} \to \mathcal{S}$ whose generic fiber is just ϕ_μ. The point is that $\mathcal{T}$ can be described quite sharply in terms of the monodromy ζ. First of all, the special fiber $\mathcal{T}_s$ is a connected curve made of several irreducible components meeting each other transversally and with no other singularity than these crossing points between distinct irreducible components.

The irreducible components of $\mathcal{T}$ fall in two parts. Those that are mapped onto $\mathcal{V}$ by φ which we call $\mathcal{V}_i$ for $1 \leq i \leq I$ and those that are mapped onto $\mathcal{W}$ which we call $\mathcal{W}_j$ for $1 \leq j \leq J$. The components $\mathcal{V}_i$ correspond to the orbites $\mathcal{O}_i$ of $(\zeta_1, \zeta_2, \zeta_4\zeta_3)$ and the restriction of φ to $\mathcal{V}_i$ is a covering ramified over three points with monodromy $(\zeta_1, \zeta_2, \zeta_4\zeta_3)|_{\mathcal{O}_i}$ where the $|$ means restriction to $\mathcal{O}_i$. Similarly the components $\mathcal{W}_i$ correspond to orbites $\mathcal{Q}_i$ of $(\zeta_2\zeta_1, \zeta_3, \zeta_4)$. Two components $\mathcal{V}_i$ and $\mathcal{W}_j$ intersect if they share a cycle of $\zeta_2\zeta_1 = (\zeta_4\zeta_3)^{-1}$ and these cycles are in bijection with the crossing points O_k for $1 \leq k \leq K$ on $\mathcal{T}_s$. These points are the points above O.

If the crossing point O_k of $\mathcal{V}_i$ and $\mathcal{W}_j$ is associated to a cycle of length ℓ_k in $\zeta_2\zeta_1$ then the local ring at O_k is isomorphic to $\mathbb{C}[[\mu, w, t]]/(wt - \mu^{\frac{e}{\ell_k}})$ where t and w are local parameters at O_k on $\mathcal{V}_i$ and $\mathcal{W}_j$ respectively. The ratio $\theta_k = \frac{e}{\ell_k}$ is called the thickness of the intersection point O_k. Note that θ_k is an integer because we assumed $(\zeta_2\zeta_1)^e = 1$. Note also that θ_k depends on the local parameter μ in the sense that if we replace μ by $\mu^{\frac{1}{o}}$ for some integer o (base change) the thickness θ_k is multiplied by o.

The points mapped to U_1 and U_2 by ϕ_μ correspond to cycles of ζ_1 and ζ_2 and their Zarisky closures in $\mathcal{T}$ cross the $\mathcal{V}_i$'s while the points mapped to U_3 and U_4 by ϕ_μ correspond to cycles of ζ_3 and ζ_4 and their Zarisky closures in $\mathcal{T}$ cross the $\mathcal{W}_j$'s.

In case the generic genus is zero, the special fiber $\mathcal{T}_s$ is a r-pointed tree of projective lines as in [12] and $\mathcal{T}$ is a deformation of it over $Spec(A)$.

We show two examples of such a situation. These examples will be two degeneracies of the covering ϕ_T studied before.

Assume first that ζ is the monodromy c_5 studied in section 4, corresponding to point C on figure 5. Both $(\zeta_1, \zeta_2, \zeta_4\zeta_3)$ and $(\zeta_2\zeta_1, \zeta_3, \zeta_4)$ have a single orbite. Therefore we have a single component $\mathcal{V}_1$ above $\mathcal{V}$ and a single component $\mathcal{W}_1$ above $\mathcal{W}$. Since $\zeta_2\zeta_1$ is a full 7-cycle we have $\ell_1 = 7$. On the other hand, the order e of the braid $t_{1,2}$ acting on ζ is also 7. Thus the thickness of the unique point O_1 above O is just $\theta_1 = 1$.

We draw the corresponding situation on figure 11.

Assume now that ζ is the monodromy d_1 given in definition 2.

$$\begin{aligned}
\zeta_1 &= [1],[2],[3,4],[5,6,7], \\
\zeta_2 &= [1,2,3,4], \\
\zeta_3 &= [1,5], \\
\zeta_4 &= (\zeta_3\zeta_2\zeta_1)^{-1}.
\end{aligned}$$

This monodromy corresponds to point D in figure 5. Now $(\zeta_1, \zeta_2, \zeta_4\zeta_3)$ has two orbites $\mathcal{O}_1 = \{1, 2, 3, 4\}$ and $\mathcal{O}_2 = \{5, 6, 7\}$ while $(\zeta_2\zeta_1, \zeta_3, \zeta_4)$ has two orbites $\mathcal{Q}_1 = \{1, 2, 3, 5, 6, 7\}$ and $\mathcal{Q}_2 = \{4\}$.

We have three points O_1, O_2 and O_3 above O corresponding to the three cycles $(1, 2, 3)$, $(5, 6, 7)$ and (4) of $\zeta_2\zeta_1$.

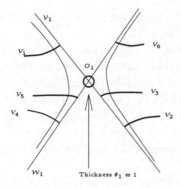

Figure 11: The special fiber at point $C \in \mathcal{H}$

The order e of braid action is 3 thus the thicknesses are $\theta_1 = 1$, $\theta_2 = 1$, $\theta_3 = 3$.

We draw the corresponding picture on figure 12.

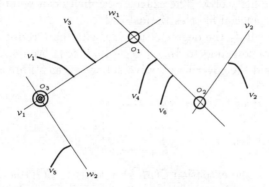

Figure 12: The special fiber at point $D \in \mathcal{H}$

Thicknesses are very useful to compute the order of vanishing (μ-adic valuation) of cross-ratios. For example, in figure 12 the thickness θ_1 of O_1 is nothing but the valuation of the cross-ratio $[V_4, V_1, V_6, V_3]$.

More generally, recall that associated to each component $\mathcal{K}$ of a stable tree of projective lines there is a projection $P_{\mathcal{K}}$ of the full curve onto this component. There is also a unique *median* component $\mathcal{L}_\delta$ associated to any triple of non-singular points $\delta = (\delta_1, \delta_2, \delta_3)$. This component is the unique one on which the three points have pairwise distinct projections.

If we have four points V_1, V_2, V_3, V_4 crossing the special fiber at non-singular distinct points, then there are only two possibilities.

Either the intersections of these points with the special fiber have distinct projections on some component of it. In this case the cross-ratio $\rho = [V_1, V_3, V_2, V_4]$ is a unit in $A = \mathbb{C}[[\mu]]$ and so is $\rho - 1$. This is the case for V_1,

V_3, V_4, V_5 on figure 12 since their instersections with the special fiber have distinct projections on $\mathcal{V}_1$.

If the intersections of V_1, V_2, V_3, V_4 don't project on distinct points on any component, then there are two components $\mathcal{K}_1$ and $\mathcal{K}_2$ with associated projections $P_{\mathcal{K}_1}$ and $P_{\mathcal{K}_2}$ such that for example $P_{\mathcal{K}_1}(V_1), P_{\mathcal{K}_1}(V_2), P_{\mathcal{K}_1}(V_3)$ are pairwise distinct while $P_{\mathcal{K}_1}(V_3) = P_{\mathcal{K}_1}(V_4)$ and $P_{\mathcal{K}_2}(V_3), P_{\mathcal{K}_2}(V_4), P_{\mathcal{K}_2}(V_1)$ are pairwise distinct while $P_{\mathcal{K}_2}(V_1) = P_{\mathcal{K}_2}(V_2)$. In this case, the cross-ratio $[V_1, V_3, V_2, V_4]$ has μ-adic valuation equal to the distance between $\mathcal{K}_1$ and $\mathcal{K}_2$ which is defined as the sum of the thicknesses of all intersections between $\mathcal{K}_1$ and $\mathcal{K}_2$.

In particular, we see that the multiplicity of this cross-ratio is bounded by e times the number of intersections. And there are no more intersections than the degree d_ϕ of the covering ϕ_μ.

We deduce that if we have a genus zero covering ϕ of degree d_ϕ of $\mathbb{P}_1$ ramified over four points U_1, U_2, U_3, U_4 and if we pick V_1, V_2, V_3, V_4 above U_1, U_2, U_3, U_4 respectively, then the cross-ratio $\rho = [V_1, V_3, V_2, V_4]$ only vanishes when $\lambda = [U_1, U_3, U_2, U_4]$ vanishes and the multiplicity of ρ is at most d_ϕ times the multiplicity of λ. The exact multiplicity can be obtained from the geometry of the special fiber as we just did.

Let us now call d_Λ the degree of the Hurwitz map from the moduli space $\mathcal{H}$ for a family of coverings to $M_{0,4} = \mathbb{P}_1 - \{0, 1, \infty\}$. We represent below this Hurwitz map and the covering of universal curves as a fibration above it.

$$
\begin{array}{ccc}
\mathcal{H} & \xleftarrow{\ \Gamma\ } & \mathcal{T} \\
{\scriptstyle \Lambda}\big\downarrow & & \big\downarrow{\scriptstyle \Phi} \\
M_{0,4} & \longleftarrow & M_{0,5}
\end{array}
$$

The degree of the extension $\mathbb{C}(M_{0,4}) = \mathbb{C}(\lambda) \subset \mathbb{C}(\mathcal{H})$ is d_Λ and from the calculation above we deduce that the degree of the extension $\mathbb{C}(\rho) \subset \mathbb{C}(\mathcal{H})$ is bounded above by $d_\Lambda \times d_\phi$.

We thus have a bound for the degree of coefficients appearing in some algebraic model for ϕ.

This bound is indeed very pessimistic since it assumes that we always have many orbites and thick intersections between them. In practice one rather expects degrees like d_Λ/d_ϕ. In our example we have $d_\phi = 7$ and $d_\Lambda = 48$ so that our estimate gives 48×7. If we look at formulae in section 5 we find that all quantities are rational fractions of degree $6 = 48/7$ of the parameter.

In all cases, the extensive combinatorial study of all degenerations of a genus zero covering gives the divisor of zeros and poles of any cross-ratio between branched points. One can then cook a multiplicative combination of these cross-ratios with smallest possible degree. This amounts to finding the shortest vector in the linear space generated by the divisors of all cross-ratios.

Incidentally we obtain a criterion for the Hurwitz space $\mathcal{H}$ to be rational : it is rational if there exists a (combination of) cross-ratio with a single zero.

This may be checked easily on the monodromy ζ in some cases.

The method presented here applies to any genus zero covering of the sphere minus r points. When r is bigger than 4 one has to consider maximally degenerate covers. The local ring at the corresponding points in the moduli space $\bar{M}_{0,r}$ is generated by cross-ratios. The ramification at the minimal primes corresponding to these cross-ratios is computed in terms of braid action and the order of vanishing of cross-ratios of branched points along the corresponding divisors can be evaluated as before. This gives a method for computing the *partial* degrees of any coefficient in the algebraic model.

7 Conclusion

We have shown how the computation of a family of coverings can efficiently reduce to the computation of degenerate coverings of the sphere minus three points. The most degenerate will be the better. Reciprocally, if we want to compute a covering of the sphere minus three points, we may find it as a special fiber in a family of higher dimension. The latter family may well be easy to compute provided it admits an(other) simple special fiber. We also have given a method for computing the degree of a coefficient c in some algebraic model (the degree of $\mathbb{C}(\mathcal{H})/\mathbb{C}(c)$) in terms of the monodromy ζ. These degrees depend on the geometry of the degenerate coverings in the family.

References

[1] A.O.L. Atkin and H.P.F. Swinnerton-Dyer. Modular forms over noncongruence subgroups. In *Proceedings of symposia in pure mathematics*, volume 19. AMS, 1971.

[2] Bryan Birch. Noncongruence subgroups, covers and drawings. In Leila Schneps, editor, *The Grothendieck Theory of Dessins d'Enfants*, volume 200 of *Lecture Notes in Math.* Cambridge University Press, 1994.

[3] G. Boccara. Nombre de représentations d'une permutation comme produit de deux cycles de longueur donnée. *Discrete Math.*, 29:105–134, 1980.

[4] G. Boccara. Cycles comme produit de deux permutations de classes données. *Discrete Math.*, 38:129–142, 1982.

[5] Kevin Coombes and David Harbater. Hurwitz families and arithmetic Galois groups. *Duke mathematical journal*, 52:821–839, 1985.

[6] J.-M. Couveignes and L. Granboulan. Dessins from a geometric point of view. In L. Schneps, editor, *The Grothendieck theory of dessins d'enfants*. Cambridge University Press, 1994.

[7] Jean-Marc Couveignes. Quelques revêtements définis sur $\mathbb{Q}$. *Manuscripta mathematica*, 94-4, 1997.

[8] P. Debès and M. D. Fried. Nonrigid constructions in Galois Theory. *Pacific Journal of Math*, 163(1):81–122, 1994.

[9] M. D. Fried and H. Völklein. The inverse Galois problem and rational points on moduli spaces. *Math. Ann.*, 290:771–800, 1991.

[10] Michael D. Fried. Fields of definition of function fields and Hurwitz families–Groups as Galois groups. *Comm. Alg.*, 5:17–82, 1977.

[11] W. Fulton. Hurwitz schemes and irreducibility of moduli of algebraic curves. *Ann. Math.*, 90:542–575, 1969.

[12] L. Gerritzen, F. Herrlich, and M. van der Put. Stable n-pointed trees of projective lines. *Ind. Math.*, 50:131–163, 1988.

[13] I.P. Goulden and D.M. Jackson. The combinatorial relationship between trees and certain connection coefficients for the symmetric group. *European J. Combin.*, 13:357–365, 1992.

[14] Louis Granboulan. Construction d'une extension régulière de $\mathbb{Q}(t)$. *Experimental Math.*, 5:1–13, 1996.

[15] D. Harbater. Galois coverings of the arithmetic line. *Lecture Notes in Math.*, 1240:165–195, 1987.

[16] A. Hurwitz. Über Riemann'sche Flächen mit gegebenem Verzweigungspunkten. *Math. Annalen*, 39:1–61, 1891.

[17] D.M. Jackson. Some combinatorial problems associated with products of conjugacy classes of the symmetric group. *J. Combin. Theory Ser. A*, 2:363–369, 1988.

[18] Henri Lebesgue. *Leçons sur les constructions géométriques*. Gauthier-Villars, 1950.

[19] H. Lenstra, A. Lenstra, and L. Lovasz. Factoring polynomials with rational coefficients. *Math. Ann.*, 261:515–534, 1982.

[20] G. Malle and B. H. Matzat. Action of Braids. In *Inverse Galois Theory*, chapter 3. Preprint. University of Heidelberg, 1993.

[21] B.H. Matzat. *Konstructive Galoistheorie*. Springer, 1987.

[22] Leila Schneps. Dessins d'enfant on the Riemann sphere. In Leila Schneps, editor, *The Grothendieck Theory of Dessins d'Enfants*, volume 200 of *Lecture Notes in Math*. Cambridge University Press, 1994.

[23] Helmut Völklein. Moduli spaces for covers of the Riemann sphere. *Israel J. of Math*, 85:407–430, 1994.

[24] Helmut Völklein. *Groups as Galois groups – an Introduction*. Cambridge Studies in Advanced Math. Cambridge Univ. Press, 1996.

[25] Helmut Wielandt. *Finite permutation groups*. Academic Press, 1964.

Some Arithmetic Properties
of Algebraic Covers

PIERRE DÈBES

ABSTRACT. Consider a Galois extension F/K and an algebraic cover f: $X \to B$ a priori defined over F. The cover f may have several models (and possibly none) over each given subfield E of F. How do these models compare to each other? Are there better models than others? We establish here a structure result for the set of all various models which can be used to investigate these questions. The structure, which is of cohomological nature, yields an interesting arithmetical tool: K-covers can be 'twisted' to provide other K-models with possibly better properties. One application is concerned with the Beckmann-Black problem. E. Black conjectures that each Galois extension E/K is the specialization of a Galois branched cover of $\mathbb{P}^1$ defined over K with the same Galois group G. We show the conjecture holds when G is abelian and K is an arbitrary field; this was known for number fields from results of S. Beckmann (1992) and E. Black (1995). Other applications include discussions of existence for a given cover, of a "good" model, a stable model, a model with a totally rational fiber, etc. Also we clarify an inaccurracy in a result of Fried about field of moduli and extensions of constants. Finally, we continue our study of local-global principles for covers: if a cover is defined over each $\mathbb{Q}_p$, does it follow it is defined over $\mathbb{Q}$? Here we consider the case of Galois covers of a general base space B.

1. Introduction

This paper is organized as follows. §2 is devoted to the structure result mentioned in the abstract for the set of all K-models of a given cover f : $X \to B$ a priori defined over a Galois extension of K. There are two versions. In the first one (§2.4) we assume that the base space B of the cover satisfies a certain condition introduced in [DeDo1] and called the (Seq/Split) condition. That condition, which is recalled in §2.3, holds for example if the base space has unramified K-rational points. The general case of the structure result is given in §2.5. §2.1-§2.3 review some basics relative to the arithmetic of covers.

§2 is used in §3 to study two questions about the realization of covers with some arithmetical constraint on some fiber. The first one (§3.1) is known as the Beckmann-Black problem: given a field K, is every finite Galois extension E/K the specialization of a Galois branched cover of $\mathbb{P}^1$

which is defined over K and has the same Galois group? Beckmann and Black answered positively when G is abelian and K is a number field. We extend this result to arbitrary fields K. The second question (§3.2) is about the existence for a given K-cover of an unramified totally K-rational fiber. We comment on this property: in particular we recapitulate what is known about it and review several situations where it revealed helpful.

§4 is about field of moduli (§4.2) and extension of constants (§2.5.1). Given a K-cover f_K, consider, on one hand, the Galois closure $\widehat{f_K}$ of f_K over K and, on the other hand, the Galois closure $\widehat{f}$ of $f_K \otimes_K K_s$ over the separable closure K_s of K. Th.4.1 (§4.3) relates the extension of constants of f_K in $\widehat{f_K}$ and the field of moduli of $\widehat{f}$ as G-cover. Two consequences are given (§4.4). The first one is a criterion for the field of moduli of $\widehat{f}$ to be a field of definition (as G-cover). The second one clarifies an inacurracy in a result of Fried about extensions of constants in a cover with centerless group. A counter-example to Fried's original statement is given in §4.1.

The theme of §5 is the local-to-global principle for covers: if a cover is defined over each $\mathbb{Q}_p$, does it follow it is defined over $\mathbb{Q}$? The state of the question is recalled in §5.1. Essentially the local-to-global principle holds for G-covers; and for mere covers, it holds under some additional hypotheses on the monodromy group and is conjectured not to hold in general. We consider the case of mere covers that are Galois over $\overline{\mathbb{Q}}$. The local-to-global principle is known to hold then if the base space B satisfies the (Seq/Split) condition. Here we consider the general case, that is, we do not assume the (Seq/Split) condition. Our main result appears in §5.3. Our approach uses the notion of Galois covers given with the action of a subgroup of the automorphism group. These objects, called SG-covers, and which generalize both mere covers and G-covers, are introduced in §5.2.

NOTATION — Given a Galois extension E/k, its Galois group is denoted by $G(E/k)$. Given a field k, we denote by k_s a separable closure of k and by $G(k)$ the absolute Galois group $G(k_s/k)$.

2. Structure result for models of a cover

2.1. Mere covers and G-covers. The main topic of this paper is the arithmetic of covers. There are classically two situations. One is concerned with not necessarily Galois covers — traditionally called "mere covers" — while the other one considers "G-covers", *i.e.*, Galois covers given with the Galois action.

Given a field K, a regular projective geometrically irreducible variety B defined over K, mere covers $f : X \to B$ over K correspond (*via* an equivalence of categories) to finite separable regular field extensions $K(X)/K(B)$ while G-covers of B of group G over K correspond to regular Galois extensions $K(X)/K(B)$ given with an isomorphism of the Galois group $G(K(X)/K(B))$ with G.

We denote the variety B with the reduced ramification divisor D removed by B^* and the K-arithmetic fundamendal group of B^* by $\Pi_K(B^*)$ or simply by Π_K when the context is clear. Degree d mere covers of B over F, unramified over B^* correspond to transitive representations $\Psi : \Pi_F(B^*) \to S_d$ such that the restriction to $\Pi_{F_s}(B^*)$ is transitive. G-covers of B of group G over F correspond to surjective homomorphisms $\Phi : \Pi_F(B^*) \to G$ such that $\Phi(\Pi_{F_s}(B^*)) = G$.

We freely use these notions in the sequel; see [DeDo1;§2] for more details. In the rest of this section we fix a field K and a variety B as above.

2.2.   Descent of the field of definition of [G-]covers.

As in [DeDo1], we frequently use the word "[G-]cover" for the phrase "mere cover [resp. G-cover]". Suppose given a Galois extension F/K and a [G-]cover $f : X \to B$ *a priori* defined over F and such that the ramification divisor D is defined over K. In the mere cover situation, we will always assume that the Galois closure over F of the mere cover is, as G-cover, defined over F. This insures that the group of the cover (*i.e.*, the Galois group of the Galois closure) is the same over F as over F_s. This is of course not restrictive in the absolute situation, *i.e.*, when F is separably closed.

A K-model of the [G-]cover f is a [G-]cover $f_K : X_K \to B$ over K such that the [G-]cover over F obtained from f_K by extension of scalars from K to F, which we denote by $f_K \otimes_K F$, is isomorphic to f over F. A [G-]cover f is said to be defined over K if it has a K-model f_K. A significant problem is to study the descent of the field of definition of the [G-]cover f, and, more generally, to find its k-models for $K \subset k \subset F$. We introduce some notation that makes it possible to handle these questions simultaneously in both the mere cover and G-cover situations.

Let $\Psi : \Pi_F(B^*) \to S_d$ [resp. $\Phi : \Pi_F(B^*) \to G$] the representation of Π_F corresponding to the mere cover [resp. G-cover] $f : X \to B$. In both cases let G denote the group of the cover. Then set

$$N = \begin{cases} G & \text{in the G-cover case} \\ \mathrm{Nor}_{S_d} G & \text{in the mere cover case} \end{cases}$$

$$C = \mathrm{Cen}_N G = \begin{cases} Z(G) & \text{in the G-cover case} \\ \mathrm{Cen}_{S_d} G & \text{in the mere cover case} \end{cases}$$

where $Z(G)$ is the center of G and $\mathrm{Nor}_{S_d} G$ and $\mathrm{Cen}_{S_d} G$ are respectively the normalizer and the centralizer of G in S_d. Finally regard N as a subgroup of S_d where d is the degree of f: in the mere cover case, an embedding $N \hookrightarrow S_d$ is given by definition; in the G-cover case, embed $N = G$ in S_d by the regular representation of G.

Then, in both the mere cover and G-cover situations, the [G-]cover $f : X \to B$ corresponds to the representation $\phi : \Pi_F(B^*) \twoheadrightarrow G \subset N$ and the following holds [DeDo1].

Proposition 2.1 — *(a) The K-models of f correspond to the homomorphisms $\Pi_K(B^*) \to N$ that extend the homomorphism $\phi : \Pi_F(B^*) \to G \subset N$. In particular, the [G-]cover f can be defined over the field K if and only if the homomorphism $\phi : \Pi_F(B^*) \to G \subset N$ extends to an homomorphism $\Pi_K(B^*) \to N$,*

(b) Two [G-]covers over F are isomorphic if and only if the corresponding representations ϕ and ϕ' of Π_F are conjugate by an element φ in the group N, that is, $\phi'(x) = \varphi\phi(x)\varphi^{-1}$ for all $x \in \Pi_F(B^*)$

A representation $\Pi_K(B^*) \to N$ extending $\phi : \Pi_F(B^*) \to G \subset N$ will be called a K-model of ϕ. From (a), K-models of the representation ϕ correspond to K-models of the [G-]cover f.

2.3. Fibers of a cover.

2.3.1. Condition (Seq/Split). Fix a Galois extension F/K, a divisor D of B with simple components defined over K and assume that the exact sequence of fundamental groups

$$1 \to \Pi_F(B^*) \to \Pi_K(B^*) \to G(F/K) \to 1$$

splits. This condition was introduced and called (Seq/Split) in [DeDo1].

Consider the special case $F = K_s$ and the base space B has K-rational points off the branch point set D. Then condition (Seq/Split) classically holds: indeed each unramified K-rational point t_o provides a section $s_{t_o} :$ $G(K) \to \Pi_K$ [1].

2.3.2. Arithmetic action of $G(F/K)$ on a fiber. Suppose given a degree d mere cover $f_K : X_K \to B$ over K unramified over B^* and let $\phi_K : \Pi_K(B^*) \to \mathrm{Nor}_{S_d} G$ be the associated representation. Given an unramified K-rational point t_o on B, denote the compositum of all fields of definition over K of points in the fiber $f_K^{-1}(t_o)$ by K_{f_K,t_o}; equivalently, K_{f_K,t_o} is the compositum of all residue fields at t_o of the Galois closure of the extension $K(X)/K(B)$. We call the field K_{f_K,t_o} the *splitting field of f_K at t_o*.

Proposition 2.2 [Del;Prop.2.1] — *For each $\tau \in G(K)$, the element $(\phi_K s_{t_o})(\tau)$ is conjugate in S_d to the action of τ on the fiber $f_K^{-1}(t_o)$. Furthermore, the splitting field K_{f_K,t_o} of f_K at t_o corresponds via Galois theory to the homomorphism $\phi_K s_{T_o} : G(K) \to N$; that is, it is the fixed field in K_s of $Ker(\phi_K s_{t_o})$ and the Galois group of the extension $K_{f_K,t_o}/K$ is the image group of $\phi_K s_{t_o}$.*

---

[1] On the other hand, condition (Seq/Split) does not always hold: an example in which it does not is given in [DeEm].

Return to the general case: let $s : G(F/K) \to \Pi_K$ denote a section to the map $\Pi_K \twoheadrightarrow G(F/K)$ (not necessarily of the form s_{t_o}). Each element of Π_K induces a permutation of the different embeddings of the function field $K(X_K)$ in a separable closure $(K(B))_s$ of $K(B)$. This set of embeddings $K(X_K) \hookrightarrow (K(B))_s$ can be viewed as the geometric generic fiber of the cover. By analogy with the case $s = s_{t_o}$, for each $\tau \in G(F/K)$, the element $(\phi_K s)(\tau)$ is called the *arithmetic action of τ on the generic fiber associated with the section* s [DeDo1;§2.9]. Furthermore, the fixed field in F of $Ker(\phi_K s)$ is denoted by $K_{f_K,s}$ and called the *splitting field* of f at s; the Galois group of the extension $K_{f_K,s}/K$ is the image group of $\phi_K s$.

2.3.3. *Remarks on K-points versus K-sections.*

(a) Assume $F = K_s$ and call K-sections on B^* the sections to the map $\Pi_K(B^*) \to G(K)$. For simplicity, assume $\text{char}(K) = 0$. It is unclear how big is the set of K-sections on B^* that do not come from K-rational points on B. This set need not be empty. Take for K a non PAC field of cohomological dimension ≤ 1 (*e.g.* $K = \mathbb{Q}^{ab}$); there is a smooth projective K-curve B with no K-rational points but from condition $\text{cd}(K) \leq 1$ there are K-sections on B^*.

(b) For some purposes, K-sections can be as useful as K-rational points. For example, assume a finite group G can be realized as the Galois group of a regular Galois extension $E/K(B)$; let $\phi_K : \Pi_K \twoheadrightarrow G$ be the associated representation. Assume B is a K-rational variety and K is hilbertian. Using Prop.2.2, the hilbertian property can be rephrased to assert that there exists an unramified K-rational point $t_o \in \mathbb{P}^1$ such that the composed map $\phi_K s_{t_o} : G(K) \to G$ is onto, thus yielding a realization of G as Galois group over K. In fact any K-section such that $\phi_K s : G(K) \to G$ is onto would be just as good. So the question arises whether there is a weaker assumption on B than "B is K-rational" that guarantees that if K is hilbertian, there exists a K-section s such that $\phi_K s : G(K) \to G$ is onto.

(c) Consider the Fried-Völklein/Pop theorem: a field that is countable, PAC and hilbertian necessarily has a pro-free absolute Galois group of countable rank. At some point, the proof uses the existence of K-rational points on some K-variety. If K-rational points could be replaced in the proof by K-sections, then the assumption PAC could be replaced by "$\text{cd}(K) \leq 1$". Thus the Fried-Völklein conjecture — $\text{cd}(K) \leq 1$ and K hilbertian implies $G(K)$ pro-free of countable rank, for countable fields – would follow. And the Shafarevich conjecture — $\mathbb{Q}^{ab}$ pro-free —, which is a special case, would follow as well.

2.4. **Structure result under (Seq/Split).** Assume that condition (Seq/Split) holds and let $s : G(F/K) \to \Pi_K$ be a section to the map $\Pi_K \twoheadrightarrow G(F/K)$. Let $f : X \to B$ be a [G-]cover defined over F, unramified over B^* and $\phi : \Pi_F(B^*) \to G \subset N$ be the associated representation. The

following notation is used below. Given two maps α, β, the composed map is denoted by $\alpha\beta$ (when it exists); for maps from a set S to a group G, the product map, sending each $s \in S$ to $\alpha(s)\beta(s)$, is denoted below by $\alpha \cdot \beta$.

Proposition 2.3 — *Assume f is defined over K. Let $\phi_K^o : \Pi_K \to N$ be the representation associated with some K-model f_K^o of f and set $\varphi^o = \phi_K^o s$. Then the set of all K-models ϕ_K of ϕ is in one-one correspondence with the 1-cochain set $Z^1(\mathrm{G}(F/K), C, \varphi^o)$ (precisely defined in the proof).*

More precisely, the K-models of ϕ are those maps $\phi_K : \Pi_K \to N$ which equal ϕ on Π_F and equal $\theta \cdot \varphi^o$ on $\mathrm{G}(F/K)$ (via s), for some $\theta \in Z^1(\mathrm{G}(F/K), C, \varphi^o)$.

Furthermore, if s' is a section to the map $\Pi_K \to \mathrm{G}(F/K)$, then $s' = \sigma \cdot s$ for some $\sigma \in Z^1(\mathrm{G}(F/K), \Pi_F, s)$ and the arithmetic action of $\mathrm{G}(F/K)$ on the generic fiber of ϕ_K associated with the section s' is given by

$$\varphi = \phi\sigma \cdot \theta \cdot \varphi^o$$

Proof. The field K is the field of moduli of the [G-]cover f relative to the extension F/K (since f is defined over K). ¿From [DeDo1] (see Main Theorem (III)), we have the following. Let $\overline{\varphi} : \Pi_K \to N/C$ be the representation of Π_K modulo C given by the field of moduli condition [DeDo1;§2.7]. The K-models of f correspond in a one-one way to the liftings $\varphi : \mathrm{G}(F/K) \to N$ of the map $\overline{\varphi}s$. More precisely, to each given lifting $\varphi : \mathrm{G}(F/K) \to N$ corresponds a K-model of the representation ϕ, namely the one that equals ϕ on Π_F and equals φ on $\mathrm{G}(F/K)$ (via s). This K-model has the further property that the action $\varphi : \mathrm{G}(F/K) \to N \subset S_d$ is the arithmetic action of $\mathrm{G}(F/K)$ on the generic fiber associated with the section s.

The map φ^o is a lifting of $\overline{\varphi}s$. Therefore the set of all liftings exactly consists of those maps $\varphi : \mathrm{G}(F/K) \to N$ of the form $\varphi = \theta \cdot \varphi^o$ where θ is any element of $Z^1(\mathrm{G}(F/K), C, \varphi^o)$, i.e., is any map $\mathrm{G}(F/K) \to C$ satisfying the cocycle condition

$$\theta(uv) = \theta(u)\ \theta(v)^{\varphi^o(u)}$$

Finally, let s' be a section to the map $\Pi_K \to \mathrm{G}(F/K)$. Then $\sigma = s's^{-1}$ satisfies the cocycle condition

$$\sigma(uv) = \sigma(u)\ \sigma(v)^{s(u)}$$

thus, lies in $Z^1(K, \Pi_{K_s}, s)$. The arithmetic action on the generic fiber of ϕ_K associated with the section s' is given by

$$\phi_K s' = \phi_K(\sigma \cdot s) = \phi\sigma \cdot \theta \cdot \varphi^o \qquad \square$$

2.5. Structure result (General Case). We fix a Galois extension F/K and a divisor D of B with simple components defined over K. We no longer assume that condition (Seq/Split) holds.

2.5.1. Extension of constants in the Galois closure ([DeDo1;§2.8]. Let $f_K : X_K \to B$ be a [G-]cover over K and let $\phi_K : \Pi_K(B^*) \to N$ be the associated representation of $\Pi_K(B^*)$. Consider the function field extension $K(X_K)/K(B)$ associated to f_K. Denote the Galois closure of the extension $K(X_K)/K(B)$ by $K(\widehat{X_K})/K(B)$; its Galois group is the *arithmetic* Galois group of f_K; denote it by $\widehat{G}$. Consider then the field $\widehat{K} = K(\widehat{X_K}) \cap F$. The extension $\widehat{K}/K$ is called the *extension of constants in the Galois closure* of f_K of f.

Denote by Λ the unique homomorphism $G(K) \to N/G$ that makes the following diagram commute.

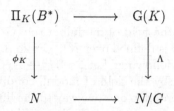

Proposition 2.4 [DeDo1;Prop.2.3] — *The homomorphism Λ corresponds to the extension of constants $\widehat{K}/K$ in the Galois closure of the model f_K of f_F. That is, $G(F/\widehat{K}) = Ker(\Lambda)$. The field $\widehat{K}$ can also be described as the smallest extension k of K such that $\phi_K(\Pi_k) \subset G$, or, equivalently, such that $kK(\widehat{X_K})/k$ is a regular extension.*

The homomorphism $\Lambda : G(K) \to N/G$ is called the *constant extension map (in Galois closure)* of f_K. For G-covers, $N/G = \{1\}$, Λ is trivial and $\widehat{K} = K$: by definition, G-covers over K do not have any extension of constants in their Galois closure.

Proposition 2.5 — *Let f be a [G-]cover over F and $\phi : \Pi_F(B^*) \to G \subset N$ be the associated representation. Assume f is defined over K. Let $\phi_K^o : \Pi_K(B^*) \to N$ be the representation associated with some K-model f_K^o of f.*

Then the set of all K-models ϕ_K of ϕ with the same constant extension map Λ as ϕ_K^o is in one-one correspondence with the 1-cochain set $Z^1(G(F/K), Z(G), \Lambda)$. More precisely, the K-models of ϕ are those maps

$\phi_K : \Pi_K \to N$ which are, for some $\Theta \in Z^1(\mathrm{G}(F/K), Z(G), \Lambda)$, of the form $(\Theta \mathrm{P}) \cdot \phi_K^o$, where P is the natural surjection $\mathrm{P} : \Pi_K \twoheadrightarrow \mathrm{G}(F/K)$.

Proof. From [DeDo1], a K-model ϕ_K with constant extension map Λ is any homomorphism $\phi_K : \Pi_K \to N$ extending $\phi : \Pi_F \to N$ and inducing the map $\Lambda : \mathrm{G}(F/K) \to N/G$. Furthermore, ϕ_K should also necessarily induce the representation $\overline{\varphi} : \Pi_K \to N/C$ given by the field of moduli condition. Consequently $\tilde{\theta} = \phi_K \cdot (\phi_K^o)^{-1}$ has values in $C \cap G = Z(G)$ and factors through $\mathrm{G}(F/K)$. Prop.2.5 immediately follows. □

3.   Covers with prescribed fibers

In this section, the base space B is the projective line $\mathbb{P}^1$ and $F = K_s$. In particular, condition (Seq/Split) holds. We retain this conclusion from Prop.2.3. If a K-model of a [G-]cover f (*a priori* defined over K_s) is known, then other K-models can be obtained by "twisting" by elements of a 1-cochain set $Z^1(\mathrm{G}(K), C, -)$. More precisely, let $t_o \in \mathbb{P}^1(K)$ be an unramified point and s_{t_o} the associated section $\mathrm{G}(K) \to \Pi_K$. The representations $\phi_K : \Pi_K \to N$ associated with K-models of f are completely determined by their restriction to Π_{K_s} (which is given) and their restriction $\varphi = \phi_K \mathrm{s}_{t_o}$ to $\mathrm{G}(K)$. If one K-model is known that has $\varphi = \varphi^o$, others are obtained by replacing φ^o by $\theta \cdot \varphi^o$, for any $\theta \in Z^1(\mathrm{G}(K), C, \varphi^o)$. This arithmetical twist is an important ingredient of the results of this section.

3.1.   The Beckmann-Black problem. In [Be2], S. Beckmann asks the following question: is every finite Galois extension $E/\mathbb{Q}$ the specialization of a Galois branched cover of $\mathbb{P}^1$ which is defined over $\mathbb{Q}$ and has the same Galois group? A finite group G is said to have the *lifting property* (over $\mathbb{Q}$) when the answer is "Yes" for every Galois extension $E/\mathbb{Q}$ of group G. She shows that finite abelian groups and symmetric groups have the lifting property. This problem has also been considered by E. Black who conjectured that each finite group has the lifting property over each field K (instead of $\mathbb{Q}$) and proposed a cohomological approach. Her main result is that over a hilbertian field K, a semi-direct product of a finite cyclic group A with a group H having the lifting property also has the lifting property if $(|H|, |A|) = 1$ and $(\mathrm{char}(K), |A|) = 1$ [Bl2]. That includes the case of abelian groups and also gives new examples of groups with the lifting property, *e.g.* the dihedral groups D_n of order $2n$ when n is odd (see also [Bl1]). Using our terminology, E. Black's conjecture can be reformulated as follows.

Conjecture 3.1 (E. Black) — *Let K be an arbitrary field, G be a finite group and E/K be a Galois extension of group G. Then there exists a G-cover $f : X_K \to \mathbb{P}^1$ of group G defined over K and some unramified point $t_o \in \mathbb{P}^1(K)$ such that the splitting field extension $K_{f_K, t_o}/K$ of f_K at t_o (see §2.3.2) is K-isomorphic to E/K.*

The main result of this section is the following one, which improves on the initial results of Beckmann and Black in that the field K is here an arbitrary field.

Theorem 3.2 — *The Black conjecture holds if G is an abelian group and K is an arbitrary field. In particular, the conjecture holds if $G(K)$ is abelian (e.g. K is finite).*

Proof. Here is our strategy. We realize G as the group of a [G-]cover $f : X \to \mathbb{P}^1$ defined over K. The splitting field extension $K_{f,t_o}/K$ of f at the given point t_o is some extension of K. Then we twist the K-model in such a way that the splitting field extension at t_o equals the given extension E/K.

More specifically, suppose given a finite abelian group G and an extension E/K of group G. Realize it as the Galois group of a G-cover $f_K^o : X_K \to \mathbb{P}^1$ over K with at least one unramified point $t_o \in \mathbb{P}^1(K)$. This is easy if K is infinite. Indeed, take any regular Galois extension $F/K(T)$ of group G (such extensions exist (*e.g.* [MatMa;Ch.4,Th.2.4])); there exists K-rational points on $\mathbb{P}^1$ different from the branch points of the extension $F/K(T)$. However this is more difficult if K is a finite field, especially when the order of G is divisible by the characteristic of K. Nevertheless, this is possible: see [De3] where it is proved that each finite abelian group G can be realized as the Galois group of a G-cover defined over K and unramified over each element of a finite subset $D \subset \mathbb{P}^1(\overline{K})$ given in advance.

Next set $s = s_{t_o}$. Let $\varphi^o : G(K) \to G$ be the arithmetic action of $G(K)$ on the fiber above t_o. The given extension E/K corresponds to some surjective homomorphism $\varphi : G(K) \twoheadrightarrow G$. For abelian G-covers, the set $Z^1(K, C, \varphi^o)$ involved in Prop.2.3 is merely $\mathrm{Hom}(G(K), G)$ (since here $C = Z(G) = G)$). Consequently, the cover $f_K^o \otimes_K K_s$ has another K-model f_K (as G-cover) such that the arithmetic action on the fiber above t_o equals φ: indeed take $\theta = \varphi(\varphi^o)^{-1}$ in Prop.2.3. This concludes the proof of the first part of Th.3.2. The second part readily follows. $\square$

3.2. Models with a totally rational fiber.

In Th.3.2, the extension E/K can be more generally any Galois extension of group $H \subset G$ (instead of $H = G$). In particular for $H = \{1\}$ we obtain that each finite abelian group is the Galois group a G-cover of $\mathbb{P}^1$ over K with a totally K-rational unramified fiber $f_K^{-1}(t_o)$. Furthermore, the point t_o can be prescribed in advance in $\mathbb{P}^1(K)$. This had been noticed in [De1] and [Des].

Recall that, given a cover $f_K : X_K \to \mathbb{P}^1$ and a point $t_o \in \mathbb{P}^1(K)$ not a branch point, the fiber $f_K^{-1}(t_o)$ is said to be *totally K-rational* if it consists only of K-rational points on X_K. Below we comment on this condition; in particular we review some situations where the existence of a totally rational fiber revealed helpful.

3.2.1. Existence results. It is known that a finite group G is the Galois group a G-cover of $\mathbb{P}^1$ over K with a totally K-rational fiber in the following situations:

- K is an arbitrary field and G is an abelian group (see above).

- K is an ample field and G is an arbitrary group: this result is essentially due to Harbater and Pop (see *e.g.* [DeDes]). Recall a field K is called *ample* if every smooth K-curve with at least one K-rational point has infinitely many K-rational points. Algebraically closed fields, separably closed fields, more generally PAC fields are ample. Local fields are ample too. The fields $\mathbb{Q}^{tr}$ [resp. $\mathbb{Q}^{tp}$] of all totally real [resp p-adic] algebraic numbers are other typical examples of ample fields.

It is unclear whether this existence result holds for any field K and any group G. That would follow from a generalization of E. Black's conjecture, where the extension E/K would be allowed to be any Galois extension of group contained in the given group G.

3.2.2.   Field of definition of G-covers.   A mere cover over K which is Galois over K_s and has a totally K-rational fiber is then automatically defined over K as [G-]cover [De1].

3.2.3.   Patching covers.   D. Harbater showed that, over a complete valued field K, G-covers with a totally K-rational fiber can be patched and glued to provide a G-cover still defined over K and of group the group generated by the groups of the initial covers. This patching and gluing result is an essential tool in the proof of the Regular Inverse Galois Problem over a complete valued field [Ha1] and also in the proof of Abhyankar's conjecture [Ha2].

3.2.4. Stable models. I proved in [De1] that a [G-]cover f_K defined over a Galois extension K of k and which has a totally K-rational unramified fiber, is then *stable* over k. That is, the field of moduli of f_K relative to the extension K/k equals the field of moduli of $f = f_K \otimes_k k_s$ relative to the extension k_s/k. D. Harbater and I combined this Stability Criterion with a Good Models result due to S. Beckmann [Be1] to prove the following result [DeHa]. A G-cover of $\mathbb{P}^1$ over $\overline{\mathbb{Q}_p}$ is defined over its field of moduli (relative to the extension $\overline{\mathbb{Q}_p}/\mathbb{Q}_p$), except possibly if p is a *bad* prime, *i.e.*, if p divides the order of the group of the cover or if the branch points of the cover coalesce modulo p. The result actually holds in a more general situation where $\mathbb{Q}_p$ is replaced by the fraction field of a henselian discrete valuation ring, with a perfect residue field of cohomological dimension ≤ 1. M. Emsalem recently extended this result to mere covers [Em].

3.2.5. The Beckmann-Black problem. In [De2] I prove, for ample fields, a *mere* form of the conjecture where the Galois cover is required to be defined over K but only as mere cover. Existence of a G-cover defined over

K with a totally K-rational unramified fiber is also a key ingredient of the proof.

4. Field of moduli and extension of constants

4.1. On a result of M. Fried. In [Fr;Prop.2], M. Fried states that if $f : X \to \mathbb{P}^1$ is a mere cover defined over K with group a centerless group G then the arithmetic Galois group $\widehat{G}$ (see §2.5.1) satisfies: $\widehat{G} \cap \mathrm{Cen}_{S_d} G = \{1\}$ (where d is the degree of f). It seems that this statement is not exactly correct. Here is an argument showing why. Here again, we use the structure result (Prop.2.3) and the derived arithmetical 'twist' on K-models of a cover.

We use the function field viewpoint. Start with a regular Galois extension $E/K(T)$ of group G and with a totally K-rational fiber above some K-rational unramified point t_o: if K is an ample field, this can be achieved for any group G. Let $\phi_K : \Pi_K \to G$ be the representation of Π_K associated with the extension $E/K(T)$ and ϕ be the restriction of ϕ_K to Π_{K_s}: for $x \in \Pi_{K_s}$ and $\tau \in G(K)$, we have, using Prop.2.2,

$$\phi_K(x \mathrm{s}_{t_o}(\tau)) = \phi(x)$$

Let $C = \mathrm{Cen}_{S_d} G$ be the centralizer of G in its regular representation and $\varphi : G(K) \to C$ be any homomorphism. ¿From Prop.2.3, the representation ϕ has a K-model $\phi_K^{\varphi} : \Pi_K \to \mathrm{Nor}_{S_d} G$ which equals ϕ on Π_{K_s} and equals φ on $(G(K)$ (*via* s) (since here $\varphi \in Z^1(G(K), C, 1) = \mathrm{Hom}(G(K), C))$. Let $E^{\varphi}/K(T)$ be the field extension associated with the "twisted" representation ϕ_K^{φ}: this extension is still Galois over K_s but no longer over K. The arithmetic Galois group is the group generated by G and $\varphi(G(K))$. This contradicts Prop.2 of [Fr] since elements in $\varphi(G(K))$ centralize G.

In the proof of Prop.2 of [Fr], it seems that an inaccuracy occurs when the author says (after display (2.6)): "Therefore $F^o(Y) \subset F^o(Y^o) \subset \widehat{F(Y)}$". Indeed $Y^o \to X$ is obtained by descent from a cover $\widehat{Y} \to X$ which is defined only up to $\widehat{F}$-isomorphism. Thus the containments above hold only up to $\widehat{F}$-isomorphism (but not up to F-isomorphism).

However it is possible to modify Fried's statement so it holds true. The rectified version is the second consequence of Th.4.1 (Cor.4.3 below).

4.2. Field of moduli [DeDo1;§2]. We recall below some basics about fields of moduli. Fix a Galois extension F/K. Given a mere cover [resp. G-cover] $f : X \to B$ a priori defined over F, for each $\tau \in G(F/K)$, let $f^{\tau} : X^{\tau} \to B^{\tau}$ denote the corresponding conjugate [G-]cover. Consider the subgroup $M(f)$ [resp. $M_G(f)$] of $G(F/K)$ consisting of all elements $\tau \in G(F/K)$ such that the covers [resp., the G-covers] f and f^{τ} are isomorphic over F. Then the *field of moduli* of the cover f [resp., the G-cover f] (relative to the extension F/K) is defined to be the fixed field $F^{M(f)}$ [resp.$F^{M_G(f)}$]

of $M(f)$ [resp. $M_G(f)$] in F. The field of moduli of a [G-]cover is contained in each field of definition k such that $K \subset k \subset F$. So it is the smallest field of definition provided that it *is* a field of definition.

Assume $\phi : \Pi_F(B^*) \to G \subset N$ is the representation corresponding to the [G-]cover $f : X \to B$ over F. Then K is the field of moduli of the [G-]cover f relative to the extension F/K if and only if the ramification divisor D is defined over K and the following condition — called the *field of moduli condition* —, holds.

(FMod)   For each $U \in \Pi_K(B^*)$, there exists $\varphi_U \in N$ such that

$$\phi(x^U) = \varphi_U \phi(x) \varphi_U^{-1} \qquad \text{(for all } x \in \Pi_F(B^*))$$

4.3. Statement of the result. Let $f_K : X_K \to B$ be a mere cover over K and $\phi_K : \Pi_K(B^*) \to G \subset S_d$ be the associated representation. Set $f = f_K \otimes_K F$ and let $\widehat{f} : \widehat{X} \to B$ be the Galois closure over F of the mere cover $f : X \to B$. Recall from §2.2 that we assume that $\widehat{f}$ is defined over F as G-cover, in other words, that $F\widehat{K(X_K)}$ is a regular extension of F; this assumption is not restrictive if $F = K_s$ (*i.e.*, in the absolute situation). The aim of this section is the following result.

Theorem 4.1 — *Let $\widehat{K}/K$ be the extension of constants in the Galois closure of f_K and $\widehat{G}$ be the arithmetic Galois group. Let K_G be the field of moduli of $\widehat{f}$ as G-cover (relative to the extension F/K). Then $\widehat{K}$ is an extension of K_G of degree*

$$[\widehat{K} : K_G] = [\mathrm{Cen}_{S_d} G \cap \widehat{G} : Z(G)]$$

4.4. Consequences. In this paragraph, we suppose given a mere cover $f : X \to B$ over F and its Galois closure $\widehat{f} : \widehat{X} \to B$ over F and let $\phi : \Pi_F(B^*) \to G$ be the representation corresponding to $\widehat{f}$. Let K_G be the field of moduli of $\widehat{f}$ as G-cover (relative to the extension F/K).

Corollary 4.2 — *Assume that f is defined over K and that $\mathrm{Cen}_{S_d} G = Z(G)$. Then $\widehat{f} : \widehat{X} \to B$ is defined over its field of moduli K_G (as G-cover).*

Proof. Let f_K be a K-model of f as mere cover. Apply Th.4.1 to get $\widehat{K} = K_G$. Now $\widehat{K}$ is a field of definition of the G-cover $\widehat{f}$ (Prop.2.4). $\square$

Corollary 4.3 — *Assume that $Z(G) = \{1\}$. Then f has a unique K_G-model $f_{K_G} : X_{K_G} \to B$ with no extension of constants in its Galois closure (up to K_G-isomorphism). Furthermore the arithmetic Galois group $\widehat{G}$ of each K-model $f_K : X_K \to B$ of f_{K_G} satisfies $\widehat{G} \cap \mathrm{Cen}_{S_d} G = \{1\}$.*

§4.1 shows the last conclusion may fail if f_K is a model of f that is not a model of f_{K_G}.

Proof. The assumption $Z(G) = \{1\}$ insures that the Galois cover $\widehat{f}$ is defined over its field of moduli K_G [DeDo1;Cor.3.2]. That is, the homomorphism $\phi : \Pi_F \to G$ extends to an homomorphism $\phi_{K_G} : \Pi_{K_G} \to G$. The representation $\psi : \Pi_F \to S_d$ associated with the mere cover f is obtained by composing ϕ with some embedding $i : G \hookrightarrow S_d$. The homomorphim ψ extends to an homomorphism $\Pi_{K_G} \to S_d$, namely the homomorphism $i\phi_{K_G}$. By construction, the associated K_G-model of f has no extension of constants in its Galois closure. The uniqueness of such a K_G-model of f follows from Prop.2.5.

Suppose given a K-model f_K of f_{K_G}, i.e., an extension $\psi_K : \Pi_K \to S_d$ of ψ_{K_G}. It follows from

$$\psi_K(\Pi_{K_G}) = \psi_{K_G}(\Pi_{K_G}) \subset G$$

and the definition of $\widehat{K}$ (§2.5.1) that $\widehat{K} \subset K_G$. On the other hand, $\widehat{K}$ is a field of definition of $\widehat{f}$ as G-cover. Therefore $K_G \subset \widehat{K}$. Whence $\widehat{K} = K_G$. Thus Th.4.1 yields

$$\mathrm{Cen}_{S_d} G \cap \widehat{G} = Z(G) = \{1\} \qquad \square$$

4.5. Proof of Theorem 4.1. The field $\widehat{K}$ is a field of definition of the G-cover $\widehat{f}$. Therefore $\widehat{K}$ contains K_G which is the field of moduli of the G-cover $\widehat{f}$.

Let $\Lambda : \mathrm{G}(F/K) \to N/G$ be the constant extension map of f_K (in Galois closure) where as usual $N = \mathrm{Nor}_{S_d} G$. From Prop.2.4, $\mathrm{G}(F/\widehat{K}) = Ker(\Lambda)$ and for each field k such that $K \subset k \subset \widehat{K}$, the kernel of the restriction of Λ to $\mathrm{G}(F/k)$ is $\mathrm{G}(F/k) \cap \mathrm{G}(F/\widehat{K}) = \mathrm{G}(F/\widehat{K})$. Conclude that

(1) $[\widehat{K} : k] = |\Lambda\,(\mathrm{G}(F/k))|$

Denote the quotient group $G \cdot (\mathrm{Cen}_{S_d} G \cap \widehat{G})/G$ by Γ; it is a subgroup of $\widehat{G}/G$. We claim that $\Lambda\,(\mathrm{G}(F/K_G)) \subset \Gamma$. Indeed, since K_G is the field of moduli of the G-cover $\widehat{f}$, for each $\mathrm{u} \in \Pi_{K_G}$, there exists some $\varphi_\mathrm{u} \in G$ such that

$$\phi(x^{\mathsf{U}}) = \varphi_{\mathsf{U}}\phi(x)\varphi_{\mathsf{U}}^{-1} \qquad (x \in \Pi_F)$$

where $\phi : \Pi_F \to G$ is the restriction of ϕ_K to Π_F. Now the above formula also holds with φ_{U} replaced by $\phi_K(\mathsf{U})$. Conclude that $\phi_K(\mathsf{U})$ lies in G, up to an element in $\mathrm{Cens}_{S_d} G$ (which also obviously lies in $\widehat{G} = \phi_K(\Pi_K)$). This proves the claim since the map Λ is the map induced modulo G by ϕ_K over $G(F/K)$. Formula (1) then yields

$$(2) \qquad\qquad\qquad [\widehat{K} : K_G] \leq |\Gamma|$$

Let k be the fixed field in F of $\Lambda^{-1}(\Gamma)$. We claim that $K_G \subset k$. Indeed, let $\tau \in G(F/k)$, *i.e.*, $\Lambda(\tau) \in \Gamma$. Pick an element $\mathsf{U} \in \Pi_K$ mapping to τ *via* the map $\Pi_K \to G(F/K)$. The element $\phi_K(\mathsf{U})$ can be written $\phi_K(\mathsf{U}) = g \cdot c$ with $g \in G$ and $c \in \mathrm{Cens}_{S_d} G$. Thus we obtain

$$\phi(x^{\mathsf{U}}) = \phi(x)^g \qquad (x \in \Pi_F)$$

This shows that $\tau \in G(K_G)$ and proves the claim. Using formula (1), we obtain

$$(3) \qquad\qquad [\widehat{K} : K_G] \geq [\widehat{K} : k] = |\Lambda(\Lambda^{-1}(\Gamma))| = |\Gamma|$$

which together with (2) completes the proof (the equality $|\Lambda(\Lambda^{-1}(\Gamma))| = |\Gamma|$ holds since $\Gamma \subset \widehat{G}/G$ and that $\widehat{G}/G$ is the image group of Λ). $\qquad\square$

5. The local-to-global principle

5.1. The problem. We retain the notation of previous sections and assume further that K is a number field and $F = \overline{\mathbb{Q}}$. Assume that the [G-]cover $f : X \to B$ can be defined over each completion K_v of K. Does it follow that the cover can be defined over K? We say that the *local-to-global* principle holds when the answer is "Yes". In his thesis, E. Dew conjectured that the local-to-global principle for G-covers of $\mathbb{P}^1$ over number fields. This was proved in [De1] except possibly for number fields that are exceptions to Grunwald's theorem (the field $\mathbb{Q}$ is not exceptional). This result was extended to G-covers of a general base space B in [DeDo1]. The case of mere covers was then considered in [DeDo2]: the local-to-global principle was shown to hold under some additional assumptions on the group G of the cover and the monodromy representation $G \hookrightarrow S_d$ (with $d = deg(f)$). Here we will consider the special case of mere covers that are Galois over $\overline{\mathbb{Q}}$.

Recall that the field of moduli of a cover embeds in each field of definition. Therefore if a cover is defined over each K_v, then its field of moduli is K (relative to the extension $\overline{K}/K$). Hence if the field of moduli is a field of definition, then the local-to-global principle holds. From [DeDo1] that is the case if the cover is Galois over $\overline{\mathbb{Q}}$ and condition (Seq/Split) (see §2.3.1) holds. The goal of this section is to investigate the problem when the cover is Galois over $\overline{\mathbb{Q}}$ but condition (Seq/Split) is not assumed.

5.2. SG-covers. Our treatment uses SG-covers which are more general objects than [G-]covers. They were originally introduced in [DeDo1;Final Note] (under a different name). Here we will only consider *Galois* SG-covers. Given a group G and a subgroup $\Gamma \subset G$, a *Galois* SG-*cover of fixed subgroup* $\Gamma \subset G$ over K is a mere cover $f : X \to B$ over K which is Galois over $\overline{K}$ and is given with an isomorphism $G \simeq \mathrm{G}(\overline{K}(X)/\overline{K}(B))$ and an embedding $\Gamma \hookrightarrow \mathrm{Aut}(K(X)/K(B))$ such that the following square diagram commutes

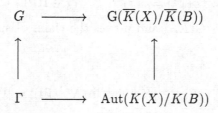

For $\Gamma = \{1\}$, SG-covers are mere covers; for $\Gamma = G$, SG-covers are G-covers. A SG-cover $f : X \to B$ over $\overline{K}$ is defined over K if the mere cover f together with the automorphisms in Γ can be defined over K.

From [DeDo1;Final Note], isomorphism classes of SG-covers of fixed subgroup $\Gamma \subset G$ $f : X \to B$ over $\overline{K}$ correspond to surjective homomorphisms $\phi : \Pi_{\overline{K}}(B^*) \to G$ regarded modulo conjugation by elements of the group

(1) $$N = Nor_{S_d} G \cap Cen_{S_d}(\Gamma^*)$$

where

- the embedding $G \subset S_d$ is given by the (free transitive) action of G on the d conjugates of a primitive element of the extension $\overline{K}(X)/\overline{K}(B)$, and
- Γ^* is the image of Γ *via* the classical anti-isomorphism $* : G \to Cen_{S_d} G$: identify each $g \in G$ with the element in S_d induced by the left-multiplication by g; then the map $*$ send g on the right-multiplication by g.

For $\Gamma = \{1\}$ (mere cover case), we have $N = Nor_{S_d}(G)$; for $\Gamma = G$ (G-cover case), we have $N = G$. Thus in general, we have $G \subset N \subset Nor_{S_d}(G)$.

As in both the mere cover and G-cover situations, we have the following assertions (which generalize Prop.2.1). The group N is the one given in (1).

(2) *(a) A SG-cover* $f : X \to B$ *of fixed subgroup* $\Gamma \subset G$ *over* K_s *corresponds to a representation* $\phi : \Pi_{K_s}(B^*) \twoheadrightarrow G \subset N$

(b) The K-models of f (as SG-cover) correspond to the homomorphisms $\Pi_K(B^*) \to N$ *that extend the homomorphism* $\phi : \Pi_{K_s}(B^*) \to G$. *In particular, the SG-cover f can be defined over the field K if and only if the homomorphism* $\phi : \Pi_{K_s}(B^*) \to G \subset N$ *extends to an homomorphism* $\Pi_K(B^*) \to N$,

(c) Two [G-]covers over K_s *are isomorphic if and only if the corresponding representations* ϕ *and* ϕ' *are conjugate by an element* φ *in the group* N, *that is,* $\phi'(x) = \varphi\phi(x)\varphi^{-1}$ *for all* $x \in \Pi_{K_s}(B^*)$

The map * can be described more intrinsecally. Denote by G_+ the group

$$G_+ = G \times^s \mathrm{Aut}(G)$$

The group G_+ is isomorphic to $\mathrm{Nor}_{S_d}(G)$ (*e.g.* [DeDo1;Prop.3.1]). Next use the embedding $G \hookrightarrow G_+$ sending g to $(g,1)$ to identify G with a subgroup of G_+ (still denoted by G). For each $g \in G$, denote the conjugation by g by c_g ($c_g(h) = ghg^{-1}$). Then the anti-isomorphism * is the map

$$\begin{cases} G \to G_+ \\ g \to g^* = (g, c_g^{-1}) \end{cases}$$

Some groups play a central role in [DeDo1] and [DeDo2]: they are N (defined in (1)), $C = \mathrm{Cen}_N(G)$, N/C and $CG/G \simeq C/Z(G)$ [2].

Proposition 5.1 — *Denote the subgroup of* $\mathrm{Aut}(G)$ *of all automorphisms* χ *that are trivial on* Γ *by* $\mathrm{Aut}_\Gamma(G)$. *Then we have.*

(a) $N = \{(\gamma, \chi) \in G_+ | \gamma \in G, \chi \in \mathrm{Aut}_\Gamma(G)\}$.

(b) $N/G = \mathrm{Aut}_\Gamma(G)$.

(c) $C = (\mathrm{Cen}_G(\Gamma))^*$.

(d) CG/G *is isomorphic to* $\mathrm{Cen}_G(\Gamma)/Z(G)$, *which embeds in* $\mathrm{Aut}_\Gamma(G)$.

(e) $Z(G)$ *is a direct summand of* C *if and only if* $Z(G)$ *is a direct summand of* $\mathrm{Cen}_G(\Gamma)$.

(f) $Z(G) \subset Z(N)$ *if and only if for each* $\chi \in \mathrm{Aut}_\Gamma(G)$, *we have* $\chi|_{Z(G)} = 1$.

Proof. By definition, $N = G_+ \cap \mathrm{Cen}_{G_+}(\Gamma^*) = \mathrm{Cen}_{G_+}(\Gamma^*)$ and $C = \mathrm{Cen}_N(G)$; (a) and (c) follow straightforwardly. The projection $G_+ \to$

---

[2] $Z(G)$ should *a priori* be replaced by $C \cap G$, but $C \cap G = Z(G)$ in general.

$\mathrm{Aut}(G)$ mapping each $(u, \chi) \in G_+$ to $\chi \in \mathrm{Aut}(G)$ factors through G_+/G to yield the isomorphism $G_+/G \simeq \mathrm{Aut}(G)$; (b) and (d) follow immediately. The anti-isomorphism $*$ sends $\mathrm{Cen}_G(\Gamma)$ onto $(\mathrm{Cen}_G(\Gamma))^*$ and $Z(G)$ onto itself. This provides an isomorphism $\mathrm{Cen}_G(\Gamma)/Z(G) \simeq (\mathrm{Cen}_G(\Gamma))^*/Z(G)$ and proves (e). Finally (f) readily follows from (a) and the definitions.   □

5.3.   The local-to-global principle for Galois covers. [DeDo2] studies the local-to-global problem for mere covers, *i.e.*, SG-covers of subgroup $\Gamma = \{1\}$. But the paper was written in a more general setting. Namely the objects we considered were representations $\phi : \Pi_{K_s} \twoheadrightarrow G \subset N$. In [DeDo2] we were mainly interested in mere covers, which correspond to $N = \mathrm{Nor}_{S_d}(G)$, and G-covers, which correspond to $N = G$. But from (2) above, [DeDo2] applies more generally to SG-covers provided that N is understood as in (1), C as $\mathrm{Cen}_N(G)$. In this context, the second Theorem in §1 of [DeDo2] rewrites as follows (using Prop.5.1).

Theorem 5.2   — *Assume that $K = \mathbb{Q}$, or more generally, that K is a number field for which the special case of Grunwald's theorem does not occur. Then the local-to-global principle holds for SG-covers of fixed subgroup $\Gamma \subset G$ satisfying simultaneously the five conditions below.*

(i) For each $\chi \in \mathrm{Aut}_\Gamma(G)$, we have $\chi|_{Z(G)} = 1$,

(ii) $Z(G)$ is a direct summand of $\mathrm{Cen}_G(\Gamma)$

(iii/1) $Z(\mathrm{Cen}_G(\Gamma)/Z(G))$ is a direct summand of $\mathrm{Cen}_G(\Gamma)/Z(G)$.

(iii/2) $Z(\mathrm{Cen}_G(\Gamma)/Z(G)) \subset Z(\mathrm{Aut}_\Gamma(G))$.

(iii/3) $\mathrm{Inn}(\mathrm{Cen}_G(\Gamma)/Z(G))$ has a complement in $\mathrm{Aut}(\mathrm{Cen}_G(\Gamma)/Z(G))$.

These five conditions hold for example if $\mathrm{Cen}_G(\Gamma) = Z(G) \subset \Gamma$.

The local-to-global principle also holds for SG-covers satisfying simultaneously conditions (i), (ii) above and the following condition

(iii)' $\mathrm{Cen}_G(\Gamma)/Z(G)$ has a complement in $\mathrm{Aut}_\Gamma(G)$.

Originally we were interested by the local-to-global principle for Galois mere covers. Th.5.2 above can be used as follows. Suppose given a Galois mere cover defined over $\mathbb{Q}_p$ for each p. If in addition there is a subgroup $\Gamma \subset G$ such that for each p, the mere cover has a model defined over $\mathbb{Q}_p$ along with the automorphisms in Γ, then f satisfies the local assumption of the local-to-global principle not only as mere cover but also as SG-cover and Th.5.2 may be used. The bigger Γ the stronger the hypothesis. If $\Gamma = G$, *i.e.*, is as big as can be, then the hypothesis does insure that the mere cover is defined over $\mathbb{Q}$ (it is even defined over $\mathbb{Q}$ as G-cover). The weakest hypothesis is for $\Gamma = \{1\}$. Although no example is known, it is most likely that the local-to-global principle does not hold in general.

We state some intermediate special cases. In both statements below, $K = \mathbb{Q}$, or more generally, K is a number field for which the special case of Grunwald's theorem does not occur.

Corollary 5.3 — *Let Γ be a subgroup of a group G such that $\mathrm{Cen}_G(\Gamma) \subset \Gamma \cap Z(G)$ (e.g. $\mathrm{Cen}_G(\Gamma) = \{1\}$). Then the local-to-global principle holds for all SG-covers with fixed subgroup $\Gamma \subset G$.*

Corollary 5.4 — *Let Γ be a subgroup of a group G such that $Z(G) \subset \Gamma$, $Z(G)$ is a direct summand of $\mathrm{Cen}_G(\Gamma)$ and $\mathrm{Cen}_G(\Gamma)/Z(G)$ has a complement in $\mathrm{Aut}_\Gamma(G)$. Then the local-to-global principle holds for all SG-covers with fixed subgroup $\Gamma \subset G$.*

Indeed, assumptions in Cor.5.3 guarantee that $\mathrm{Cen}_G(\Gamma) = Z(G) \subset \Gamma$. Note that if they hold, then necessarily $Z(G) = \mathrm{Cen}_G(\Gamma) = Z(\Gamma)$. As to Cor.5.4, the assumptions guarantee here that conditions (i), (ii) and (iii)' of Th.5.2 hold. Assumptions of both Cor.5.3 and Cor.5.4 hold if $\Gamma = G$, *i.e.*, in the G-cover situation. Thus both these results generalize Th.3.8 of [DeDo1].

References

[Be1] S. Beckmann, *On extensions of number fields obtained by specializing branched coverings*, J. reine angew. Math., **419**, (1991), 27-53.

[Be2] S. Beckmann, *Is every extension of $\mathbb{Q}$ the specialization of a branched covering?*, J. Algebra, **164**, (1994), 430–451.

[Bl1] E. Black, *Arithmetic lifting of dihedral extensions*, J. Algebra, **203**, (1998), 12–29.

[Bl2] E. Black, *On semidirect products and arithmetic lifting property*, J. London Math. Soc., (to appear).

[De1] P. Dèbes, *Covers of $\mathbb{P}^1$ over the p-adics*, in *Recent developments in the Inverse Galois Problem*, Contemp. Math., **186**, (1995), 217–238.

[De2] P. Dèbes, *Galois covers with prescribed fibers: the Beckmann-Black problem*, preprint, (1998).

[De3] P. Dèbes, *Regular realization of abelian groups with controlled ramification*, in Proc. of the Finite Field conference in Seattle (Summer 1997), Contemp.. Math., (to appear).

[DeDes] P. Dèbes and B. Deschamps, *The Inverse Galois problem over large fields*, in *Geometric Galois Action*, London Math. Soc. Lecture Note Series, Cambridge University Press, (1997), 119–138.

[DeDo1] P. Dèbes and J-C. Douai, *Algebraic covers: field of moduli versus field of definition*, Annales Sci. E.N.S., 4ème série, **30**, (1997), 303–338.

[DeDo2] P. Dèbes and J-C. Douai, *Local-global principle for algebraic covers*, Israel J. Math., **103**, (1998), 237–257.

[DeEm P. Dèbes and M. Emsalem, *On fields of moduli of curves*, J. Algebra, (to appear).

[DeHa] P. Dèbes and D. Harbater, *Field of definition of p-adic covers*, J. fur die reine und angew. Math., **498**, (1998), 223–236.

[Des] B. Deschamps, *Existence de points p-adiques sur des espaces de Hurwitz pour tout p*, in *Recent developments in the Inverse Galois Problem*, Contemp. Math., **186**, (1995), 239–247.

[Em] M. Emsalem, *On reduction of covers of arithmetic surfaces*, in Proc. of the Finite Field conference in Seattle (Summer 1997), Contemp. Math., (to appear).

[Fr] M. Fried, *Fields of definition of function fields and Hurwitz families–Groups as Galois groups*, Comm. in Alg., **5/1** (1977), 17–82.

[Ha1] D. Harbater, *Galois covering of the arithmetic line*, Lecture Notes in Math. **1240**, (1987), 165–195.

[Ha2] D. Harbater, *Abhyankar's conjecture on Galois groups over curves*, Invent. math., **117**, (1994), 1–25

[MatMa] B. H. Matzat and G. Malle, *Inverse Galois theory*, IWR preprint series, Heidelberg, (1993-96).

UNIV. LILLE, MATHÉMATIQUES, 59655 VILLENEUVE D'ASCQ CEDEX, FRANCE.
E-mail address: Pierre.Debes@univ-lille1.fr

Curves with infinite K-rational geometric fundamental group

Gerhard Frey, Ernst Kani and Helmut Völklein

1   Rational Geometric Fundamental Groups

Let K be a finitely generated field with separable closure K_s and absolute Galois group G_K. By a curve C/K we always understand a smooth geometrically irreducible projective curve. Let $F(C)$ be its function field and let $\Pi(C)$ be the Galois group of the maximal unramified extension of $F(C)$. We have the exact sequence

$$1 \to \Pi_g(C) \to \Pi(C) \to G_K \to 1 \qquad\qquad (\star)$$

where $\Pi_g(C)$ is the geometric (profinite) fundamental group of $C \times \mathrm{Spec}(K_s)$ (i.e. $\Pi_g(C)$ is equal to the Galois group of the maximal unramified extension of $F(C) \otimes K_s$).

This sequence induces a homomorphism ρ_C from G_K to $Out(\Pi_g(C))$ which is the group of automorphisms modulo inner automorphisms of $\Pi_g(C)$. It is well known that ρ_C is an important tool for studying C. For instance, it determines C up to $K-$isomorphisms if the genus of C is at least 2 and K is a number field or even a $\mathfrak{p}-$adic field (see [Mo]).

So it is of interest to find quotients $\bar{\Pi}(C)$ of $\Pi(C)$ such that the induced map of G_K is not the identity but the induced representation $\bar{\rho_C}$ becomes trivial. We give a geometric interpretation of such quotients. For this we assume that the sequence $(\star)$ is split and choose a section s which induces a homomorphism σ from G_K to $Aut(\Pi_g(C))$. Let U be a normal subgroup of $\Pi(C)$ contained in $\Pi_g(C)$. The representation ρ_C becomes trivial modulo U if and only if the map $\sigma \bmod U$ has its image inside of $Inn(\Pi_g(C)/U)$.

Let Z be the center of $\Pi_g(C)/U$ and $\bar{\Pi} = (\Pi_g(C)/U)/Z$. Our condition on U implies that there is exactly one group theoretical section $\bar{s}$ from G_K to $(\bar{\Pi}(C)/U)/Z$ inducing the trivial action on $\bar{\Pi}$ and so $\bar{\Pi}$ occurs as Galois group of an unramified regular extension of $F(C)$ in a natural way.

Thus, to find center free infinite factors of Π_g on which ρ_C becomes trivial is equivalent with finding infinite regular Galois coverings of C. Choosing as base point a geometric point of C we can say that "$\bar{\Pi}$ is a factor of the geometric fundamental group of C over K".

Remark 1.1 If $\bar{s}$ can be extended to a (group theoretical) section into $\Pi(C)/U$ then again it induces the trivial action on $\Pi_g(C)/U$ and hence this group occurs as Galois group of a regular unramified extension of $F(C)$. This is always true if K is a finite field or if Π/U is abelian.

But there is a difficulty: The sections $\bar{s}$ corresponding to different quotients of Π_g may not be compatible (i.e. the composite of the corresponding coverings are not regular) and so we do not have a "maximal unramified regular covering" of C over K.

So we specialize, assume that C has a K–rational point P and choose a splitting s_P of $(\star)$ corresponding to P by identifying G_K with the decomposition group of an extension of the place corresponding to P in $F(C)$ to its separable closure . Now the finite quotients of $\Pi_g(C)$ on which $s_P(G_K)$ operates trivially correspond to unramified Galois coverings C' of C on which the decomposition group of P operates trivially. Hence P has K-rational extensions to C' and the choice of s_P corresponds to the choice of such an extension P'. (If we make another choice, then the corresponding section is replaced by a conjugate in $\Pi(C)$.) We can regard C' as etale covering of C with respect to the base points P resp. P'.

Taking the limit we get the **K-rational geometric fundamental group of C with base point P** :

$$\Pi_g(C, P) := \Pi_g(C)/\langle (s_P(\sigma) - 1)\Pi_g(C)\rangle_{\sigma \in G_K}.$$

Remark 1.2 The construction of unramified coverings with base points is also of interest for coding theory:
Assume that $g(C) \geq 1$. Since

$$g(C') - 1 = [C' : C](g_C - 1)$$

and $\#\{P' \in C'(K)\}$ divisible by $[C' : C]$ we get (if $C'(K) \neq \emptyset$) that

$$\frac{|C'(K)|}{g(C') - 1} \geq \frac{1}{(g_C - 1)}$$

and this means that C' lies in a "good family" of coverings and so it is desirable to have infinite towers in this restricted sense over curves C (of small genus).

If K is a finite field such towers can be constructed by using class field theory of curves over any finite field (cf. Serre[Se1]) or over fields of square order by using Shimura curves (cf. [Ih]). By Ihara's method one can even obtain curves of genus 2 which are optimal. The curve given by $y^2 = x^6 - 1$ is such an example.
All these known examples are defined over finite fields and are of very special type.

In this paper we shall use either quartic coverings of the projective line or quadratic coverings of elliptic curves with enough ramification points (instead of quadratic coverings of the projective line as in the class field tower constructions) to find for every genus $g \geq 3$ curves with infinite geometric fundamental group defined over $\mathbb{Q}(i)$ or over $\mathbb{F}_q(i)$ (where i is a forth root of unity) and we find even parametric families of such curves over every ground field containing i (cf. Theorem 5.22). [1]

More precisely we get the

Result: *Let K be a field of characteristic $\neq 2, 3$ which contains a fourth root of unity. Take $a = \frac{2b^2}{1+b^4}$ with $b \in K^*$, $b^4 \neq 1$.*
Let C be a cyclic covering of degree 4 of the projective line given by an equation

$$s^4 = (1 + b^4)t(t^2 - 1)(t - a)g(t)$$

with a polynomial $g(t)$ such that $g(0)g(1)g(-1) \neq 0$ and $g(a) = 1$. Then C has an infinite K-rational geometric fundamental group with base point P_a corresponding to the place $t = a$.

We apply this and get the

Consequence: *Let b and a be as above.*
Let E_b be the elliptic curve defined over K by the equation

$E_b :$ $\qquad\qquad y^2 = (1+b^4)x(x^2-1)(x-a).$

Then for every natural number $g \geq 3$ there exist quadratic coverings C of E_b defined over K of genus $g_C = g$ with infinite K-rational geometric fundamental group with base point lying over P_a, the point of E_b corresponding to $x = a$.

Example: *Let K, b and a be as above. The curve*

$$C_a : y^4 = (1 + b^4)x(x^2 - 1)(x - a)$$

has an infinite K-rational geometric fundamental group with base point $(a, 0)$.

We thus have examples of curves of genus g with an infinite K-rational geometric fundamental group for every $g \geq 3$. No such example can exist for curves of genus $g = 1$, as will be explained in the next section. This leaves only the case $g = 2$. Here (cf. Example 4.13) we shall find very special curves with an infinite tower of regular unramified Galois coverings but we cannot decide there whether there are rational points in the tower.

There are other interesting aspects related to curves of genus 2 in the context of the questions discussed in this paper which will be investigated in more detail in the paper [FKV].

---

[1] We remark that we do not have any example of a curve defined over $\mathbb{Q}$ with infinite $\mathbb{Q}$-rational geometric fundamental group.

2 Abelian coverings

In this section we shall give a short review of the special case of abelian unramified coverings.

The largest abelian factor of $\Pi_g(C)$ which appears as the Galois group of a regular unramified extension of $F(C)$ is $\Pi_g(C)^{ab}/\langle (s_P(\sigma) - 1)\Pi_g(C)^{ab}\rangle_{\sigma \in G_K}$ and so it is independent of the choice of P or of the choice of the splitting. (But the coverings will depend on the choices.) Moreover, one sees that every abelian group which can be realized as Galois group of a regular unramified covering of C can be realized in such a way that a given point $P \in C(K)$ splits completely (cf. [V] or [Se]). So we do not have to distinguish between extensions with or without K-rational base points as long as we only want to describe the occurring Galois groups.

Let us begin by looking at elliptic curves. Here we have the following finiteness result:

Theorem 2.1 *Let K be a finitely generated field. Then there is a number $N = N(K)$ such that for all elliptic curves E defined over K we have $|E(K)_{tor}| \leq N$.*

Proof. We use induction with respect to the transcendence degree of K over its prime field K_0.

If K is a finite field the claim is obvious. If K is a number field the theorem of Merel gives a bound N depending on $[K : \mathbb{Q}]$ only (cf. [Me]).

Let K be transcendental of degree $d_K \geq 1$ over K_0. We fix a subfield $K_1 \subset K$ such that K_1 is finitely generated over K_0 and such that K is a function field of one variable over K_1 of genus g_K.

Let E/K be an elliptic curve with a K-rational point P of order n. Let K_2 be the smallest extension field of K_1 in K over which E is defined and over which P is rational.

If K_2 is equal to K_1 we can use the induction hypothesis, otherwise K_2 is a function field over K_1 with a K_1-rational embedding of the function field of the modular curve $X_1(n)$ into K_2 and hence into K.

Thus, g_K bounds the genus of $X_1(n)$ which is of size $O(n^2)$ (cf. Miyake[Mi], if $(n, \mathrm{char}(K)) = 1$ and Igusa[Ig], for $n = \ell^r$, $\ell = \mathrm{char}(K)$), and so n is bounded.

Corollary 2.2 *There is a universal bound $n = n(K)$ such that for all regular unramified Galois covers ϕ of elliptic curves E over K we have $\deg(\phi) \leq n$.*

Proof. If ϕ is an unramified Galois covering of E, then the covering curve E' is an elliptic curve too, ϕ is an isogeny and the Galois group G_K acts on the kernel of ϕ trivially, i.e. $\mathrm{Ker}(\phi) \leq E'(K)$. Using the universal bound for torsion points obtained in the theorem the corollary follows.

Next we consider the case that the genus of C is larger than 1, and study as before the existence of regular unramified abelian Galois coverings of C.

Proposition 2.3 *If $\phi : C_1 \to C$ is a regular abelian unramified covering, then $\deg \phi$ is bounded by a number depending on K and on C.*

Proof. By Serre [Se] we know that there is an abelian variety A/K such that ϕ induces an isogeny

$$\phi_* : A \to J_C$$

with $\mathrm{Ker}(\phi_*) \subset A(K)$ and $|\mathrm{Ker}(\phi_*)| = \deg(\phi)$. If K is finite we have $|A(K)| = |J_C(K)|$, and so we are done. If K is a number field or a field of transcendence degree larger than 1 over its prime field we can use reduction theory to get the result.

Remark: Since at present we do not have an analogue of Merel's theorem for abelian varieties, we can only find a bound depending on the genus and on the conductor of C, and not only on K, even if K is a number field.

3 Coverings of $\mathbb{P}^1$ with given ramification type

3.1 The Construction Principle

In the last section we had seen that for a given curve C over a finitely generated field K, the degree of the abelian unramified regular extensions of C/K is bounded.

The situation changes if we look for non-abelian coverings. Indeed, in the next section we shall give examples of families of curves of arbitrary genus $g \geq 3$ defined over a finitely generated field K which have infinite towers of unramified regular Galois coverings. However, it should be remarked that the dimension of these families is small compared with the dimension of the moduli space of curves of genus g and so it can be conjectured that for "general curves", the degree of a regular unramified Galois covering is bounded by a number $N(g, K)$.

The construction of these curves is based on the following principle. We begin with the projective line $\mathbb{P}^1$ and choose a finite G_K-invariant set S_0 of points of $\mathbb{P}^1$. Then we look for Galois coverings C' of $\mathbb{P}^1$ with a *simple* Galois group G (in our examples G is a projective linear group). We require that this covering is unramified outside of S_0 and that the ramification in points $P_i \in S_0$ has an order e_{P_i} which is small and prime to $\mathrm{char}(K)$. We take a Galois covering C_0 of $\mathbb{P}^1$ disjoint to C' which ramifies in all P_i with an order which is a multiple of e_{P_i}. By Abhyankar's Lemma it follows that $C' \times_{\mathbb{P}^1} C_0/C_0$ is unramified with Galois group G and, because of the simplicity of G, regular.

To find C' we follow [FV] and use Hurwitz spaces which are moduli spaces for the ramification cycle configuration inside of G; this boils down to proving the existence of K-rational points on these spaces. This Diophantine problem leads to interesting relations between arithmetic geometry and group theory. Sometimes group theory or, to be more precise, the theory of rigidity implies the existence of rational points on the corresponding Hurwitz spaces. This is the case in our examples which are related to Thompson tuples, and so in this case the existence of such coverings is obtained by pure group theory (and Riemann's existence theorem).

However, this method fails to produce K-rational points on these coverings and so arithmetic-geometric methods have to be invoked in order to prove the existence of such points (which is necessary for our considerations of K-rational fundamental groups).

In the case of the Thompson tuples, such a method is readily available, for behind their very construction (cf. [V3]) lies a deep arithmetic reason, viz. *the existence of a family of abelian varieties with good reduction.*

3.2 Rigidity

For the convenience of the reader we shall now give a very short overview over rigid systems of conjugacy classes. We shall assume that char$(K) = 0$, but actually it is enough to avoid fields with characteristics dividing the order of the occurring Galois groups and we shall use this fact in our examples.

We want to describe the regular Galois coverings of the projective line. For this, we recall the situation over the complex numbers over which we can use as crucial tool the Riemann existence theorem.

We have the following invariants for a finite Galois extension $L/\mathbb{C}(x)$: its Galois group G, the finite set $\underline{P} = \{P_1, ..., P_r\}$ of branch points (which lie in $\mathbb{P}^1(\mathbb{C}) = \mathbb{C} \cup \{\infty\}$) and for each branch point P_i the associated conjugacy class C_i of G consisting of the distinguished inertia group generators over P_i. Define the **ramification type** of $L/\mathbb{C}(x)$ to be the class of the triple $(G, \underline{P}, \mathbf{C})$, where $\mathbf{C} = (C_1, ..., C_r)$. We say that two such triples are in the same class if the sets $\underline{P}$ are the same and if there is an isomorphism between the groups that identifies the conjugacy classes associated with the same point of $\underline{P}$.

The extension $L/\mathbb{C}(x)$ is defined over a subfield K of $\mathbb{C}$ if there is a Galois extension $L_K/K(x)$, regular over K, which becomes equivalent to L when tensored with $\mathbb{C}$.

$\mathbf{C}$ is **weakly rigid** if any two generating systems $\sigma_1, ..., \sigma_r$ of G with $\sigma_i \in C_i$ and $\sigma_1 \cdots \sigma_r = 1$ are conjugate under an automorphism of G and if there is one such generating system. By Riemann's existence theorem, $\mathbf{C}$ is weakly rigid if and only if there is exactly one Galois extension of $\mathbb{C}(x)$ of type $(G, \underline{P}, \mathbf{C})$ (up to equivalence), see [V], Th. 2.17. This uniqueness can be exploited to define the covering of $\mathbb{P}^1$ over a smaller field. To do this, we sharpen the group theoretical conditions for the ramification type as follows.

C is **quasi-rigid** if it is weakly rigid and every automorphism of G fixing each C_i is inner.

Now suppose **C** is quasi-rigid and that $L/\mathbb{C}(x)$ is a corresponding extension. Let L' be the fixed field of the center $Z(G)$ of G. Then the extension $L'/\mathbb{C}(x)$ is defined over the field generated over $\mathbb{Q}$ by all branch points and by all roots of unity of order $\mathrm{ord}(\sigma)$, $\sigma \in C_1 \cup ... \cup C_r$. The exact minimal field of definition of L' may be even smaller and can be determined using the branch cycle argument (see [V], 3.2 and 3.3.2). This construction principle for Galois extensions of rational function fields ("rigidity criterion") has successfully been used to realize given groups as Galois groups.

4 Construction of infinite towers of unramified curve covers

4.1 Thompson tuples and Belyi triples

Let q be a power of the prime p, and $n \geq 3$. We denote by $\mathbb{F}_q^*$ the multiplicative group of the field $\mathbb{F}_q$ with q elements. For any $\sigma \in \mathrm{GL}_n(q)$, let $c_1, ..., c_n$ be the eigenvalues of σ (counted with multiplicities) and $\chi_\sigma(T) = \prod_i (T - c_i)$ its characteristic polynomial (normalized to be monic). We call σ a **perspectivity** if it has an eigenspace of dimension $n - 1$.

Definition 4.1 Let $r = n + 1$ (respectively, $r = 3$). A *Thompson tuple* (respectively, a *Belyi triple*) in $\mathrm{GL}_n(q)$ is an r-tuple $(\sigma_1, ..., \sigma_r)$ such that $\sigma_1, ..., \sigma_r$ generate an irreducible subgroup of $\mathrm{GL}_n(q)$, their product satisfies $\sigma_1 \cdots \sigma_r = 1$ and σ_i is a perspectivity for all $i = 1, ..., r$ (respectively, for $i = 3$).

In [V1] one finds

Theorem 4.2 *Suppose that $(\sigma_1, ..., \sigma_r)$ and $(\sigma_1', ..., \sigma_r')$ are Thompson tuples (respectively, Belyi triples) in $\mathrm{GL}_n(q)$ with $\chi_{\sigma_i} = \chi_{\sigma_i'}$ for all i. Then there is an element $g \in \mathrm{GL}_n(q)$ with $\sigma_i' = g^{-1}\sigma_i g$ for all i. Thus $\sigma_1, ..., \sigma_r$ are weakly rigid generators of $G = \langle \sigma_1, ..., \sigma_r \rangle$, and they are quasi-rigid generators if the normalizer of G in $\mathrm{GL}_n(q)$ equals $\mathbb{F}_q^* G$.*

We can apply the theory of rigid ramification types presented in the last section to obtain (cf. [V], Theorems 2.17, 3.26 and 7.9):

Corollary 4.3 *Let* $(\sigma_1, ..., \sigma_r)$ *be a Thompson tuple or a Belyi triple in* $GL_n(q)$. *Let* C_i *be the conjugacy class of* σ_i *in* $G = \langle \sigma_1, ..., \sigma_r \rangle$, *and let* $P_1, ..., P_r$ *be distinct points of* $\mathbb{P}^1(\mathbb{C})$. *Put* $\underline{P} = \{P_1, ..., P_r\}$, $C_{P_i} = C_i$, *and* $\mathbf{C} = (C_P)_{P \in \underline{P}}$. *Then we have:*

(a) There is a unique Galois extension $L'/\mathbb{C}(x)$ *of ramification type* $[G, \underline{P}, \mathbf{C}]$.

(b) Suppose that the normalizer of G *equals* $\mathbb{F}_q^* G$. *If* $L = Fix(Z(G))$ *denotes the fixed field of the center* $Z(G)$ *of* G, *then the extension* $L/\mathbb{C}(x)$ *is defined over any subfield* $K \subset \mathbb{C}$ *which contains all the (finite)* $P_i \in \underline{P}$ *and all roots of unity of order* $ord(\sigma_i)$. *In particular, there is a regular Galois extension* L_0 *of* $K(x)$ *with Galois group* $\bar{G} = G/Z(G)$ *with ramification type* $[\bar{G}, \underline{P}, \bar{\mathbf{C}}]$, *where* $\bar{C}_P$ *is the image of* C_P *in* $\bar{G}$.

Corollary 4.4 *Suppose in addition that* M *is a regular extension of* $K(x)$ *which is linearly disjoint to the above extension* L_0 *over* K, *and which has the property that for each* $i = 1, ..., r$ *and each place* m *of* M *lying above the place corresponding to* P_i *of* $\mathbb{C}(x)$, *the ramification index of* m *is a multiple of* $ord(\sigma_i)$. *Then the composite* $L_0 \cdot M$ *is an unramified Galois extension of the function field* M *and its Galois group is isomorphic to* $\bar{G}$.

We therefore see that to find such field extensions L, it is enough to find Thompson tuples or Belyi triples. These will now be investigated in more detail. The first result in this direction (taken from [V1]) shows that they can be characterized by their characteristic polynomials:

Theorem 4.5 *Let* $f_1, ..., f_r \in \mathbb{F}_q[T]$ *be monic polynomials of degree* n *with* $\prod_i f_i(0) = (-1)^{rn}$.

(a) There exists a Belyi triple $(\sigma_1, \sigma_2, \sigma_3)$ *in* $GL_n(q)$ *with* $\chi_{\sigma_i} = f_i$ *if and only if* $f_3(T) = (T - c)^{n-1}(T - d)$ *for elements* $c, d \in \mathbb{F}_q^*$ *satisfying* $abc \neq 1$ *for all* $a, b \in \bar{\mathbb{F}}_q$ *with* $f_1(a) = f_2(b) = 0$.

(b) There exists a Thompson tuple $(\sigma_1, ..., \sigma_r)$ *in* $GL_n(q)$ *with* $\chi_{\sigma_i} = f_i$ *if and only if* $f_i(T) = (T - a_i)^{n-1}(T - b_i)$ *for elements* $a_1, ..., a_r, b_1, ..., b_r \in \mathbb{F}_q^*$ *satisfying* $a_1 \cdots a_r \neq 1$ *and* $b_j \prod_{i \neq j} a_i \neq 1$ *for all* $j = 1, ..., r$.

From Theorem 4.2 we see that the tuples $(\sigma_1, ..., \sigma_r)$ are classified by their characteristic polynomials χ_{σ_i}, and so the group $G = \langle \sigma_1, ..., \sigma_r \rangle$ depends only on the χ_{σ_i} (up to conjugation). In the case of Thompson tuples there is a complete classification of the related groups (cf. [V1] and [V4]). In the following Theorem 4.6 we shall state only that part of this classification which will be relevant for us below.

In order to state the result, we require the following terminology. A Thompson tuple $(\sigma_1, ..., \sigma_{n+1})$ is called **normalized** if $\chi_{\sigma_i} = (T - 1)^{n-1}(T - b_i)$ for $i = 1, ..., n$.

Theorem 4.6 *Let n be odd and ≥ 3. Suppose $(\sigma_1, ..., \sigma_{n+1})$ is a normalized Thompson tuple in $\mathrm{GL}_n(q)$ generating the group G with $\chi_{\sigma_i}(T) = (T - 1)^{n-1}(T - b_i)$ for certain $b_i \in \mathbb{F}_q$ for $i = 1, \cdots, n$. If $n = 3$ assume that some b_i has multiplicative order > 3. Suppose that $\mathbb{F}_q = \mathbb{F}_p(a_{n+1}, b_1, \cdots b_{n+1})$, and assume that if $q = q_0^2$ is a square, then not all the a_i, b_i have norm 1 over $\mathbb{F}_{q_0}$. Then*

$$\mathrm{SL}_n(q) \leq G \leq \mathrm{GL}_n(q).$$

4.2 Curves with infinite towers of unramified regular Galois extensions

We now apply the above results to construct unramified Galois G-covers of curves with group $G = \mathrm{PSL}_n(p)$ for suitable values of n and p. In order to be able to specify precisely which values are suitable here, we first introduce the following notation.

Notation. Let $n \geq 3$ be an odd natural number, and let $d_1, ..., d_{n+1} > 3$ be integers. Then we denote by $\mathbb{P}(n, d_1, \cdots, d_{n+1})$ the set of all prime numbers p satisfying the following conditions:

(i) $(n, p - 1) = 1$;

(ii) there are elements $b_1, \ldots, b_{n+1} \in \mathbb{F}_p^*$ with $b_1 \cdots b_{n+1} = 1$ such that each b_i has multiplicative order d_i in $\mathbb{F}_p^*$.

A trivial but useful fact is

Lemma 4.7 *If $d_1, \ldots, d_n$ are prime to n and $d_{n+1} = lcm(d_1, \ldots, d_n)$, then every prime p with $p \not\equiv 1 \bmod l$ for all primes $l \mid n$ and $p \equiv 1 \bmod d_{n+1}$ is in $\mathbb{P}(n, d_1, \cdots, d_{n+1})$. Moreover, there is a constant c (depending only on n and $d_1, \ldots, d_{n+1}$) such that for all primes $l \geq c$ there are infinitely many $p \in \mathbb{P}(n, d_1, \cdots, d_{n+1})$ such that l does not divide the order of $\mathrm{GL}_n(p)$.*

From now on, suppose that the integers $d_1, ..., d_{n+1}$ satisfy the conditions of this lemma, and that K is a field of characteristic 0 containing a primitive d^{th} root of unity, where $d = lcm(d_1, \ldots, d_{n+1}, 2) = lcm(d_{n+1}, 2)$. Moreover, let $p \in \mathbb{P}(n, d_1, \cdots, d_{n+1})$ and choose $b_i \in \mathbb{F}_p^*$ such that condition (ii) holds. Then by Theorem 4.5 there is a normalized Thompson tuple $(\sigma_1, ..., \sigma_{n+1})$ in $\mathrm{GL}_n(p)$ corresponding to these b_i with $a_{n+1} = -1$. The order of σ_i equals d_i for $i \leq n$ and it equals $lcm(d_{n+1}, 2) = d$ for $i = n + 1$.

By Theorem 4.6 the group $G = \langle \sigma_1, ..., \sigma_{n+1} \rangle$ contains $\mathrm{SL}_n(p)$. Thus the group $\bar{G} = G/Z(G)$ is isomorphic to the simple group $\mathrm{PSL}_n(p)$ (because we assumed $(n, p - 1) = 1$).

It remains to choose the ramification points.

Since the automorphism group of $\mathbb{P}^1$ acts sharply triple transitive we can assume without loss of generality that $P_{n-1} = 0$, $P_n = 1$ and $P_{n+1} = \infty$. The other ramification points are then denoted by $t_1, \cdots, t_{n-2} \in K$.

Using Corollary 4.3 we see that for every $p \in \mathbb{P}(n, d_1, \cdots, d_{n+1})$, we get a regular Galois extension F'_p of $K(X)$ with Galois group $\mathrm{PSL}_n(p)$ ramified only in $0, 1, \infty, t_1, \ldots, t_{n-2}$ with ramification indices dividing d.

Since for different primes p the Galois groups of the F''_p's are non-isomorphic simple groups, the composite field

$$F'_{\mathbb{P}(n, d_1, \cdots, d_{n+1})} = \prod_{p \in \mathbb{P}(n, d_1, \cdots, d_{n+1})} F'_p$$

has the same regularity and ramification properties.

Next, let M_d be any regular extension field of $K(X)$ such that every extension of the places $0, 1, \infty, t_1, \cdots, t_{n-2}$ is ramified with an order divisible by d. (For instance, take M_d as a cyclic cover of $K(x)$ of degree d totally ramified at $0, 1, \infty, t_1, \cdots, t_{n-2}$. In this case the genus of M_d is equal to $(n-1)(d-1)/2$.) Using Corollary 4.4, we get that $F'_{\mathbb{P}(n, d_1, \cdots, d_{n+1})} \cdot M_d$ is a regular unramified extension of M_d with Galois group equal to

$$\prod_{p \in \mathbb{P}(n, d_1, \cdots, d_{n+1})} \mathrm{PSL}_n(p).$$

Thus, if we take $K = \mathbb{Q}(\zeta_d, t_1, \ldots, t_{n-2})$, where the $t_1, \ldots, t_{n-2}$ are algebraically independent over $\mathbb{Q}$, then we obtain:

Theorem 4.8 *Let $n \geq 3$ be an odd number and let $d_1, \cdots, d_n > 3$ be integers prime to n, $d_{n+1} := \mathrm{lcm}(d_1, \cdots, d_n)$ and $d = \mathrm{lcm}(2, d_{n+1})$. Then $\mathbb{P}(n, d_1, \cdots, d_{n+1})$ is an infinite set and there is an $(n-2)$-dimensional rational family $\mathcal{C}$ of curves of genus $(n-1)(d-1)/2$ defined over $\mathbb{Q}(\zeta_d)$ such that for all $p \in \mathbb{P}(n, d_1, \cdots, d_{n+1})$ there is an unramified regular Galois cover $\mathcal{C}_p$ with Galois group $\mathrm{PSL}_n(p)$ and the composite of these covers is an infinite unramified regular Galois tower.*

On the other hand, if we choose the t_i's to be algebraic over $\mathbb{Q}$, we obtain:

Theorem 4.9 *Let K be a number field. Let n be odd and at least equal to 3 and let d be as in the previous theorem. Then there are infinitely many curves of genus $(d-1)(n-1)/2$ defined over K with infinite towers of unramified regular Galois covers defined over $K(\zeta_d)$.*

As was mentioned above, the Hurwitz space theory and the theory of Thompson tuples works also over fields of characteristic l as long as l is prime to the order of the Galois group. Using the second part of Lemma 4.7 we see that Theorem 4.8 holds if we replace $\mathbb{Q}$ by $\mathbb{F}_l$ for almost all primes l. Hence we have

Corollary 4.10 *Let n and d be as above, and let l be a sufficiently large prime such that there are infinitely many primes p with $p^i - 1$ prime to l for $i = 1, \ldots, n$. Then there exist curves of genus $(d-1)(n-1)/2$ defined over $K = \mathbb{F}_l(\zeta_d)$ which have an infinite tower of unramified regular Galois covers defined over K.*

Now we discuss the simplest possibility and get the following example which will play an important role in Section 5.3.

Example 4.11 Let $n = 3$ and $d_1 = \ldots = d_4 = d = 4$, and let K be a field of characteristic ≥ 5 which contains a fourth root of unity $\zeta_4 = i$. Furthermore, let $t \in K$ be such that $t \neq 0, 1$.
Then for every $p \equiv 5 \bmod 12$ we get an extension of $\mathbb{P}^1$ defined over K whose Galois group is equal to $PSL_3(p)$ and which is ramified of order 4 in $\{0, 1, \infty, t\}$.
Fix an integer $g_0 \geq 1$. If $g_0 = 1$, let C_0 be an elliptic curve which covers $\mathbb{P}^1$ with ramification points $\{0, 1, \infty, t\}$, and if $g_0 > 1$, let C_0 be a hyperelliptic curve of genus g_0 whose set of Weierstrass points contains $\{0, 1, \infty, t\}$. Let C be a quadratic extension of C_0 ramified in the extensions of $\{0, 1, \infty, t\}$. Then C has genus $g := g_0 + \delta/2$, where δ is the number of ramified primes in the cover C/C_0.
For example, if we choose C as a cyclic cover of degree 4 over $\mathbb{P}^1$ which is ramified in the Weierstrass points of C_0 with ramification order 4 and in s additional points of $\mathbb{P}^1$ with ramification order 2, then $g = 3g_0 + s$.

Hence we get

Corollary 4.12 *Let K be a field of characteristic $\neq 2, 3$ which contains a primitive fourth root of unity $i = \zeta_4$. Let $t \in K$, $t \neq 0, 1$ and let C_0 be either an elliptic curve ($g_0 = 1$) or a hyperelliptic curve of genus g_0 defined over K with at least four K-rational points of order 2 respectively, Weierstrass points equal to $\{0, 1, \infty, t\}$.*
Then for all non-negative integers s there are quadratic covers of C_0 defined over K of genus $3g_0 + s$ with an infinite tower of regular unramified Galois covers over K.

If we want to obtain examples as above for curves of genus 2, then we have to use Belyi triples, but in that case we find only isolated examples. Here is one of them:

Example 4.13 Let Φ_m be the cyclotomic polynomial of order m, and let $(\sigma_1, \sigma_2, \sigma_3)$ be a Belyi triple in $GL_3(p)$ with characteristic polynomials $\chi_{\sigma_1} = (T + 1)\Phi_6$, $\chi_{\sigma_2} = (T - 1)\Phi_3$, and $\chi_{\sigma_3} = (T - 1)^2(T + 1)$. Then the images $\bar{\sigma}_1, \bar{\sigma}_2, \bar{\sigma}_3$ generate the simple group $PSL_3(p)$ for $p \equiv -1 \bmod 3$, and there is a corresponding extension $L_p/\mathbb{Q}(x)$ branched at $0, \infty, 1$. Take $x = t^6$ and let

$M/\mathbb{Q}(t)$ be the genus 2 function field extension branched where t is a sixth root of 1, i.e. M corresponds to the curve with equation

$$y^2 = t^6 - 1.$$

Then $L_p \cdot M/M$ is unramified with group $\mathrm{PSL}_3(p)$ and defined over $\mathbb{Q}$. It follows that the curve $y^2 = t^6 - 1$ has an infinite tower of regular unramified Galois extensions over $\mathbb{Q}$ and over $\mathbb{F}_l$ for all primes $l \geq 5$.

We observe that although we obtain in this way the example of Ihara mentioned in the first section, we have actually realized different Galois groups by this method. At present, however, we cannot answer the question concerning the existence of rational points in the associated infinite tower.

Further examples (for $\mathrm{PSL}_3(p)$ again) are obtained by taking χ_{σ_3} as above, and $\chi_{\sigma_1} = (T-1)\Phi_3$, $\chi_{\sigma_2} = (T+1)\Phi_4$, or alternatively, $\chi_{\sigma_1} = (T+1)\Phi_3$, $\chi_{\sigma_2} = (T-1)\Phi_4$. We thank G. Malle for pointing out these examples.

4.3 Geometric Interpretation of Thompson tuples

In the last section, the examples of almost unramified covers of $\mathbb{P}^1$ were constructed by using group theory. This construction did not provide any method for determining the existence of rational points in the covers, nor did it explain the arithmetical "reason" behind the existence of such examples. However, both these aspects are greatly elucidated by using the results of [V3] to relate the Thompson tuples to torsion points of Jacobians of curves which are cyclic covers of the projective line.

The basic idea here is the following. If $\phi : C \rightarrow \mathbb{P}^1_K$ is a cyclic covering with group $Z = \mathrm{Aut}(\phi)$ defined over a field K and $\lambda : Z \rightarrow \mathbb{F}^*_q$ is a faithful character, then the Galois group G_K acts $\mathbb{F}_q$-linearly on the λ-component $J_C[p]_\lambda$ of the group $J_C[p]$ of p-torsion points of the Jacobian variety J_C of C. Now it turns out that if the ramification points $t_1, \ldots, t_{n+2} \in K$ of ϕ are in "sufficiently general position", then the image of the Galois group G_K in $\mathrm{GL}_n(J_C[p]_\lambda)$ is (essentially) generated by a Thompson tuple, and conversely, every Thompson tuple arises in this fashion.

In order to make this connection between cyclic coverings and Thompson tuples more precise, it is useful to introduce the following alternate description of Thompson tuples, using the *Braid group*.
For this, let $s = n+2$, and let $Q_1, \ldots, Q_{s-1}$ be the standard generators of the *Artin braid group* $\mathcal{B}_s$ on s strings. Mapping Q_i to the transposition $(i, i+1)$ yields a homomorphism $\kappa : \mathcal{B}_s \rightarrow S_s$. The kernel of κ is called the *pure braid group*, and is denoted by $\mathcal{B}^{(s)}$. It is generated by the elements

$$Q_{ij} = Q_{j-1} \cdots Q_{i+1} \, Q_i^2 \, Q_{i+1}^{-1} \cdots Q_{j-1}^{-1}, \quad 1 \leq i < j \leq s.$$

The braid group $\mathcal{B}_s$ acts on tuples $(g_1, \ldots, g_s)$ of elements of any group by the rule that Q_i maps $(g_1, \ldots, g_s)$ to

$$(g_1, \ldots, g_{i-1}, \ g_{i+1}, \ g_{i+1}^{-1} g_i g_{i+1}, \ldots, g_s)$$

Lemma 4.14 *(a) If $\underline{\zeta} = (\zeta_1, \ldots, \zeta_{n+2})$ is an $(n+2)$-tuple of elements $\zeta_i \in \mathbb{F}_q^*$ satisfying the conditions*

$$\prod_{j=1}^{n+2} \zeta_j = 1, \quad \zeta_i \neq 1, \text{ for all } i, \tag{1}$$

then there is a unique (up to conjugation) normalized Thompson tuple $\underline{\sigma} = \underline{\sigma}_\zeta = (\sigma_1, \ldots, \sigma_{n+1})$ in $\mathrm{GL}_n(q)$ such that

$$\begin{aligned} \chi_{\sigma_i}(T) &= (T-1)^{n-1}(T - \zeta_i^{-1}\zeta_{n+2}^{-1}), \quad 1 \leq i \leq n, \\ \chi_{\sigma_{n+1}}(T) &= (T - \zeta_{n+2})^{n-1}(T - \zeta_{n+1}^{-1}). \end{aligned}$$

Moreover, every normalized Thompson tuple $\underline{\sigma}$ in $\mathrm{GL}_n(q)$ is of the form $\underline{\sigma} = \underline{\sigma}_\zeta$, for a unique $(n+2)$-tuple $\underline{\zeta}$ satisfying (1).

(b) In the situation of (a) there is a homomorphism

$$\Phi_\zeta : \mathcal{B}^{(s)} \to \mathrm{GL}_n(q)$$

such that for $i = 1, \ldots, s-1$ the elements $\tau_i := \Phi_\zeta(Q_{is})$ are perspectivities with $\chi_{\tau_i} = (T-1)^{n-1}(T - \zeta_i^{-1}\zeta_s^{-1})$ and $\tau_1 \cdots \tau_{s-1} = \zeta_s^{-1} \cdot$ id. Thus $(\tau_1, \ldots, \tau_{s-2}, \zeta_s \tau_{s-1})$ is a Thompson tuple associated with $\underline{\zeta}$.

Proof. (a) Put $s = n + 2$. Recall from Theorem 4.5 that a normalized Thompson tuple $(\sigma_1, \ldots, \sigma_{n+1})$ is characterized by the characteristic polynomials of the elements σ_i and so by the s-tuples $(b_1, \cdots, b_{n+1}, a) \in (\mathbb{F}_q)^{n+2}$ which satisfy the additional properties that $\prod_{i=1}^{n+1} b_i \cdot a^{n-1} = 1$ and that $a \neq 1$, $b_{n+1} \neq 1$ and $a \cdot b_i \neq 1$ for $1 \leq i \leq n$.

We associate to $(\sigma_1, \ldots, \sigma_{n+1})$ the s−tuple $(\zeta_1, \cdots, \zeta_{n+1}, \zeta_s)$ with $\zeta_i = (ab_i)^{-1}$ for $1 \leq i \leq n$, $\zeta_{n+1} = b_{n+1}^{-1}$ and $\zeta_s = a$. This tuple has the property that all ζ_i are different from 1 and that $\prod_{i=1}^s \zeta_i = 1$.

Conversely, if $\zeta_1, \ldots, \zeta_{n+2} \in \mathbb{F}_q^*$ satisfying condition (1) are given, then $a := \zeta_{n+2}$, $b_{n+1} := \zeta_{n+1}$ and $b_i := \zeta_i^{-1}\zeta_{n+2}^{-1}$, for $1 \leq i \leq n$ satisfy the above conditions and hence give rise to a Thompson tuple $\underline{\sigma}_\zeta = (\sigma_1, \ldots, \sigma_{n+1})$ with the indicated characteristic polynomials.

(b) For the definition of Φ_ζ see [V4] and [MSV]. (It is essentially the *Gassner representation* of $\mathcal{B}^{(s)}$ associated with ζ). The first assertion follows from [V4], Lemma 3 and relation (5). The relation $\tau_1 \cdots \tau_{s-1} = \zeta_s^{-1} \cdot$ id follows from the definition of Φ_ζ and the fact that the element

$$Q_{1s} \cdots Q_{s-1,s} = Q_{s-1} \cdots Q_2 Q_1^2 Q_2 \cdots Q_{s-1}$$

acts by sending $(g_1, \ldots, g_s)$ to $(g_1, \ldots, g_s)^{g_s}$. The τ_i generate an irreducible group by [V1], Lemma 3.3.

We thus see that ζ-tuples give rise to Thompson tuples and conversely. On the other hand we can also attach to such tuples an (essentially unique) cyclic covering $\phi : C \to \mathbb{P}^1$ which is ramified at $s = n + 2$ prescribed points $t_1, \ldots, t_s \in K$ as follows.

Lemma 4.15 *Suppose $\zeta_1, \ldots, \zeta_s \in \mathbb{F}_q^*$ satisfy $\zeta_1 \cdot \ldots \cdot \zeta_s = 1$ and $d_i :=$ ord$(\zeta_i) \neq 1$, $1 \leq i \leq s$. Let K be a field containing a primitive d-th root of unity where $d = \mathrm{lcm}(d_1, \ldots, d_s)$ and suppose that $t_1, \ldots, t_s \in K$ are s distinct points. Then there is a cyclic covering $\phi = \phi_{\underline{t}, \zeta} : C \to \mathbb{P}^1_K$ and a faithful character $\lambda : Z := \mathrm{Aut}(\phi) \to \mathbb{F}_q^*$ such that ϕ is unramified outside of $\{t_1, \ldots, t_s\}$ and is ramified at t_i in such a way that the distinguished inertia group generator z_i at t_i satifies $\lambda(z_i) = \zeta_i$, for $1 \leq i \leq s$. Furthermore, all automorphisms of ϕ are defined over K.*

Proof. This is elementary and well-known, but in order to be able to discuss the examples below it is useful to make this precise. Thus, fix a primitive d-th root $\zeta \in \mathbb{F}_q$ (which exists by hypothesis). Then there exist unique numbers $m_1, \ldots, m_s$ with $0 < m_i < d$ such that $\zeta_i = \zeta^{m_i}$. Now consider the cyclic covering $\phi : C \to \mathbb{P}^1$ defined by the equation

$$y^d = (x - t_1)^{m_1} (x - t_2)^{m_2} \cdots (x - t_s)^{m_s}.$$

Since $(m_1, \ldots, m_s) = 1$, this is a covering of degree d and since $m_1 + \ldots + m_s \equiv 0 \bmod d$, it follows that ϕ is unramified at infinity and hence also outside of $\{t_1, \ldots, t_s\}$. Now let σ be the unique generator of $Z := \mathrm{Aut}(\phi)$ such that $\sigma(y) = e^{2\pi i/d} y$. Then $z_i = \sigma^{m_i}$ is the distinguished inertia group generator at t_i, and so if we let $\lambda : Z \to \mathbb{F}_q^*$ be the unique homomorphism such that $\lambda(\sigma) = \zeta$, then $\lambda(z_i) = \zeta_i$, as desired.

Now let J_C denote the Jacobian of the curve C of Lemma 4.15, and let p be a prime not dividing d. Then the group $Z = \mathrm{Aut}(\phi)$ acts on the group $J_C[p]$ of ($\bar{K}$-rational) p-torsion points of J_C, and so $J_C[p]$ is the direct sum of its λ-components $J_C[p]_{\lambda^k} = J_C[p] \otimes_{\mathbb{F}_p[Z]} V_{\lambda^k}$, where $V_{\lambda^k} \simeq \mathbb{F}_{q_k}$ denotes the representation space of the linear characters $\lambda^k : Z \to \mathbb{F}_{q_k}^*$ of Z. Since the automorphisms $z \in Z$ are defined over K, the Galois action commutes with the Z-action, and so G_K acts $\mathbb{F}_{q_k}$-linearly on each space $J_C[p]_{\lambda^k}$ (which has a natural $\mathbb{F}_{q_k}$-vector space structure). Since $J_C[p]_\lambda$ has dimension $n = s - 2$ over $\mathbb{F}_q$ (see [V3], or equation (3) below), this yields a homomorphism $G_K \to GL_n(q)$.

Theorem 4.16 *Let $\zeta = (\zeta_1, \ldots, \zeta_s)$ be a ζ-tuple in $\mathbb{F}_q$ satisfying condition (1) with the property that $\mathbb{F}_q = \mathbb{F}_p(\zeta_1, \ldots, \zeta_s)$, and let $\underline{\sigma}_\zeta = (\sigma_1, \ldots, \sigma_{s-1})$ be the associated Thompson tuple in $GL_n(q)$, where $n = s - 2$. In addition, let*

$t_1, \ldots, t_{s-1}$ *be distinct points in* k, *and let* $t_s = X$ *be transcendental over* k. *Put* $K = k(X)$, *and choose accordingly the cyclic covering*

$$\phi = \phi_{\underline{t}, \zeta} : C \to \mathbb{P}^1_K$$

and character $\lambda : Z \to \mathbb{F}_q^*$ *as in the previous Lemma. Let* $S = J_C[p]_\lambda$ *be the associated* λ-*component of the* p-*torsion points of* J_C, *and let* $N = N_{\lambda, Z'}$ *be the fixed field of the kernel of the homomorphism*

$$\rho_{\lambda, Z'} : G_K \to \mathrm{GL}_n(q)/Z'$$

induced by the action of G_K *on* $S \simeq \mathbb{F}_q^n$, *where* $Z' \leq Z(\mathrm{GL}_n(q))$ *is any subgroup containing* $\lambda(Z)$.

(a) The Galois group of the extension $\bar{k}N/\bar{k}(X)$ *is the subgroup*

$$G' = \langle \sigma'_1, \ldots, \sigma'_{s-1} \rangle \leq \mathrm{GL}_n(q)/Z'$$

generated by the images σ'_i *of the Thompson tuple in* $\mathrm{GL}_n(q)/Z'$. *Moreover, the ramification type of this extension is* $[G', \{t_1, \ldots, t_{s-1}\}, \mathbf{C}']$, *where* $\mathbf{C}' = (C'_1, \ldots, C'_{s-1})$ *and* C'_i *denotes the class of* σ'_i *in* G'.

(b) If G' *is self-normalizing in* $\mathrm{GL}_n(q)/Z'$, *then* N *is regular over* k, *and hence* $\mathrm{Gal}(N/k(X)) \simeq \mathrm{Gal}(\bar{k}N/\bar{k}(X)) \simeq G'$.

Proof. The analogous result for the case that the branch points $t_1, \ldots, t_s$ are algebraically independent over k was proved in [V3], Theorem A. So we have to carry through a specialization argument which essentially boils down to replacing the pure braid group $\mathcal{B}^{(s)}$ by its normal subgroup generated by $Q_{1,s}, \ldots, Q_{s-1,s}$.

As in the proof of Theorem A of [V3], let $\mathcal{O}_s$ be the space of subsets of $\mathbb{C}$ of cardinality s, and $\mathcal{H}(\mathbf{C})$ the Hurwitz space related to covers of type ζ; and let $\mathbf{p}$ be a point of $\mathcal{H}(\mathbf{C})$ over the point $\bar{\mathbf{p}} = \{t_1, \ldots, t_s\}$ of $\mathcal{O}_s$ such that the corresponding cover of $\mathbb{P}^1$ has ramification of type ζ_i over t_i. We can proceed as in that proof till to step 4.

There we specialize: $\mathcal{O}$ has to be replaced by the curve $\tilde{\mathcal{L}}$ defined as follows: Let $\mathcal{L}$ be the curve on $\mathcal{O}_s$ consisting of all $\{t_1, \ldots, t_{n+1}, t\}$, $t \in \mathbb{C} \setminus \{t_1, \ldots, t_{n+1}\}$, and $\tilde{\mathcal{L}}$ the (absolutely) irreducible component of the inverse image of $\mathcal{L}$ in $\mathcal{H}_s(G)^{(\mathbf{A}_1)}$ that contains $\mathbf{p}_1$; here $\mathbf{A}_1 = V \cdot Z'$. The fundamental group $\pi_1(\mathcal{O}_s, \bar{\mathbf{p}})$ can be identified with the braid group $\mathcal{B}_s$ in such a way that $Q_{1,s}, \ldots, Q_{s-1,s}$ correspond to loops around $t_1, \ldots, t_{s-1}$ on $\mathcal{L}$ that generate $\pi_1(\mathcal{L}, \bar{\mathbf{p}})$. Let H be the stabilizer of $\tilde{\mathcal{L}}$ in $\mathbf{A}_0/\mathbf{A}_1 \cong \mathrm{GL}_n(q)/Z'$. Let $\Phi'_\zeta : \mathcal{B}^{(s)} \to \mathrm{GL}_n(q)/Z'$ be the composition of Φ_ζ (see Lemma 4.14) with the canonical map $\mathrm{GL}_n(q) \to \mathrm{GL}_n(q)/Z'$. As in [V5], Lemma 2.2 we see that H is conjugate in $\mathrm{GL}_n(q)/Z'$ to the group G' generated by $\Phi'_\zeta(Q_{1,s}), \ldots, \Phi'_\zeta(Q_{s-1,s})$. We may assume $H = G'$.

Then similarly as in Step 4, the Galois action of $G_{\bar{k}(X)}$ on the subspace $S \cong \mathbb{F}_q^n$ of the p-division points of the Jacobian of C induces modulo Z' the group $H = G'$. Furthermore, if we let $N' = \bar{k}N$ (which is the fixed field of the kernel of the map $G_{\bar{k}(X)} \to G'$), then the extension $N'/\bar{k}(X)$ is ramified where t equals some t_i, and $\Phi'_\zeta(Q_{i,s})$ is a corresponding distinguished inertia group generator. This (together with Lemma 4.14) proves part (a) of the theorem.

(b) Note that $G_{\bar{k}(X)}$ is normal in $G_{k(X)}$. Thus, if G' is self-normalizing in $\mathrm{GL}_n(q)/Z'$, then the image of $G_{k(X)}$ in $\mathrm{GL}_n(q)/Z'$ also equals G', and so the corresponding extension $N/k(X)$ is regular over k.

Remark 4.17 Let $K(S_\lambda)$ denote the extension of K obtained by adjoining the coordinates of the p-torsion points in $S_\lambda \subset J_C[p]$ to K. Then clearly $K(S_\lambda) = N_{\lambda,1}$ is the fixed field of the kernel of the representation

$$\rho_\lambda : G_K \to \mathrm{Aut}(S_\lambda) \simeq \mathrm{GL}_n(q)$$

on S_λ, and hence is a Galois extension of K which contains the field(s) $N = N_{\lambda,Z'}$ of Theorem 4.16. Moreover, $K(S_\lambda)$ is cyclic over N with Galois group $\mathrm{Gal}(K(S_\lambda)/N) \leq Z'$. We shall see below in Theorem 5.14 that the extension $K(S_\lambda)/N_{\lambda,Z'}$ is always an *unramified* extension.

The consequence of Theorem 4.16 is that the almost unramified families of covers of $\mathbb{P}^1$ constructed in the previous section are obtained by adjoining Galois invariant subspaces of torsion points of families of abelian varieties to $K(X)$. This imposes strong conditions on these abelian varieties.

In the next section we shall discuss some consequences of these conditions in more detail.

5   A family of curves with infinite geometric fundamental group

5.1   The abelian variety J_C^{new}

In order to interpret the results of the previous section geometrically, it is useful to introduce the following "new part" J_C^{new} of the Jacobian J_C of a cyclic covering $\pi : C \to C'$ of curves, which is (partially) analogous to the new part of the Jacobian of the modular curve $X_0(N)$.

Definition 5.1 Let $\pi : C \to C'$ be a cyclic covering of curves defined over an arbitrary field K, and let $Z = \mathrm{Aut}(\pi) \simeq \mathbb{Z}/N\mathbb{Z}$ denote its covering group.

For each subgroup $H \leq Z$ let $\pi_H : C \to C_H = C/H$ denote the quotient map. Then we call the abelian subvariety

$$J_C^{old} := \sum_{\substack{H \leq Z \\ H \neq 1}} \pi_H^* J_{C/H}$$

the *old part* of J_C, and its orthogonal complement (with respect to the canonical polarization on J_C) the *new part* J_C^{new}.

For our purposes it is important to observe that J_C^{new} and J_C^{old} can be expressed in terms the following idempotent ε_{new} of the group ring $\mathbb{Q}[Z]$:

$$\varepsilon_{new} = \sum_{d|N} \mu(d)\varepsilon_d,$$

where, for a subgroup $H \leq Z$ of order d,

$$\varepsilon_d = \varepsilon_H = \frac{1}{d}\sum_{h \in H} h \in \mathbb{Q}[Z].$$

Lemma 5.2 *If* $\chi : Z \to \mathbb{C}^*$ *is a fixed character of order* $N = |Z|$, *then*

$$\varepsilon_d = \sum_{\substack{k=0 \\ d|k}}^{N-1} \varepsilon_{\chi^k} \quad and \quad \varepsilon_{new} = \sum_{\substack{k=1 \\ (k,N)=1}}^{N-1} \varepsilon_{\chi^k}, \tag{2}$$

where $\varepsilon_{\chi^k} = \frac{1}{N}\sum_{g \in Z} \chi(g)^k g^{-1} \in \mathbb{C}[Z]$ *denotes the primitive idempotent associated to the character* χ^k. *In particular,* ε_{new} *is a symmetric idempotent of* $\mathbb{Q}[Z]$ *and*

$$\varepsilon_d \cdot \varepsilon_{new} = 0, \quad for \ all \ d|N, \ d \neq 1.$$

Proof. Since the χ^k form a dual basis of the group ring $\mathbb{C}[Z]$, it is enough to verify (2) after evaluating both sides by χ^k, for $k = 0, \ldots, N-1$. Now for any class function ψ on Z we have $\psi(\varepsilon_H) = (1_H, \psi_{|H})$, $\psi(\varepsilon_{\chi^k}) = (\chi^k, \psi)$. Since for a subgroup H of order d we have $\chi_{|H}^k = 1_H \iff \mathrm{Ker}(\chi^k) \geq H \iff d|k$, it follows that $\chi^k(\varepsilon_d) = 1$ if $d|k$ and $\chi^k(\varepsilon_d) = 0$ otherwise. From this, the first formula of (2) is obvious and the second follows readily, using the fact that $s := \sum_{d|(k,N)} \mu(d) = 1$, if $(k, N) = 1$ and $s = 0$ otherwise.

Thus, by (2) we see that ε_{new} is a sum of pairwise orthogonal idempotents, and hence is also an idempotent. Furthermore, ε_{new} is symmetric since all ε_d are symmetric.

Finally, if $d \neq 1$, then (2) shows that ε_d and ε_{new} have no common components ε_{χ^k}, and hence are orthogonal.

Corollary 5.3 *Put $\tilde{\varepsilon}_{new} = N\varepsilon$ and $\tilde{\varepsilon}_{old} = N - \tilde{\varepsilon}$. Then $\tilde{\varepsilon}_{new}, \tilde{\varepsilon}_{old} \in \mathrm{End}(J_C)$ and we have*

$$J_C^{old} = \tilde{\varepsilon}_{old}(J_C), \quad J_C^{new} = \tilde{\varepsilon}_{new}(J_C).$$

Furthermore, $J_C = J_C^{old} + J_C^{new}$, and $J_C^{old} \cap J_C^{new} \leq J_C[N]$; in particular, $J_C \sim J_C^{old} \times J_C^{new}$.

Proof. By definition, $N\varepsilon_{new} \in \mathbb{Z}[Z] \subset \mathrm{End}(J_C)$, and so $\tilde{\varepsilon}_{new}, \tilde{\varepsilon}_{old} \in \mathrm{End}(J_C)$; note that $\tilde{\varepsilon}_{old} = N(1 - \varepsilon_{new}) = \sum_{1 \neq d | N} \mu(d) N\varepsilon_d$. Put $\tilde{\varepsilon}_d = d\varepsilon_d \in \mathbb{Z}[Z]$. Then $\tilde{\varepsilon}_d(J_C) = \pi_{H_d}^*(J_{C/H_d})$, so it follows from the definition that $\tilde{\varepsilon}_{old}(J_C) \subset \sum_{1 \neq d | N} \pi_{H_d}^*(J_{C/H_d}) = J_C^{old}$. On the other hand, by Lemma 5.2 we have $\tilde{\varepsilon}_{old}\tilde{\varepsilon}_d = N\tilde{\varepsilon}_d$, if $d \neq 1$, so $\pi_{H_d}^*(J_{C/H_d}) = \tilde{\varepsilon}_d(J_C) = \tilde{\varepsilon}_{old}(\tilde{\varepsilon}_d(J_C)) \subset \tilde{\varepsilon}_{old}(J_C)$ and hence $J_C^{old} = \tilde{\varepsilon}_{old}(J_C)$. By construction, $\tilde{\varepsilon}_{new} + \tilde{\varepsilon}_{old} = N$, $\tilde{\varepsilon}_{new} \cdot \tilde{\varepsilon}_{old} = 0$, so $J_C^{new} := \tilde{\varepsilon}_{new}(J_C)$ is the orthogonal complement of J_C^{old}, and hence the isogeny relation follows.

Corollary 5.4 *Suppose $p \neq \mathrm{char}(K)$ is a prime with $p \equiv 1 \bmod N$. Then there exist primitive characters $\lambda : Z \to \mathbb{F}_p^*$ and $\tilde{\lambda} : Z \to \mathbb{Z}_p^*$ of order N and we have natural identifications (of Z-modules and G_K-modules)*

$$J_C^{new}[p] = \bigoplus_{\substack{1 \leq k < N \\ (k,N)=1}} J_C[p]_{\lambda^k}, \quad \text{and} \quad T_p(J_C^{new}) = \bigoplus_{\substack{1 \leq k < N \\ (k,N)=1}} T_p(J_C)_{\tilde{\lambda}^k},$$

where $T_p(A) = \varprojlim A[p^n]$ denotes the Tate-module of an abelian variety A. In particular, $\dim J_C^{new} = \frac{1}{2}\phi(N)(2g_{C'} - 2 + r)$, where $r = \#\{P \in C'(\bar{K}) : e_P(\pi) > 1\}$ denotes the number of ramified primes.

Proof. By the hypothesis on p, both $\mathbb{F}_p$ and $\mathbb{Z}_p$ contain a primitive N-th root of unity and so λ and $\tilde{\lambda}$ exist. Thus, if we replace χ by λ and $\tilde{\lambda}$ in (2), the analogous formulae of (2) hold in $\mathbb{F}_p[Z]$ and in $\mathbb{Z}_p[Z]$. From this the above decomposition follows because $J_C^{new}[p] = \varepsilon_{new}(J_C[p])$ and $J_C[p]_{\lambda^k} = \varepsilon_{\lambda^k}(J_C[p])$, and similarly, $T_p(J_C^{new}) = \varepsilon_{new}T_p(J_C)$ and $T_p(J_C)_{\tilde{\lambda}^k} = \varepsilon_{\tilde{\lambda}^k}T_p(J_C)$. (Note that this is also a decomposition of G_K-modules since the Z-action commutes with the G_K-action.)

To work out the dimension of J_C^{new}, we use the well-known fact (cf. Serre[Se2], p. 106) that the character h_1 of the representation of Z on $T_p(J_C) \otimes \mathbb{Q}_p$ is given by

$$h_1 = 2 \cdot 1_Z + (2g_{C'} - 2)\mathrm{reg}_Z + a_Z$$

where $a_Z = \sum_{P' \in C'} a_{P'}$, and $a_{P'} = (\mathrm{reg}_Z - 1_{D_P}^*)$ denotes the (tame) Artin character at P'. (Here D_P is the decomposition group of any $P \in \pi^{-1}(P')$, and $*$ denotes induction.) From this it follows easily (by Frobenius reciprocity) that if $k \not\equiv 0 \bmod N$, then the rank of $T_p(J_C)_{\tilde{\lambda}^k}$ $(= \dim_{\mathbb{F}_p}(J_C[p]_{\lambda^k}))$ is

$$(h_1, \tilde{\lambda}^k) = (2g_{C'} - 2) + r_k, \text{ where } r_k := (a_Z, \tilde{\lambda}^k) = \#\{P' \in C' : e_{P'} \nmid k\}. \quad (3)$$

Since $\mathrm{rank}(T_p(J_C^{new})) = 2\dim(J_C^{new})$, the asserted dimension formula follows.

Using the abelian variety J_C^{new}, we can now give the following geometric interpretation of the main result (Theorem 4.16) of the previous section.

Theorem 5.5 *Let $K = k(t)$, where t is transcendental over k and k contains the N-th roots of unity. Suppose that $t_1,\ldots,t_{n+1}$ are $n+1$ distinct elements in k, and put $t_{n+2} = t$. In addition, suppose that $m_1,\ldots,m_{n+2}$ are integers with $1 \leq m_i < N$ such that $\gcd(m_1,\ldots,m_{n+2},N) = 1$, $m_1 + \ldots + m_{n+2} \equiv 0$ mod N and $m_i \neq N - m_{n+2}$, for $1 \leq i \leq n+1$. Let $\pi : C \to \mathbb{P}^1$ denote the cyclic covering of degree N defined by the equation*

$$y^N = c(x - t_1)^{m_1} \cdots (x - t_{n+2})^{m_{n+2}}, \tag{4}$$

where $c \in k^*$, and let $A = J_C^{new}$ denote the new part of the Jacobian J_C of C. Then $\dim A = \frac{1}{2}\phi(N)n$, and for any prime $p \equiv 1$ mod N we have:

(a) The group $A[p]$ of p-torsion points of A is the direct sum of G_K-invariant subspaces $S_i = J_C[p]_{\lambda^i}$, where $(i, N) = 1$, and $\lambda : Z = \mathrm{Aut}(\pi) \to \mathbb{F}_p^*$ is a primitive character. Moreover, G_K acts irreducibly on each S_i, and hence semi-simply on $A[p]$.

(b) Let L_i (respectively, $\tilde{M}_i$ and M_i) denote the fixed field of the kernel of the action of G_K on S_i (respectively, on $S_i/\lambda^i(Z)$ and on $\mathbb{P}(S_i)$), so $L_i \supset \tilde{M}_i \supset M_i$. Then $L_i/\tilde{M}_i$ is a cyclic extension of order dividing N, and $\tilde{M}_i/K$ (respectively, M_i/K) is a Galois extension whose geometric part $\tilde{G} := \mathrm{Gal}(\bar{k}\tilde{M}_i/\bar{k}K) \leq \mathrm{GL}_n(p)/\lambda^i(Z)$ (respectively, $\bar{G} := \mathrm{Gal}(\bar{k}M_i/\bar{k}K) \leq \mathrm{PGL}_n(p)$) is generated by the image of the Thompson tuple attached to the covering. In addition, $\tilde{M}_i$ (and hence also M_i) is ramified only at $t_1,\ldots,t_{n+1}$ with ramification order dividing N.

(c) The field $K(A[p])$ (which is generated over K by the coordinates of all p-torsion points of A) is an abelian extension of the compositum $\tilde{M}$ of all the $\tilde{M}_i$'s of degree dividing $N^{\phi(N)}$. Moreover, $\tilde{M}/K$ is ramified only at $t_1,\ldots,t_{n+1}$ with ramification order dividing N.

(d) If $(n, p-1) = 1$, (and $N \nmid 6$ if $n = 3$), then M_i is regular over k and

$$\mathrm{Gal}(M_i/K) = \mathrm{PGL}_n(p) = \mathrm{PSL}_n(p).$$

Proof. (a) The first assertion follows directly from the decomposition of Corollary 5.4. Next, choose a "canonical generator" σ of Z as in the proof of Lemma 4.15. Then for any i with $(i, N) = 1$, the hypotheses on the m_j's guarantee that the $(n + 2)$-tuple $\zeta_1 = \lambda^i(\sigma^{m_1}),\ldots,\zeta_{n+2} = \lambda^i(\sigma^{m_{n+2}})$ satisfies condition (1) of Lemma 4.14 and so by Theorem 4.16(a) (with $Z' = Z(\mathrm{GL}_n(p))$) it follows that the image of the Galois group $G_{\bar{k}K}$ on $\mathbb{P}(S_i)$ is generated by a Thompson tuple and hence $G_K \geq G_{\bar{k}K}$ acts irreducibly on S_i.

(b) Since $\mathrm{Gal}(L_i/\tilde{M}_i) = \mathrm{Gal}(L_i/K) \cap \lambda^i(Z) \leq \lambda^i(Z)$ (viewed as subgroups of $\mathrm{Aut}_{\mathbb{F}_p}(S_i) \simeq \mathrm{GL}_n(p)$), the first assertion is clear. The other assertions follow directly from Theorem 4.16(a) by taking $Z' = \lambda^i(Z)$ (respectively, $Z' = Z(\mathrm{GL}_n(p))$).

(c) Since $A[p]$ is the direct sum of the S_i's (cf. part (a)), we have $K(A[p]) = \prod L_i$, and so both assertions follow from part (b).

(d) The hypothesis implies that n is odd and that $\mathrm{PSL}_n(p) = \mathrm{PGL}_n(p)$. Thus, from Theorem 4.6 it follows that the image $\bar{G}$ of the Thompson tuple in $\mathrm{PGL}_n(p)$ generates the whole group. In particular, $\bar{G}$ is self-normalizing, and so M_i is regular over k by Theorem 4.16(b).

Remark 5.6 *We shall see below in Theorem 5.14 that the extension L_i/M_i is always unramified; in particular, $K(A[p])$ is unramified over $\tilde{M}$.*

By applying a variant of the Serre-Tate criterion of potentially good reduction (see below), we obtain

Theorem 5.7 *If C/K is as in Theorem 5.5, then the new part $A = J_C^{new}$ of its Jacobian J_C has potentially good reduction over K. In other words, there is a finite extension K_0 of K such that $A_{K_0} = A \otimes K_0$ extends to an abelian scheme $\mathcal{A}$ over the projective curve T_0 with function field K_0, i.e. for each geometric point $P \in T_0(\bar{K})$ the fibre of $\mathcal{A}$ over P is an abelian variety.*

Proof. By Theorem 5.5 (c), we see that for all primes $p \equiv 1 \mod N$, the ramification degrees of the extension $K(A[p])/K$ divide $N^{\phi(N)+1}$, and so the assertion follows from the following criterion:

Proposition 5.8 (Serre-Tate Criterion) *Let K be a field with a discrete valuation v, and let A be an abelian variety over K. Then the following conditions are equivalent:*

(i) A has potentially good reduction at v;

(ii) the ramification degree of v in $K(A[m])$ is bounded for all m prime to $\mathrm{char}(\kappa(v))$;

(iii) there is a constant c such that the ramification degree $e_v(K(A[p])/K) \leq c$, for infinitly many primes p.

Proof. This is well-known, but for convenience of the reader we present the proof.

(i) $\Rightarrow$ (ii): This follows from the criterion of Néron-Ogg-Shafarevich for good reduction; cf. Serre-Tate[ST], Theorem 2.

(ii) $\Rightarrow$ (iii): Trivial.

(iii) $\Rightarrow$ (i): By replacing K by a suitable finite extension K', we may assume without loss of generality that A has semi-stable reduction at v. (For example, we could take $K' = K(A[p])$, for any prime $p \geq 3$, $p \neq \mathrm{char}(\kappa(v))$; cf. Grothendieck[Gro], Prop. 4.7). In addition, we may assume that K is henselian with separably closed residue field. Then the implication (iii) $\Rightarrow$ (i) follows once we have verified:

Claim: If A does not have good reduction at v, then $p|e_v(K(A[p])/K)$, for every prime $p \nmid \#(\Phi(A_K))$ ($p \neq \mathrm{char}(\kappa(v))$), where $\Phi(A_K)$ denotes the (finite) group of components of the Néron model of A/K with respect to v.

To see this, we first note that for p as above, $\#(A[p](K)) = \#(A^0[p](\kappa(v))) = p^{t+2a}$, where t denotes the toric rank and a the abelian rank of the connected component A^0 of the identity of the reduction of the Néron model. (Note that $t + a = d := \dim A$, and that $t > 0$ since A has bad reduction.) On the other hand, for $K' = K(A[p])$ we clearly have $\#(A[p](K')) = p^{2d}$, and so $p^t | \#\Phi(A_{K'})$. Thus, $p | \#(\Phi(A_{K'})/\Phi(A_K))$. On the other hand, the quotient group $\Phi(A_{K'})[p]/\Phi(A_K)[p]$ has exponent $e_v(K'/K)$; this follows from [Gro], Th. 11.5, together with formula (10.3.5). Thus $p|e_v(K'/K)$, as asserted.

Remark. In the above implication (iii) $\Rightarrow$ (i) we needed several deep results from Grothendieck's article [Gro]. However, all these can be avoided if in condition (iii) we can assume in addition that $K(A[p])/K$ is *tamely rami-fied* for the primes p in question. (This is always the case if $|\mathrm{GL}_{2d}(p)|$ is prime to the residue characteristic $\mathrm{char}(K)$, but for the above application we need to assume only that $(\mathrm{char}(K), N) = 1$ since we already know that the ramification degree divides a power of N.)

In that case the implication (iii) $\Rightarrow$ (i) can be proved as follows:

Again, it is enough to prove this in the case that K is henselian with sepa-rably closed residue field. If L denotes the compositum of the $K(A[p])$'s for the primes in question, then the hypothesis (and tame ramification) implies that L is a finite (cyclic) extension of K. By construction, $L(A[p]) = L$ for infinitely many p's, so by the Néron-Ogg-Shafarevich criterion for good reduction (cf. [ST], Theorem 1), A has good reduction over L and hence potentially good reduction over K.

Let us now come back to Theorem 5.7. By imposing further restrictions on the m_i's, we can conclude that all of J_C has potentially good reduction:

Corollary 5.9 *Suppose that C/K is as above and satifies in addition the condition that $m_i \not\equiv -m_{n+2} \bmod d$, for every $d|N$ with $e_{n+2} \nmid d \neq 1$, where $e_{n+2} = \frac{N}{(m_{n+2}, N)}$. Then J_C has potentially good reduction over K.*

Proof. Induct on the number of the divisors of N. If N is prime, then $J_C = J_C^{new}$ and so the assertion follows from the theorem. If N is composite, then by the theorem it is enough to show that J_C^{old} has potentially good reduction or, equivalently, that every $J_d := J_{C/H_d}$ has potentially good reduction (for $d|N$, $d \neq 1$). This is clear if $e := e_{n+2}|d$, for then $C_d := C_{H_d}$ is unramified at t_{n+2} and hence the covering is already defined over K. Thus, assume $e \nmid d$.

Then the ramification of $C_d \to \mathbb{P}^1$ satisfies the same hypotheses as $C \to \mathbb{P}^1$, and so by the induction hypothesis we have that J_{C_d} has good reduction.
t 5uMzuyN

5.2 Rational points on $K(\mathbb{P}(S_i))$

One of the advantages of the above geometric description of the extensions generated by Thompson tuples is that it is possible to determine whether or not the extension field has a rational point. This is based on the following (well-known) fact.

Proposition 5.10 *Let A be an abelian variety over K which has potentially good reduction with respect to a discrete valuation v on K. Suppose that v is unramified in $K' = K(A[m])$, where $m \geq 3$ is an integer which is relatively prime to the characteristic of the residue field $k = \kappa(v)$ of v. Then:*

(i) the reduction A_v of A at v is an abelian variety over k;

(ii) if v' denotes any extension of v to K', then its residue field is $\kappa(v') = k(A_v[m])$.

Proof. The first assertion follows from [ST], Theorem 2, Corollary 2(b). To prove the second, let $k' = \kappa(v')$. Then the reduction map $r_{v'} : A(K') \to A_{v'}(k')$ induces an isomorphism $A[m](K') \simeq A_{v'}[m](k')$, so all m-torsion points of A_v are rational over k'. Thus $k(A_v[m]) \subset k'$. To prove that equality holds, let $\sigma \in \mathrm{Gal}(k'/k(A_v[m]))$. Then σ lifts uniquely to an automorphism $\tilde{\sigma} \in D_{v'}(K'/K)$, where $D_{v'}(K'/K) \leq \mathrm{Gal}(K'/K)$ denotes the decomposition group of v'. Since the action of $\tilde{\sigma}$ on $A[m]$ is (via $r_{v'}$) the same as that of σ on $A_{v'}[m] = A_v[m]$, this means that $\tilde{\sigma}$ acts trivially on $A[m]$. But then $\tilde{\sigma}$ and hence σ are both trivial, so $k' = k(A_v[m])$.

Corollary 5.11 *In the above situation, suppose that $S \subset A[m]$ is a G_K-invariant subspace. Let $\bar{S}$ denote the image of S in $A_{v'}[m]$. Then $\kappa(v'_{|K(S)}) = k(\bar{S})$ and $\kappa(v'_{|K(\mathbb{P}(S))}) = k(\mathbb{P}(\bar{S}))$. In particular, if $m = p$ is a prime, then v splits completely in $K(\mathbb{P}(S))$ if and only if every $\bar{k}$-isogeny of A_v with kernel in $\bar{S}$ is k-rational.*

Proof. By the same argument as in the proof of the proposition, we see that $\mathrm{Gal}(k'/k(\bar{S})) \simeq D_{v'}(K'/K) \cap \mathrm{Gal}(K'/K(S)) = D_{v'}(K'/K(S))$, and so the first assertion follows. To prove the second, we note that the group $\mathrm{Gal}(K(S)/K(\mathbb{P}(S)))$ consists of those $\sigma \in \mathrm{Gal}(K(S)/K)$ which act diagonally on S. Since $\mathrm{Gal}(k(\bar{S})/k(\mathbb{P}(\bar{S})))$ is defined similarly, we see that $\mathrm{Gal}(k(\bar{S})/k(\mathbb{P}(\bar{S}))) \simeq D_{v'_{|K(S)}}(K(S)/K(\mathbb{P}(S)))$, and so the second assertion follows.

To prove the last assertion, we observe that by the above result v splits completely in $K(\mathbb{P}(S)) \Longleftrightarrow k(\mathbb{P}(\bar{S})) = k \Longleftrightarrow$ every $\sigma \in \mathrm{Gal}(k(\bar{S})/k)$ acts diagonally on $\bar{S} \Longleftrightarrow \sigma(\mathbb{F}_p \bar{s}) = \mathbb{F}_p \bar{s}$, for all $\bar{s} \in \bar{S}$ and $\sigma \in \mathrm{Gal}(k(\bar{S})/k)$ $\Longleftrightarrow$ every isogeny $\varphi : A_v \to A'$ with kernel $\mathrm{Ker}(\varphi) = \mathbb{F}_p \bar{s}$ (where $\bar{s} \in \bar{S} - \{0\}$) is k-rational $\Longleftrightarrow$ every $\bar{k}$-isogeny of A_v with kernel in $\bar{S}$ is k-rational.

The main difficulty in applying the above criterion to our situation is the fact that we need to guarantee that v is unramified in $K(A[p])$ for some p. Although we already know by Theorem 5.5 the ramification behaviour of the extension $K(\mathbb{P}(S_i))/K$, there remains the problem of understanding that of $K(S_i)/K(\mathbb{P}(S_i))$. This will be analyzed next by using the following criterion.

Proposition 5.12 *Let A/K be an abelian variety and let v be a discrete valuation of K. Moreover, let $I = I_{v'}$ denote the inertia group of an extension v' of v to $K' = K(A[p])$, where $p \neq \mathrm{char}(\kappa(v))$ is a prime. If $S \subset A[p]$ is a non-zero G_K-invariant subspace and $H = \mathrm{Gal}(K'/K(\mathbb{P}(S)))$, then $v'_{|K(S)}$ is unramified over $K(\mathbb{P}(S))$ if and only if $S^{I \cap H} \neq \{0\}$. In particular, if $S^I \neq \{0\}$, then $v'_{|K(S)}$ is unramified over $K(\mathbb{P}(S))$.*

Proof. Let $H_0 = \mathrm{Gal}(K'/K(S)) \leq H$. Then $v'_{|K(S)}$ is unramified over $K(\mathbb{P}(S)) \Longleftrightarrow I \cap H \leq H_0 \Longleftrightarrow I \cap H$ acts trivially on $S \Longleftrightarrow S^{I \cap H} = S$. Thus, if $v'_{|K(S)}$ is unramified, then clearly $S^{I \cap H} = S \neq \{0\}$. Conversely, if $S_0 := S^{I \cap H} \neq \{0\}$ and $g \in I \cap H$, then g acts diagonally on S and trivially on S_0, so g acts trivially on all of S. Thus $I \cap H \leq H_0$, and so $v'_{|K(S)}$ is unramified.

Corollary 5.13 *Suppose in addition that $S = A[p]_\lambda$ where $\lambda : Z \to \mathbb{F}_p^*$ is a character on a finite subgroup $Z \leq \mathrm{Aut}(A)$. If either $A[p]_\lambda^I \neq \{0\}$ or if $T_p(A)_{\tilde{\lambda}}^{\tilde{I}} \neq \{0\}$, where $\tilde{\lambda} : Z \to \mathbb{Z}_p^*$ is a lift of λ and $\tilde{I} \leq G_K$ denotes the absolute inertia group of an extention of v (and of v') to K^{sep}, then $v'_{|K(S)}$ is unramified over $K(\mathbb{P}(S))$.*
In particular, if the reduction A_v of the Néron model of A/K at v is not unipotent (i.e. the connected component of A_v is not an extension of additive groups), then v' is unramified over $K(\mathbb{P}(A[p]))$.

Proof. If the second hypothesis holds, then also $V := T_p(A)_{\tilde{\lambda}}^{\tilde{I}} \otimes \mathbb{F}_p \neq \{0\}$. But then $\{0\} \neq V \subset (T_p(A)_{\tilde{\lambda}} \otimes \mathbb{F}_p)^I = A[p]_\lambda^I$, which means that the second hypothesis implies the first. Now if the first hypothesis holds, then $S^{I \cap H} \supset S^I \neq \{0\}$, and so the conclusion follows directly from the proposition.
In particular, taking for $\tilde{\lambda}$ the trivial character (on $Z = \{1\}$), then by [Gro], Prop. 2.2.5 (and (2.1.11)) we have $\mathrm{rank}(T_p(A)^{\tilde{I}}) = n - \lambda(A)$, where $n = \dim(A)$ and $\lambda(A)$ denotes the unipotent rank of A_v. Thus, $T_p(A)^{\tilde{I}} = \{0\} \Longleftrightarrow A_v$ is unipotent, and so the assertion follows from the previous criterion.

The above criterion can be used in our situation in the following way.

Theorem 5.14 *In the situation and notation of Theorem 5.5, let v_i denote the place of K/k corresponding to the specialization $t \to t_i$, for $1 \leq i \leq n+1$. Then:*

(a) If v is any place of K/k with $v \notin R := \{v_1, \ldots, v_{n+1}\}$, then C and hence J_C and $A = J_C^{new}$ have good reduction at v.

(b) For each i with $1 \leq i \leq n+1$, let $\bar{C}_i$ denote the normalization of the curve

$$y^N = c(x - t_1)^{m_1} \cdots (x - t_{n+1})^{m_{n+1}} \cdot (x - t_i)^{m_{n+2}},$$

and let $\bar{\pi}_i : \bar{C}_i \to \mathbb{P}_k^1$ denote the associated cyclic covering of degree N. Then $\bar{C}_i$ is the normalization of the reduction of C at v_i, and the new part $J_{\bar{C}_i}^{new}$ of its Jacobian has dimension $\frac{1}{2}\phi(N)(n-1)$.

(c) Let J_{v_i} and A_{v_i} denote the reductions of the Néron models of J_C and $A = J_C^{new}$ at v_i, and let $J_{v_i}^0$ and $A_{v_i}^0$ denote their respective connected components of the identity. Then there are natural surjections of algebraic groups $f_i : J_{v_i}^0 \to J_{\bar{C}_i}$ and $f_i^{new} : A_{v_i}^0 \to J_{\bar{C}_i}^{new}$.

(d) Let $\tilde{I} = \tilde{I}_{v_i} \leq G_K$ denote the inertia group of an extension of v_i to K^{sep}. Then for any prime $p \equiv 1 \bmod N$ (with $p \neq \mathrm{char}(K)$) and any primitive character $\tilde{\lambda} : Z \to \mathbb{Z}_p^*$, we have that $\mathrm{rank}(T_p(J_C)_{\tilde{\lambda}}^{\tilde{I}}) \geq n - 1$.

(e) For p as above, let $\lambda : Z \to \mathbb{F}_p^*$ be a primitive character. Then for any j with $(j, N) = 1$, the extension $L_j = K(S_j)$ is everywhere unramified over $M_j = K(\mathbb{P}(S_j))$, and hence $K' = K(A[p])$ is ramified over K only at the places of R with ramification index dividing N and is unramified over $M = \prod M_j$.

Proof. Recall that C is by definition the normalization of the projective plane curve $C' \subset \mathbb{P}_K^2$ defined by equation (4). Let $\mathcal{C}'$ denote the closure of C in $\mathbb{P}_{\mathbb{P}^1}^2 = \mathbb{P}^2 \times \mathbb{P}^1$, and let $\nu : \mathcal{C} \to \mathcal{C}'$ denote its normalization (in $\kappa(\mathcal{C}') = \kappa(C)$). Then, if v is a place of K/k with $v \neq v_\infty$, the place at infinity, then the fibre $\mathcal{C}_v'$ of $\mathcal{C}'$ is the projective plane curve over $\kappa(v)$ defined by the equation

$$y^N = c(x - t_1)^{m_1} \cdots (x - t_{n+1})^{m_{n+1}} \cdot (x - \bar{t})^{m_{n+2}},$$

where $\bar{t} \in \kappa(v) \supset K$ denotes the image of t in the residue field of v. Since this curve is reduced and geometrically irreducible, it follows that the same is true for the fibres $\mathcal{C}_v$ of $\mathcal{C}$, and so the normalization $\tilde{\nu}_v : \tilde{\mathcal{C}}_v \to \mathcal{C}_v'$ of $\mathcal{C}_v'$ factors over the finite map $\nu_v : \mathcal{C}_v \to \mathcal{C}_v'$. Note that by considering a different affine model, the above argument extends to show that the last assertions are also true for $v = v_\infty$.

(a) If $v \neq v_\infty$, then the hypothesis on v means that $\bar{t} \neq t_1, t_2, \ldots t_{n+1}$. Thus, by Riemann-Hurwitz, the genus of $\tilde{\mathcal{C}}_v$ is the same as that of C, and so C has good reduction at v. Then J_C also has good reduction (cf. [BLR], 9.4/4) and hence so does any quotient A of J_C (cf. [BLR], 7.5/3). Since the argument for $v = v_\infty$ is similar, this proves assertion (a).

(b) Now suppose that $v = v_i$, so $\bar{t} = t_i$ (and $\kappa(v) = k$). Then C'_v is the projective plane curve defined by the given equation and so by what was said above, $\bar{C}_i = \tilde{C}_v$, which proves the first assertion. The second assertion follows directly from Corollary 5.4, for $\bar{\pi}_i : \bar{C}_i \to \mathbb{P}^1_k$ is a cyclic covering of degree N which is ramified at precisely $r = n + 1$ places.

(c) Let $\delta : \tilde{C} \to C$ denote a desingularization of C (which exists since the base $\mathbb{P}^1_K$ is clearly excellent). Then, since C is normal, $\bar{C}_i$ is the normalization of a component of the fibre $\tilde{C}_v$ of $\tilde{C}$ at $v = v_i$. Thus, we have a natural surjection $\mathrm{Pic}^0_{\tilde{C}_v/k} \to \mathrm{Pic}^0_{\bar{C}_i/k}$ (cf. [BLR], 9.2/13). On the other hand, by [BLR], Th. 9.5/4, $\mathrm{Pic}^0_{\tilde{C}_v/k}$ is canonically isomorphic to the connected component J^0_i of the identity of the reduction $J_i = J_{v_i}$ of the Néron model of J_C at $v = v_i$, and so we have a canonical surjection $f_i : J^0_i \to J_{\bar{C}_i}$.

Next we observe that the covering automorphisms $\sigma \in \mathrm{Aut}(\pi)$ map isomorphically onto those of $\bar{\pi}_i : \bar{C}_i \to \mathbb{P}^1_k$, and so these extend to automorphisms of J^0_i and of $J_{\bar{C}_i}$ in such a way that f_i becomes equivariant. From this it follows that if $\bar{\varepsilon}_i : J_{\bar{C}_i} \to J^{new}_{\bar{C}_i}$ denotes the projection map onto the new part of $J_{\bar{C}_i}$ (cf. Corollary 5.3), then the composition $\bar{\varepsilon}_i \circ f_i$ factors over the map $\tilde{\varepsilon}_{new,i} : J^0_i \to A^0_{v_i}$ which is induced by the universal property of Néron models; i.e. there is a homomorphism $f^{new}_i : A^0_{v_i} \to J^{new}_{\bar{C}_i}$ such that $\bar{\varepsilon}_i \circ f_i = f^{new}_i \circ \tilde{\varepsilon}_{new,i}$. Thus, since f_i and $\bar{\varepsilon}_i$ are surjective, so is f^{new}_i, which proves the assertion.

(d) By Grothendieck [Gro], Prop. 2.2.5 (and (2.2.3.3)), we have a natural identification $T_p(A^0_i) \simeq T_p(A)^I$ (after passing to the henselization of v_i). Since the action of Z commutes with the Galois action, this induces an isomorphism $T_p(A^0_i)_{\bar{\lambda}} \simeq T_p(A)^I_{\bar{\lambda}}$. Now by (c), the map f^{new}_i is surjective, hence so is the induced map on the Tate modules, and therefore $T_p(A^0_i)_{\bar{\lambda}} \to T_p(J_{\bar{C}_i})_{\bar{\lambda}}$ is surjective as well. Thus, $\dim T_p(A)_{\bar{\lambda}} \geq \dim T_p(J_{\bar{C}_i})_{\bar{\lambda}} = n - 1$, the latter by formula (3) of Corollary 5.4.

(e) Let v be a place of K/k. If $v \neq v_i$, then A has good reduction at v, so $K(A[p])$ is unramified over K at v, hence a fortiori so is $K(S_j)$ over $K(\mathbb{P}(S_j))$. On the other hand, if $v = v_i$, then in in view of (d) we can apply the criterion of Corollary 5.13 to see that $v'_{|K(S_j)}$ is unramified over $K(\mathbb{P}(S_j))$ for every extension v' of v. Thus, $L_j = K(S_j)$ is everywhere unramified over $M_j = K(\mathbb{P}(S_j))$ and hence $K(A[p]) = \prod L_j$ is unramified over $M = \prod M_j$ as well. On the other hand, it follows from Theorem 5.5 that each M_j and hence M is unramified outside R and ramified at each $v_i \in R$ of order dividing N, so the same is true for K'.

By using the above proposition, we can now make the assertion of Theorem 5.7 much more precise.

Theorem 5.15 *Let K_0 be a finite Galois extension of $K = k(t)$ such that the ramification index $e_i = e_{v_i}(K_0/K)$ at v_i is divisible by N, for $1 \leq i \leq n+1$. Then $A = J_C^{new}$ has good reduction everywhere over K_0, and hence $K_0(A[m])$ is unramified over K_0 for every integer m relatively prime to $\mathrm{char}(K)$.*

Proof. If $p \equiv 1 \bmod N$ is any prime ($p \neq \mathrm{char}(K)$), then by Abhyankar's Lemma it follows from Theorem 5.14(e) that $K_0(A[p]) = K(A[p]) \cdot K_0$ is unramified over K_0. By the criterion of Néron-Ogg-Shafarevich, this means that $A_{K_0} = A \otimes K_0$ has good reduction everywhere, and hence $K_0(A[m])$ is unramified over K_0 for all m which are prime to $\mathrm{char}(K)$.

Thus, by above theorem we can now apply Corollary 5.11 to obtain following criterion for the existence of rational points:

Corollary 5.16 *Suppose in addition that $v = v_{i_0}$ is totally ramified in K_0, and let $A_{\tilde{v}}$ denote the reduction of $A \otimes K_0$ at $\tilde{v}$, where $\tilde{v}$ denotes the (unique) extension of v to K_0. If $p \neq \mathrm{char}(K)$ is any prime and $S \subset A[p]$ is a G_K-invariant subspace such that*

($\star\star$)   *every isogeny of the reduction $A_{\tilde{v}}$ with kernel contained in $\bar{S}$ ($=$ the image of S in A_v) is rational over K,*

then every extension of $\tilde{v}$ to $M_S := K_0(\mathbb{P}(S))$ is a K-rational point of M_S.

5.3   Example

We now return to Example 4.11, but change the notation slightly to conform with that of this section. Thus, let k be a field containing a primitive fourth root of unity i (so in particular $\mathrm{char}(k) \neq 2$). As before, take $n = 3$ and (in the notation of Theorem 5.5) $N = 4, m_1 = \ldots = m_3 = 1, m_4 = 3$ and $m_5 = 2$. Moreover, let the ramification points be $t_1 = 0, t_2 = 1, t_3 = -1, t_4 = a, t_5 = t$ where $a \in k \setminus \{0, \pm 1\}$ and t is transcendental over k, and put $K = k(t)$. Then a corresponding cyclic cover $\pi : C \to \mathbb{P}^1_K$ is given by the equation

$$Y^4 = cX(X-1)(X+1)(X-a)^3(X-t)^2 \text{ with } c \in k^*. \tag{5}$$

Clearly, C has genus 4 and $\deg(\pi) = 4$. Moreover, π has a subcover of degree 2 whose quotient curve is an elliptic curve E given by the equation

$$Z^2 = cX(X-1)(X+1)(X-a) \text{ with } Z = Y^2/((X-a)(X-t)). \tag{6}$$

Note that E is a constant curve, i.e. E is already defined over k. Moreover, E maps injectively into the Jacobian J_C of C and can be identified with the old part J_C^{old} of J_C (cf. section 5.1). Its complement in J_C is therefore the new part $A = J_C^{new}$; thus $J_C \sim E \times A$ and so $\dim A = 3$.

Next we choose a cyclic extension K_0 of K of degree 4 which is totally ramified at each of the points of $R = \{0, 1, -1, a\}$. For example, we can take $K_0 = k(t, s)$ where

$$s^4 = t(t^2 - 1)(t - a)g(t), \tag{7}$$

and $g(t) \neq 0$ is any nonzero polynomial of which does not vanish at R. Thus, if T_0 denotes the smooth curve over k defined by this equation (whose function field is K_0), then by Theorem 5.15 we obtain

Proposition 5.17 *The abelian variety $A_{K_0} = A \otimes K_0$ extends to an abelian scheme $\mathcal{A}$ over the curve T_0.*

Now we want to verify that the tower of unramified covers $K_0(\mathbb{P}(S_p))$ of K_0 attached to the subspaces $S_p \subset A[p]$ for $p \equiv 1 \mod 4$ has rational points which all lie over the same base-point $\tilde{v}$ of K_0.

For this we shall use the criterion $(\star\star)$ of Corollary 5.16 applied to the place $v = v_4$ corresponding to the specialization of t to $t_4 = a$. However, in order to do this, we need to determine the structure of the reduction $A_0 := A_{\tilde{v}}$ of A_{K_0} at the unique extension $\tilde{v}$ of v to K_0. (Recall that by the above proposition we know that A_0 is an abelian variety over k.) As a first step towards this end, we shall prove:

Proposition 5.18 *Let $J_{\tilde{v}}$ denote the reduction of the Jacobian $J = J_C \otimes K_0$ at $\tilde{v}$. Then*

$$J_{\tilde{v}} \simeq J_0 \times E_d,$$

where J_0 is the Jacobian of the curve C_0 which is the normalization of the curve defined by the equation

$$\bar{Y}^4 = c\bar{X}(\bar{X} - 1)(\bar{X} + 1)(\bar{X} - a)$$

and E_d denotes the elliptic curve given by the equation

$$Y^2 = X(X^2 - d), \tag{8}$$

where $d = c/g(a)$ and g is as in (7).

Proof. First note that since $J_C \sim E \times A$, and E is a constant curve, it follows from Proposition 5.17 that J has good reduction over T_0 and in particular at $\tilde{v}$. Thus, if $\mathcal{C}$ denotes the minimal model of C at $\tilde{v}$, then $\mathcal{C}$ has semi-stable reduction (use [BLR], 9.5/4, 9.2/5 and 9.2/12). Thus, if $C_1, \ldots, C_r$ denote the irreducible components of the reduction $C_{\tilde{v}}$ of $\mathcal{C}$, then each C_i is smooth (by [BLR], 9.2/12) and we have (by [BLR], 9.5/4 and 9.2/8)

$$J_{\tilde{v}} = J_{C_1} \times \ldots \times J_{C_r}.$$

Thus, the assertion follows once we have shown that C_0 and E_d occur as components of $C_{\tilde{v}}$, for then all other components must have genus 0 (since $g(C) = 4$, $g(C_0) = 3$ and $g(E_d) = 1$). For this it is enough to show:

Claim: There are normal models C' and C'' of C over $\mathfrak{O}_{\tilde{v}}$ whose reductions $C'_{\tilde{v}}$ and $C''_{\tilde{v}}$ each have an irreducible component $C'_{\tilde{v},1}$ and $C''_{\tilde{v},1}$ with function field $\kappa(C'_{\tilde{v},1}) \simeq \kappa(C_0)$ and $\kappa(C''_{\tilde{v},1}) \simeq \kappa(E_d)$.

Indeed, if such a model C' exists, then its desingularization and hence the minimal model C have the same property (because $g(C_0) > 0$). But since all components of C are smooth, there is a component C_1 of C with $C_1 \simeq C_0$, and similarly, if such a model C'' exists, then C has a component $C_2 \simeq E_d$.

Proof of claim: a) The construction of C': Let $\tilde{C}$ denote the normal model constructed in the proof of Theorem 5.14. Its fibre $\tilde{C}_v$ at v is integral and has C_0 as its normalization (because $Y^4 = cX(X^2 - 1)(X - a)^5$ also has normalization C_0). Thus, if we let C' denote (the normalization of) $\tilde{C} \otimes \mathfrak{O}_{\tilde{v}}$, then its fibre $C'_{\tilde{v}}$ has normalization C_0, as desired.

b) The construction of C'': Let $F_0 = \kappa(C \otimes K_0)$ denote the function field of $C \otimes K_0$; thus $F_0 = K_0(X, Y) = k(t, s, X, Y)$, where X and Y are related by equation (5) and t and s by equation (7). Put $x = (X - a)/s^4$ and $y = Y/s^5$. Then $F_0 = K_0(x, y)$ and equation (5) becomes

$$y^4 = x^3(x - B)^2 g_1(x),$$

where $B = (t - a)/s^4 = (t(t^2 - 1)g(t))^{-1} \in K = k(t)$ and $g_1(x) = cX(X^2 - 1) = c(s^4x + a)((s^4x + a)^2 - 1)$.

Let $\mathfrak{A}$ denote the integral closure of $\mathfrak{B} = \mathfrak{O}_{\tilde{v}}[x]$ in F_0, and let $C'' = \mathrm{Spec}(\mathfrak{A})$. Fix an irreducible component of the reduction of C'', and let V denote the associated (normalized) valuation of F_0. We then have:

$$V(\textstyle\sum a_i x^i) = e \min(\tilde{v}(a_i)), \;\; \text{if } a_i \in K_0,$$

for some integer $e \geq 1$. Since by construction $\tilde{v}(s) = 1$, $\tilde{v}(t - a) = 4$ and $\tilde{v}(g(t)) = 0$, we see that $V(x - B) = V(g_1(x)) = 0$, and so also $V(y) = 0$. Thus, if $\bar{x}$ and $\bar{y}$ denote the images of x and y in the residue field $\kappa(V) = \mathfrak{O}_{\tilde{v}}/\mathfrak{M}_{\tilde{v}}$ of V, then the above relation specializes to

$$\bar{y}^4 = a_0 \bar{x}^3(\bar{x} - b_0)^2,$$

where $a_0 = \overline{g_1(x)} = ca(a^2 - 1)$ and $b_0 = \bar{B} = (a(a^2 - 1)g(a))^{-1}$. Since this equation is irreducible over $k(\bar{x}) = \kappa(V_{|K_0(x)})$, we see that $[k(\bar{x}, \bar{y}) : k(\bar{x})] = 4$. Thus, since $[\kappa(V) : \kappa(V_{|K_0(x)})] \leq [F_0 : K_0(x)] = 4$, it follows that $\kappa(V) = k(\bar{x}, \bar{y})$. (It also follows that $e = 1$ and that V is the unique valuation above $V_{|K_0(x)}$, i.e. that the fibre of C'' at $\tilde{v}$ is integral, but we don't need this.)

It remains to show that $k(\bar{x}, \bar{y}) = \kappa(E_d)$. For this, put $u = \bar{y}^2/(\bar{x}(\bar{x} - b_0))$ and $v = a_0 \bar{y}/u$. Then by the above relation we have $u^2 = a_0 \bar{x}$, so $k(\bar{x}, \bar{y}) = k(u, v)$. Moreover, $v^2 = a_0 \bar{y}^2/\bar{x} = u(u^2 - a_0 b_0)$, so $\kappa(V) = \kappa(E_d)$, with $d = a_0 b_0 = c/g(a)$, as desired.

Remark 5.19 It follows from the above proof that reduction $C_{\tilde{v}}$ of the minimial model $\mathcal{C}$ has the form $C_{\tilde{v}} = C_1 \cup \ldots \cup C_r$, where $C_1 \simeq C_0$, $C_r \simeq E_d$, and $C_i \simeq \mathbb{P}^1$, for $2 \leq i \leq r-1$, and each C_i meets C_{i-1} transversally in a unique point for $2 \leq i \leq r$.

Next we want to determine the abelian variety J_0 up to k-isogeny. To simplify matters, we shall assume in the following that $1-a^2$ is a square in k or, equivalently, that $a = \frac{2b}{1+b^2}$ for some $b \in k$, where $b \neq 0, \pm 1$. (In fact, $b = (1 + \sqrt{(1-a^2)})/a$.) In addition, we shall choose $c = (1+b^2)$.

Lemma 5.20 *Let C_0 be the curve defined over k by the equation*

$$Y^4 = cX(X-1)(X+1)(X-a), \quad \text{where } a = \frac{2b}{1+b^2} \text{ and } c = 1+b^2,$$

for some $b \in k\backslash\{0,\pm 1\}$. Then its Jacobian J_0 is up to an isogeny of $2-$power degree isomorphic to $E \times E_1 \times E_{b^2}$, where E is given by equation (6) and E_1 and E_{b^2} by equation (8) by putting $d = 1$ and $d = b^2$, respectively. In particular, if $b = b_1^2$ is a square in k, then J_0 is k-isogenous to $E \times E_1 \times E_1$.

Proof. Let $F = \kappa(C_0) = k(X,Y)$ denote the function field of C_0. Then $F' := k(X,Y^2) = \kappa(E)$ is the function field of E. We shall show that there are two subfields $F_1 \simeq \kappa(E_1)$ and $F_2 \simeq \kappa(E_{b^2})$ of F which, for suitable $U, V \in F$, fit into the following field diagram of quadratic extensions:

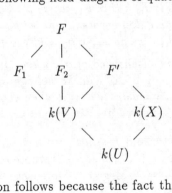

From this the assertion follows because the fact that $F/k(V)$ is a $\mathbb{Z}/2\mathbb{Z} \times \mathbb{Z}/2\mathbb{Z}$-extension implies that $J_0 = J_F \sim J_{F'} \times J_{F_1} \times J_{F_2} = E \times E_1 \times E_{b^2}$ where the indicated isogeny has 2-power degree; cf. [KR].

To find these subfields, put $U = cX(X-a)/(X^2-1)$, so $[k(X):k(U)] = 2$. Then

$$(U-1)(X^2-1) = cX(X-a) - (X^2-1) = (1-bX)^2$$

because $c - 1 = b^2$ and $ca = 2b$ by hypothesis. Thus, if we put $Z_1 = Y(1-bX)/(X^2-1)$, then

$$Z_1^4 = U(U-1)^2,$$

and so, if we put $F_1 := k(U, Z_1)$, then $[F_1 : k(U)] = 4$ and $F_1(X) = F$. In particular, since $[F : k(X)] = 4$, we see that $F_1 \cap k(X) = k(U)$ and that $[F : F_1] = 2$.

Next, let $V = Z_1^2/(U - 1)$. Then by the above equation $V^2 = U$, so $Z_1^2 = V(U - 1) = V(V^2 - 1)$. From this we see that $k(V) = k(U, Z_1^2)$, so $[F_1 : k(V)] = 2$, and hence $F_1 = k(V, Z_1) \simeq \kappa(E_1)$. In addition, it follows that $k(V, X) = F'$ because F' is the unique proper intermediate field of $F/k(V)$. In addition, let us observe that

$$(U - b^2)(X^2 - 1) = cX(X - a) - b^2(X^2 - 1) = (X - b)^2,$$

because $(c - b^2) = 1$ and $ca = 2b$ by hypothesis. Therefore, if we put $X_1 = (X - b)/(1 - bX)$ and $Z_2 = Z_1 X_1$, then $X_1^2 = (U - b^2)/(U - 1)$ and $Z_2^2 = V(U - 1) \cdot (U - b^2)/(U - 1) = V(V^2 - b^2)$. From this we obtain for $F_2 := k(V, Z_2)$ that $[F_2 : k(V)] = 2$ and that $F_2 \simeq \kappa(E_{b^2})$. We therefore see that $F_1, F_2, F', k(V)$ and $k(U)$ fit into the above field diagram as indicated, and that $F_1 \simeq \kappa(E_1)$ and $F_2 \simeq \kappa(E_{b^2})$.

We can now determine the structure of the reduction $A_0 = A_{\tilde{v}}$ of the abelian variety $A = J_C^{new}$ at $\tilde{v}$.

Proposition 5.21 *Assume again that* $a = \frac{2b}{1+b^2}$ *and* $c = 1 + b^2$ *for some* $b \in k^*$, *and let* $d = c/g(a)$. *Then:*

(a) *There is a k-isogeny $\varphi : A_0 \to A_0' := E_1 \times E_{b^2} \times E_d$ of 2-power degree, and hence A_0 is k'-isogenous to $E_1 \times E_1 \times E_1$, where $k' = k(\sqrt{b}, \sqrt[4]{d})$. Moreover, if $\sigma_0 \in \text{Aut}(A_0)$ denotes the automorphism induced by a generator $\sigma \in \text{Aut}(\pi)$ of the covering group, then we have $\varphi \circ \sigma_0 = \sigma_0' \circ \varphi$ for some $\sigma_0' \in \text{Aut}(A_0')$.*

(b) *Let $\bar{S}_p \leq A_0[p]$ be an eigenspace of σ_0, where $p \equiv 1 \mod 4$ is a prime. Then every cyclic $\bar{k}$-isogeny of A_0' with kernel contained in $\varphi(\bar{S}_p)$ is an endomorphism of A_0'. In particular, if $k' = k$, then every $\bar{k}$-isogeny of A_0 with kernel contained in $\bar{S}_p$ is defined over k.*

Proof. (a) Since $A = J_C^{new}$ and $E = J_C^{old}$, we have an isogeny $\varphi_1 : E \times A \to J = J_C$ of 2-power degree (cf. Corollary 5.3) whose reduction $\overline{\varphi}_1 : E \times A_0 \to J_{\tilde{v}}$ is a k-isogeny of the same degree. Moreover, by Proposition 5.18 we have an isomorphism $\varphi_2 : J_{\tilde{v}} \xrightarrow{\sim} J_0 \times E_d$, and by Lemma 5.20 we have a k-isogeny $\varphi_3 : J_0 \to E \times E_1 \times E_{b^2}$ of 2-power degree. We thus see that $\tilde{\varphi} = (\varphi_3 \times id_{E_d}) \circ \varphi_2 \circ \overline{\varphi}_1 : E \times A_0 \to E \times E_1 \times E_{b^2} \times E_d = E \times A_0'$ is an isogeny of 2−power degree, and so the first assertion follows (by taking $\varphi = \tilde{\varphi}_{|\{0\} \times A_0}$) once we have shown that $\tilde{\varphi}(\{0\} \times A_0) = \{0\} \times A_0'$. To prove this, it is enough to find automorphisms $\sigma_0 \in \text{Aut}(A_0)$ and $\sigma_0' \in \text{Aut}(A_0')$ with $\sigma_0^2 = [-1]_{A_0}$ and $(\sigma_0')^2 = [-1]_{A_0'}$ such that

$$\tilde{\varphi} \circ (\tilde{\sigma}_0) = (\tilde{\sigma}_0') \circ \tilde{\varphi},$$

where $\tilde{\sigma}_0 = [-1]_E \times \sigma_0$ and $\tilde{\sigma}_0' = [-1]_E \times \sigma_0'$.

Indeed, since $1+\sigma_0$ is an isogeny of A_0 (because $(1+\sigma_0)(1-\sigma_0) = 1-[-1]_{A_0} = [2]_{A_0}$), we see that $(1 + ([-1]_E \times \sigma_0))(E \times A_0) = \{0\} \times A_0$, and similarly $(1+\tilde{\sigma}_0')(E \times A_0') = \{0\} \times A_0'$, and hence $\check{\varphi}(\{0\} \times A_0) = \check{\varphi}((1+\tilde{\sigma}_0)(E \times A_0)) = (1+\tilde{\sigma}_0')\check{\varphi}(E \times A_0) = (1+\tilde{\sigma}_0')(E \times A_0') = \{0\} \times A_0'$, as desired.

To construct σ_0 and σ_0' with these properties, let σ_J denote the automorphism (of order 4) on the Jacobian J induced by $\sigma \in \mathrm{Aut}(\pi)$. Then σ_J maps the subvarieties E and A of J_C into themselves, and $(\sigma_J)_{|E} = [-1]_E$. Thus, if $\sigma_A = (\sigma_J)_{|A}$, then we have $\varphi_1 \circ ([-1] \times \sigma_A) = \sigma_J \circ \varphi_1$. Thus, if $\bar{\sigma}$ (resp. σ_0) denotes the automorphism induced by σ_J (resp. σ_A) on the reduction $J_{\bar{v}}$ (resp. on A_0), then we also have $\bar{\varphi}_1 \circ ([-1]_E \times \sigma_0) = \bar{\sigma} \circ \overline{\varphi_1}$.

Next we note that the proof of Proposition 5.18 shows that σ induces automorphisms σ_d and σ_{C_0} on E_d and on C_0 respectively, from which we see that $\varphi_2 \circ \bar{\sigma} = (\sigma_{J_0} \times \sigma_d) \circ \varphi_2$. Similarly, the proof of Lemma 5.20 shows that $\varphi_3 \circ \sigma_{J_0} = ([-1]_E \times \sigma_1 \times \sigma_{b^2}) \circ \varphi_3$, for certain automorphisms $\sigma_i \in \mathrm{Aut}(E_i)$ where $i = 1, b^2$. We thus see that if we put $\sigma_0' = \sigma_1 \times \sigma_{b^2} \times \sigma_d \in \mathrm{Aut}(A_0')$, then the above commutation relation holds.

It remains to show that $\sigma_0^2 = [-1]_{A_0}$ and that $(\sigma_0')^2 = [-1]_{A_0'}$. The latter is clear, for by construction $\sigma_0' = \sigma_1 \times \sigma_{b^2} \times \sigma_d$, where (for $i = 1, b^2, d$) the automorphism $\sigma_i \in \mathrm{Aut}(E_i)$ has order 4 and so $\sigma_i^2 = [-1]_{E_i}$ because $\mathrm{char}(k) \neq 2, 3$. To prove the former, recall that $A = \check{\varepsilon}_{new} J$, where $\check{\varepsilon}_{new} = 1 - \sigma_J^2$, so σ_J^2 acts on A like $[-1]_A$. Thus, $\sigma_A^2 = [-1]_A$ and hence by functoriality we have $\sigma_0^2 = [-1]_{A_0}$, as desired.

(b) Write $E_2 = E_{b^2}$ and $E_3 = E_d$, so $A_0' = E_1 \times E_2 \times E_3$. Let $\bar{S}_p' = \varphi(\bar{S}_p) \leq A_0'[p]$, which is an eigenspace under the action of $\sigma_0' = \sigma_1 \times \sigma_2 \times \sigma_3$.

Since $\sqrt[4]{-1} \in k$ and $j(E_i) = 1728$, we have for $i = 1, 2, 3$ that $\mathrm{End}(E_i) \supset \mathbb{Z}[\sqrt[4]{-1}]$, and equality holds unless E_i is supersingular (i.e. unless $\mathrm{char}(k) \equiv 3 \bmod 4$). Thus, since $p \equiv 1 \bmod 4$, there is an $\alpha_i \in \mathrm{End}(E_i)$ of degree p such that $\mathrm{Ker}(\alpha_i) = \bar{S}_p' \cap e_i(E_i)$, where $e_i : E_i \to A_0'$ is the embedding of E_i as the i-th factor of A_0'. Thus, $\bar{S}_p' = \mathrm{Ker}(\alpha_1) \times \mathrm{Ker}(\alpha_2) \times \mathrm{Ker}(\alpha_3)$.

Now since E_2 and E_3 are twists of E_1, there are k'-isomorphisms $f_{1i} : E_1 \to E_i$, for $i = 2, 3$, which we can choose such that $f_{1i} \circ \alpha_1 = \alpha_i \circ f_{1i}$. Thus, if $\tilde{P}_1 \in E_1(\bar{k})$ is a generator of $\mathrm{Ker}(\alpha_1)$, and $\tilde{P}_i = f_{1i}(\tilde{P}_1)$ then $\{P_1, P_2, P_3\}$ is a basis of $\bar{S}_p'$, where $P_i := e_i(\tilde{P}_i)$.

Now let h be a cyclic $\bar{k}$-isogeny of A_0' with $\mathrm{Ker}(h) \leq \bar{S}_p'$. Then $\mathrm{Ker}(h)$ is generated by $P = aP_1 + bP_2 + cP_3$, for some $a, b, c \in \mathbb{F}_p$. W.l.o.g. we may assume that $a \neq 0$ (otherwise we interchange the roles of E_1, E_2 and E_3), and hence that $a = 1$. Then the k'-endomorphism $h' : A_0' \to A_0'$, defined by the

matrix $\begin{pmatrix} (1-b)\alpha_1 & \alpha_1 f_{12}^{-1} & 0 \\ -bf_{12} & id_{E_2} & 0 \\ -cf_{13} & 0 & id_{E_3} \end{pmatrix}$ has kernel $\mathrm{Ker}(h') = \langle P_1 + bP_2 + cP_3 \rangle = \mathrm{Ker}(h)$, which proves the first assertion.

From this the second assertion is immediate. Indeed, if $k' = k$, then by what was just shown, every $\bar{k}$-subgroup of $\bar{S}_p'$ is k-rational, and hence the same is true for $\bar{S}_p$.

We can now summarize the results obtained above in the following theorem.

Theorem 5.22 *Suppose k contains a primitive fourth root of unity, and let $a = \frac{2b^2}{1+b^4}$, for some $b \in k^*$ with $b^4 \neq 1$. Let $K = k(t,s)$ be the function field of any curve of the form*

$$s^4 = t(t^2 - 1)(t - a)g(t),$$

where $g(t) \in k[t]$ is any polynomial which does not vanish at $\{0, 1, -1, a\}$, and which has been normalized in such a way that $(1 + b^4)/g(a)$ is a fourth power in k. Moreover, let $\mathfrak{p}$ be the unique place of K/k inducing the specialization $t \mapsto a$. Then the k-rational geometric fundamental group of K with base point $\mathfrak{p}$ is infinite; more precisely, for every prime $p \equiv 5 \bmod 12$ (with $p \neq \mathrm{char}(k)$), there is an unramified extension K_p/K with $\mathrm{Gal}(K_p/K) \simeq \mathrm{PSL}_3(p)$ in which $\mathfrak{p}$ splits completely.

Proof. It is clearly enough to prove the last statement. For this, let C be the curve defined over $k(t)$ by the equation $Y^2 = (1 + b^4)X(X^2 - 1)(X - a)^3(X - t)^2$, and let J denote its Jacobian and $A = J^{new}$ its new part (cf. section 5.1). Since the hypotheses on a, $c = 1 + b^2$ and g ensure that $k' := k(\sqrt{b^2}, \sqrt[4]{c/g(a)}) = k$, it follows from Proposition 5.21a) that the reduction A_0 of $A \otimes K$ at $\mathfrak{p}$ is k-isogenous to $E_1 \times E_1 \times E_1$.

Now let $p \equiv 5 \bmod 12$ be any prime (and $\neq \mathrm{char}(k)$), and let $S_p \leq A[p]$ denote an eigenspace of the automorphism σ_A (cf. proof of Proposition 5.21). By Theorem 5.5(b),(d) (and Abhyankar's Lemma) we know that $K_p := K(\mathbb{P}(S_p))$ is unramified over K and that $\mathrm{Gal}(K_p/K) = \mathrm{PSL}_3(p)$. Let $\bar{S}_p$ denote the image of S_p in $A_0[p]$ under the reduction map at $\mathfrak{p}$; clearly, $\bar{S}_p$ is identical to the set $\bar{S}_p$ as defined in Proposition 5.21(b). Thus, by that proposition, condition ($\star\star$) of Corollary 5.16 holds, and so it follows that $\mathfrak{p}$ splits completely in K_p.

Remark 5.23 The simplest examples of fields K satifying the hypothesis of Theorem 5.22 are clearly the fields $K = k(s,t)$, where

$$s^4 = (1 + b^4)t(t^2 - 1)\left(t - \frac{2b^2}{1+b^4}\right)$$

and $b \in k^*$ is any element with $b^4 \neq 1$; recall that fields of this type were studied in Lemma 5.20. Clearly, all these fields have genus 3. More generally, by choosing $g(t)$ suitably, we can obtain examples of fields of any genus $g \geq 3$ which satisfy the hypotheses of Theorem 5.22.

References

[BLR] S. BOSCH, W. LÜTKEBOHMERT AND M. RAYNAUD, Néron Models. Springer-Verlag, Berlin, 1990.

[FKV] G. FREY, E. KANI AND H. VÖLKLEIN, Curves of genus 2 with elliptic differentials and associated Hurwitz spaces. In preparation.

[FV] M. FRIED AND H. VÖLKLEIN, The inverse Galois problem and rational points on moduli spaces. Math. Annalen **290** (1991), 771-800.

[GS] A. GARCIA AND H. STICHTENOTH, On the asymptotic behaviour of some towers of function fields over finite fields. To appear in J. Number Theory.

[Gro] A. GROTHENDIECK, Modeles de Néron et monodromie. SGA 7$_I$, Exp. IX. Springer Lecture Notes **288** (1972), 313-523.

[Ig] J.-I. IGUSA, On the algebraic theory of elliptic modular functions. J. Math. Soc. Japan **20** (1968), 96–106.

[Ih] Y. IHARA, On unramified extensions of function fields over finite fields. Adv. Stud. in Pure Math. **2** (1983), 83–97.

[KR] E. KANI AND M. ROSEN, Idempotent relations and factors of Jacobians. Math. Ann. **284** (1989), 307–327.

[MSV] K. MAGAARD, K. STRAMBACH AND H. VÖLKLEIN, Finite quotients of the pure symplectic braid group, to appear in Israel J. Math.

[Me] L. MEREL, Bornes pour la torsion des courbes elliptiques sur les corps de nombres. Inventiones mathematicae **124** (1996), 434–449.

[Mi] MIYAKE, Modular Forms. Springer Verlag, Berlin, 1989.

[Mo] S. MOCHIZUKI, The Local pro-p Grothendieck Conjecture for Hyperbolic Curves. RIMS Kyoto University, Preprint 1045 (1996).

[Se] J.-P. SERRE, Groupes algébriques et corps de classes. Hermann, Paris, 1959.

[Se1] J.-P. SERRE, Sur le nombre des points d'une courbe algebrique sur un corps fini. C. R. Acad. Sci. Paris **296** (1983), série I, 397–402.

[Se2] J.-P. SERRE, Local Fields. Springer Verlag, New York, 1979.

[ST] J.-P. SERRE AND J. TATE, Good reduction of abelian varieties. Ann. of Math. **88** (1968), 492–517.

[V] H. VÖLKLEIN, Groups as Galois Groups – an Introduction, Cambr. Studies in Adv. Math. **53**, Cambridge Univ. Press 1996.

[V1] H. VÖLKLEIN, Rigid generators of classical groups, Math. Annalen **311** (1998), 421–438.

[V2]   H. VÖLKLEIN, Moduli spaces for covers of the Riemann sphere, Israel J. Math. **85** (1994), 407–430.

[V3]   H. VÖLKLEIN, Cyclic covers of P^1, and Galois action on their division points, Contemp. Math. **186** (1995), 91–107.

[V4]   H. VÖLKLEIN, Braid group action through $GL_n(q)$ and $U_n(q)$, and Galois realizations, Israel J. Math. 82 (1993), 405–427.

[V5]   H. VÖLKLEIN, Braid group action, embedding problems and the groups $PGL(n,q)$, $PU(n,q^2)$, Forum Math. **6** (1994), 513–535.

Embedding Problems and Adding Branch Points

David Harbater*

Abstract: If $Y \to X$ is a G-Galois branched cover of curves over an algebraically closed field k, and if G is a quotient of a finite group Γ, then $Y \to X$ is dominated by a Γ-Galois branched cover $Z \to X$. This is classical in characteristic 0, and was proven in characteristic p by the author [Ha6] and F. Pop [Po1] in conjunction with the proof of the geometric case of the Shafarevich Conjecture on free absolute Galois groups. The resulting cover $Z \to X$, though, may acquire additional branch points. The present paper shows how many new branch points are needed, and shows that there is some control on the positions of these branch points and on the inertia groups of $Z \to X$.

Section 1. Introduction and survey of results.

This paper concerns an aspect of the fine structure of the fundamental group of an affine curve U over an algebraically closed field k of characteristic p. In [Ra], [Ha3], it was shown which finite groups G are quotients of $\pi_1(U)$ — namely, according to Abhyankar's Conjecture, the set of such G depends only on the pair (g, n), where g is the genus of the smooth compactification X of U and n is the number of points in $X - U$. But the structure of the profinite group $\pi_1(U)$ remains a mystery, even in the case of the affine line. Moreover, the group $\pi_1(U)$ (unlike the set $\pi_A(U)$ of its finite quotients) does not depend just on (g, n) (cf. [Ha4, §1], [Ta, Thm. 3.5]), though it is unclear how it varies in moduli. In the current paper we study the structure of $\pi_1(U)$ by investigating how the finite quotients of this group fit together, and how $\pi_1(U)$ grows as additional points are deleted.

A preliminary result in this direction appeared in [Ha6], [Po1], in connection with proving the geometric case of the Shafarevich Conjecture. Namely, it was shown there that the absolute Galois group G_K of the function field K of U is a free profinite group (of rank equal to the cardinality of k). This was done by showing that every finite embedding problem for K has a proper solution, i.e. that if $\Gamma \twoheadrightarrow G$ is a surjection of finite groups, then every unramified G-Galois cover $V \to U$ of affine curves is dominated

---

* Supported in part by NSF Grant DMS94-00836.

---

by a Γ-Galois *branched* cover $W \to U$. In fact, the proof (which used patching techniques in formal or rigid geometry) showed a bit more — viz. it bounded the number of branch points of $W \to U$. That bound was not sharp, however, and here we obtain the sharp bound (Theorem 5.4 below).

More precisely, let $G = \Gamma/N$, let $p(N)$ be the subgroup of N generated by its p-subgroups (so that $\bar{N} = N/p(N)$ is the maximal prime-to-p quotient of N), and let r be the *rank* of $\bar{N}$ (i.e. the minimal number of elements in any generating set). In [Ha6, Theorem 3.5] it was shown that $W \to U$ as above can be chosen with at most $r+1$ branch points; and it was asked if it is always possible to choose $W \to U$ with at most r branch points. (It is not in general possible with only $r-1$ branch points even in characteristic 0, as topological considerations show; and that implies the same for characteristic p.) In [Po1], it was shown that this is always possible in the case that $r = 0$, thus answering [Ha6, Question 3.7]. So if N above is a *quasi-p group* (i.e. is generated by its p-subgroups, or equivalently if $\bar{N} = 1$) then the cover $W \to U$ can be chosen to be unramified.

Here we show that for arbitrary r (not just $r = 0$), the dominating cover $W \to U$ can be chosen with at most r branch points (where as above, $r = \mathrm{rk}(\bar{N})$). In fact, we show a bit more. Namely, for finite group Γ and normal subgroup N of Γ, we will define the *relative rank* of N in Γ, denoted $\mathrm{rk}_\Gamma(N)$. This will be a non-negative integer that is $\leq \mathrm{rk}(N)$ (but is often strictly less). What we will show is that in the above situation, the cover $W \to U$ can be chosen with at most $\mathrm{rk}_{\bar{\Gamma}}(\bar{N})$ branch points, where $\bar{\Gamma} = \Gamma/p(N)$. By using the $r = 0$ case, the proof of this result is reduced to the case that N is of order prime to p; and there we use methods of patching and lifting. In addition, we show that there is often control over the positions of the new branch points, and over the inertia groups of the resulting cover (cf. Props. 3.3, 3.5, 4.1, 5.1).

The results in this paper can be phrased in the language of embedding problems. This and other group-theoretic notions (along with some notions about covers) are discussed in Section 2. Then, in Section 3, we use formal patching to prove the above result in a key special case (when N has order prime to p, and one of the branch points of $Y \to X$ is tame, where $Y \to X$ is the smooth compactification of $V \to U$). In Section 4, we use a lifting result of Garuti [Ga, Theorem 2] to prove the above result in the case that $\mathrm{rk}_\Gamma(N) \leq 1$, again assuming that N has order prime to p. Section 5 combines the two special cases, and applies Pop's result [Po1] in the case $r = 0$, to prove the full result (Theorem 5.4).

I would like to thank Claus Lehr, Rachel Pries, Kate Stevenson, and the referee for helpful comments on this manuscript.

Section 2. Notions concerning covers and groups.

In this paper we work over a fixed algebraically closed field k of characteristic $p \geq 0$, and consider covers of k-curves. A *cover* will be a finite generically separable morphism $Y \to X$ of k-schemes, where X is connected. If G is a finite group, then a G-*Galois* cover consists of a cover $Y \to X$ together with a homomorphism $\rho : G \to \operatorname{Aut}(Y/X)$, such that G acts simply transitively on any generic geometric fibre of the cover, via ρ. (The top space Y is not required to be connected. For example, the trivial G-Galois cover of X is a disjoint union of copies of X indexed by the elements of G, on which G acts by the regular representation.)

Since k is algebraically closed of characteristic p, there is a primitive mth root of unity $\zeta_m \in k$ for each positive integer m not divisible by p. Here we may choose the elements ζ_m to be compatible; i.e. such that $\zeta_{mm'}^{m'} = \zeta_m$ for all m, m'. From now on, these will be fixed. For any G-Galois cover $\psi : Y \to X$ of smooth connected k-curves and any tame ramification point η lying over a branch point $\xi \in X$, the corresponding extension of complete local rings is given by $y^m = x$, for some choice of local parameters x, y. The inertia group is generated by $c : y \mapsto \zeta_m y$, and the element $c \in G$ (which is independent of the choice of local parameters) is called the *canonical generator* of the inertia group at η. (Here and just below, we follow the terminology of [St] and [HS, §§2,3].) If all the branch points of $Y \to X$ are tame, and if the branch points are given with an ordering, say $\xi_1, \ldots, \xi_r$, then we say that the cover has *description* $(c_1, \ldots, c_r)$, where c_j is a canonical generator of inertia at a point over ξ_j, and where each c_j is determined up to (individual) conjugacy. In the case that $k = \mathbf{C}$ and $\zeta_m = e^{2\pi i/m}$, the fundamental group of $U = X - \{\xi_1, \ldots, \xi_r\}$ has presentation

$$\pi_1(U) = \langle a_1, \ldots, a_g, b_1, \ldots, b_g, c_1, \ldots, c_r \mid \prod_{i=1}^{g} [a_i, b_i] \prod_{j=1}^{r} c_j = 1 \rangle, \quad (*)$$

where g is the genus of X. Here the G-Galois cover corresponding to a surjection $\phi : \pi_1(U) \to G$ has description $(\phi(c_1), \ldots, \phi(c_r))$. If p does not divide the order of G, then this also holds for an arbitrary algebraically closed field k of characteristic $p \geq 0$, via the same presentation $(*)$ of the maximal prime-to-p quotient $\pi_1(U)^{p'}$ of $\pi_1(U)$ [Gr, XIII, Cor. 2.12]. (This presentation arises via a specialization morphism between k and $\mathbf{C}$, which should be chosen so that the given roots of unity $\zeta_m \in k$ correspond to $e^{2\pi i/m} \in \mathbf{C}$. Cf. [Gr, XIII] and [GM, Thm. 4.3.2, Lemma 4.1.3].)

Consider more generally a G-Galois cover $Y \to X$ of a semi-stable k-curve X (i.e. X is connected and its only singularities are nodes). Let $\tilde{Y} \to \tilde{X}$ be the pullback of $Y \to X$ over the normalization $\tilde{X}$ of X. We say that $Y \to X$ is *admissible* if for each singular point $\eta \in Y$, the canonical generators at the two points $\eta_1, \eta_2 \in \tilde{Y}$ over η are inverses in G. A *thickening* of $Y \to X$ is a G-Galois cover of normal $k[[t]]$-curves $Y^* \to X^*$ whose closed fibre is $Y \to X$, and whose completion along the smooth locus of X is a trivial deformation of its closed fibre. If $Y \to X$ is an admissible cover, then such a $Y^* \to X^*$ is called an *admissible thickening* of $Y \to X$ if at the complete local ring at every singular point of Y the cover is given by the extension $k[[t, x_1, x_2]]/(x_1 x_2 - t^m) \hookrightarrow k[[t, y_1, y_2]]/(y_1 y_2 - t)$ for some m prime to p, where $x_i \mapsto y_i^m$ under the inclusion, and where an associated canonical generator of inertia acts by $x \mapsto \zeta_m x$, $y \mapsto \zeta_m^{-1} y$. Observe that in this situation, the singular points of the closed fibre X are isolated branch points of $Y^* \to X^*$ (and this does not contradict Purity of Branch Locus since X^* is not regular and Y^* is not flat over X^*). Since these points are branch points of the irreducible components of $Y \to X$, the process of constructing an admissible thickening can be regarded as a way of patching together these components in such a way that some of the branch points "cancel" on the general fibr! e (and cf. [HS, Thm. 7]). This ob servation will be key to the results of §3 below, and thus to the paper's main theorem, by yielding a cover with fewer branch points than would otherwise be expected.

The remainder of this section is devoted to discussing some group-theoretic notions that will be used in this paper.

If Γ is any finite group, then (following [FJ]), we define its *rank* to be the smallest non-negative integer $r = \mathrm{rk}(\Gamma)$ such that Γ has a generating set of r elements. (In the literature, this integer is also sometimes denoted by $d(\Gamma)$.) More generally, let E be a subgroup of a finite group Γ. A subset $S \subset E$ will be called a *relative generating set* for E in Γ if for every subset $T \subset \Gamma$ such that $E \cup T$ generates Γ, the subset $S \cup T$ also generates Γ. We define the *relative rank* of E in Γ to be the smallest non-negative integer $s = \mathrm{rk}_\Gamma(E)$ such that there is a relative generating set for E in Γ consisting of s elements. Thus every generating set for E is a relative generating set, and so $0 \leq \mathrm{rk}_\Gamma(E) \leq \mathrm{rk}(E)$. Also, $\mathrm{rk}_\Gamma(E) = \mathrm{rk}(E)$ if $E = 1$ or $E = \Gamma$, while $\mathrm{rk}_\Gamma(E) = 0$ if and only if E is contained in the Frattini subgroup $\Phi(\Gamma)$ of Γ.

A related notion is the following: Let G be a subgroup of a group Γ. A subset $T \subset \Gamma$ is a *supplementary generating set* for Γ *with respect to G* if $T \cup G$ generates Γ. Suppose that Γ is a finite group that is generated by two subgroups E, G. We then define the *relative rank* of $E \subset \Gamma$ *with* *respect to* G to be the smallest non-negative integer $t = \mathrm{rk}_\Gamma(E, G)$ such that there is a supplementary generating set T for Γ with respect to G, with T of cardinality t and $T \subset E$. Note that every relative generating set for E in Γ is a supplementary generating set for Γ with respect to G. So $0 \leq \mathrm{rk}_\Gamma(E, G) \leq \mathrm{rk}_\Gamma(E) \leq \mathrm{rk}(E)$.

If p is a prime number, then the *quasi-p part* of a finite group Γ is the subgroup $p(\Gamma) \subset \Gamma$ that is generated by the p-subgroups of Γ (or equivalently, by the Sylow p-subgroups of Γ). Thus $p(\Gamma)$ is a characteristic subgroup of Γ, and in particular is normal. A group Γ is defined to be a *quasi-p group* if $\Gamma = p(\Gamma)$. Thus for any finite group Γ, the subgroup $p(\Gamma)$ is a quasi-p group and $\Gamma/p(\Gamma)$ is the (unique) maximal quotient of Γ whose order is not divisible by p. (In the other case, viz. $p = 0$, we set $p(\Gamma) = 1$.)

If Π, Γ, H are groups (not necessarily finite), then an *embedding problem* for Π consists of a pair of surjective group homomorphisms $\mathcal{E} = (\alpha : \Pi \to H, f : \Gamma \to H)$. A *weak solution* to the embedding problem consists of a group homomorphism $\beta : \Pi \to \Gamma$ such that $f\beta = \alpha$. If moreover β is surjective, then it is referred to as a *proper solution* to the embedding problem. An embedding problem is *finite* if Γ is finite. The motivation for the notion of embedding problems comes from Galois theory: If $K \subset L$ is a Galois field extension with group H, and if Π is the absolute Galois group G_K of K, then Galois theory yields a corresponding surjection $\alpha : G_K \twoheadrightarrow H$. Let $f : \Gamma \twoheadrightarrow H$ be a surjective homomorphism of finite (or profinite) groups. Then a proper [resp. weak] solution to the embedding problem (α, f) corresponds to a Γ-Galois fiel! d extension of K [resp. to a *Gamma*-Galois K-algebra] containing the H-Galois extension L, such that the actions of Γ and H are compatible with the surjection $\Gamma \twoheadrightarrow H$. That is, the H-Galois extension L is embedded in a Γ-Galois extension via a solution to the embedding problem.

Observe that if $\phi : \Pi' \twoheadrightarrow \Pi$ is a surjective homomorphism of groups, then every embedding problem for Π induces an embedding problem for Π'. Namely, if $\mathcal{E} = (\alpha : \Pi \to H, f : \Gamma \to H)$ is an embedding problem for Π, then there is an *induced embedding problem* $\mathcal{E}' = (\alpha' : \Pi' \to H, f : \Gamma \to H)$ for Π', where $\alpha' = \alpha\phi$. Moreover, a weak or proper solution to the given embedding problem induces such a solution to the new problem. On the other hand, not every solution to the new problem need come from a solution to the original problem. These observations will be useful later, when considering the fundamental groups $\Pi = \pi_1(U)$ and $\Pi' = \pi_1(U')$ of two affine curves $U' \subset U$. In that context, solutions to embedding problems for Π correspond to certain unramified covers of U, whereas solutions to embedding problems for Π' are required merely to be unramified over U'.

As above, let $\mathcal{E} = (\alpha : \Pi \to H, f : \Gamma \to H)$ be an embedding problem for Π. If the exact sequence $1 \to N \to \Gamma \to H \to 1$ is split, where $N = \ker f$, then we say that $\mathcal{E}$ is a *split* embedding problem. A split embedding problem $\mathcal{E} = (\alpha, f)$ always has a weak solution, viz. $s\alpha : \Pi \to \Gamma$, where s is a section of $\Gamma \to H$. Often, finding proper solutions to embedding problems can be reduced to doing so for split embedding problems — e.g. see [FJ, §20.4], [Ha3, proofs of Thm. 5.4, Prop. 6.2], [Ha6, proof of Prop. 3.3], and [Po2, §1B(2)]. For the sake of completeness, we conclude this section with a precise statement of this reduction, in a form that can be cited later (in sections 3 and 5 below).

Proposition 2.1. *Let $\mathcal{E} = (\alpha : \Pi \to H, f : \Gamma \to H)$ be an embedding problem, and let $N = \ker f$. Suppose that $\mathcal{E}$ has a weak solution $\alpha_0 : \Pi \to \Gamma$, and let $H_0 \subset \Gamma$ be the image of α_0. Consider the semi-direct product $\Gamma_0 = N \rtimes H_0$, with respect to the conjugation action of H_0 on $N \lhd \Gamma$, and let $f_0 : \Gamma_0 \twoheadrightarrow H_0$ be the natural quotient map. If the split embedding problem $\mathcal{E}_0 = (\alpha_0 : \Pi \to H_0, f_0 : \Gamma_0 \to H_0)$ has a proper solution, then so does $\mathcal{E}$.*

Proof. Since α_0 is a weak solution to $\mathcal{E}$, we have $f\alpha_0 = \alpha$; or equivalently $\bar{\mu}\alpha_0 = \alpha$, where $\bar{\mu} : H_0 \to H$ is the restriction of $f : \Gamma \twoheadrightarrow H$ to H_0. Since f has kernel N, and since its restriction $f|_{H_0} = \bar{\mu}$ is surjective onto H (because $\bar{\mu}\alpha_0 = \alpha$ is), it follows that Γ is generated by N and H_0. Let $\mu : \Gamma_0 \to \Gamma$ be the homomorphism defined by taking the identity inclusion on each factor of $\Gamma_0 = N \rtimes H_0$. (This is a homomorphism since the conjugation action of H_0 on N in Γ_0 is the same as the conjugation action of H_0 on N in Γ.) Then μ is a surjection since N and H_0 generate Γ, and it is straightforward to check that $f\mu = \bar{\mu}f_0$. We thus obtain the following commutative diagram (where as above $\bar{\mu}\alpha_0 = \alpha : \Pi \to H$):

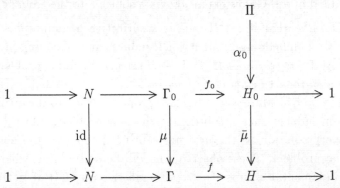

So any proper solution $\beta_0 : \Pi \twoheadrightarrow \Gamma_0$ to the split embedding problem $\mathcal{E}_0 = (\alpha_0 : \Pi \to H_0, f_0 : \Gamma_0 \to H_0)$ yields a proper solution $\beta : \Pi \twoheadrightarrow \Gamma$ to the original embedding problem $\mathcal{E}$, viz. $\beta = \mu\beta_0$. $\quad\square$

In particular, the reduction in the above proposition can be always accomplished in the case that the group Π is *projective* (which by definition [FJ, §20.4] means that every finite embedding problem for Π has a weak solution). Indeed, in that situation, the given embedding problem $\mathcal{E}$ has a weak solution α_0, and so the above hypotheses are satisfied.

Section 3. Results via patching.

This section uses formal patching methods in order to prove the main result of the paper in a special case. Namely, we consider a finite group Γ and a quotient $G = \Gamma/N$, together with a G-Galois étale cover of smooth affine k-curves $V \to U$ (where, as always, k is algebraically closed of characteristic $p \geq 0$). We consider the smooth completions X, Y of U, V, and assume that $Y \to X$ is tamely ramified at some branch point ξ. We will also assume that p does not divide the order of N. In this situation, we will show that there is a Γ-Galois cover $W \to U$ dominating $V \to U$, having at most $\mathrm{rk}_\Gamma(N)$ branch points, and with specified inertia groups over those points (Prop. 3.5). This solves a certain embedding problem (Cor. 3.6). Moreover we will obtain greater control on the number of branch points of the constructed cover and on the inertia groups over $X - U$ in the case that the embedding problem is split (under an additional assumption on normalizers). Cf. Prop. 3.3 and Cor. 3.4. A more general and more precise version of these results appears first, in Prop. 3.1 an Cor. 3.2).

Patching methods, in formal or rigid geometry, have previously been used to prove a number of results concerning fundamental groups of varieties, especially for curves in characteristic p — e.g. [Ha1], [Ha2], [Ra,§§3-5], [St], [Sa], [Ha6], [Po1], [HS]; see also [Ha5, §2]. The basic idea is to build a simpler, but possibly degenerate, cover with similar properties, and then to deform it to a family of covers whose generic member is as desired. In order to reduce the number of branch points of the cover we construct here, and thus achieve the desired sharp bound on that number, we will use a construction involving *admissible* covers; cf. §2 above and the remark following Proposition 3.3 below.

Below we preserve the terminology of Section 2, and begin with an assertion concerning the problem of modifying a cover so as to expand its Galois group. (Cf. also [Ha2, Theorem 2] for a related result.) Note that here, and in the next few results, it suffices to use the value $\mathrm{rk}_\Gamma(E, G)$, rather than having to use the possibly larger value $\mathrm{rk}_\Gamma(E)$.

Proposition 3.1. *Let Γ be a finite group generated by two subgroups G, E, where p does not divide $|E|$, and let $r \geq \mathrm{rk}_\Gamma(E, G)$. Let $V \to U$*

be a G-Galois étale cover of smooth connected affine k-curves with smooth completion $Y \to X$. Suppose that $Y \to X$ is tamely ramified over some point ξ of $B = X - U$, and that some inertia group over ξ normalizes E. Then there is a smooth connected Γ-Galois cover $W \to U$ having at most r branch points.

Moreover, if $\{e_1, \ldots, e_r\} \subset E$ is a supplementary generating set for Γ with respect to G, then the above cover can be chosen so that:

(i) The H-Galois cover $W/N \to U$ agrees with $V/(N \cap G) \to U$, where N is the normal closure of E in Γ and where $H = \Gamma/N = G/(N \cap G)$.

(ii) There are inertia groups of $W \to U$ over the branch points $\xi_1, \ldots, \xi_r$ having canonical generators $e_1, \ldots, e_r$, respectively.

(iii) Each inertia group of $Y \to X$ over any point $\chi \in B - \{\xi\}$ is also an inertia group of $Z \to X$ over χ, where Z is the smooth completion of W.

Proof. Let $R = k[[t]]$, let $\tilde{X} = X \times_k R$, and let X^* be the blow-up of $\tilde{X}$ at the closed point of $\bar{\xi} = \xi \times_k R$. Thus X^* is a regular two-dimensional scheme that is projective as an R-curve. Its closed fibre X_0 is connected and consists of two irreducible components: a proper transform that is isomorphic to X, and an exceptional divisor that is isomorphic to $\mathbf{P}_k^1$. These two components meet at the point on the proper transform corresponding to ξ on X, and to the point $s = 0$ on the projective s-line $\mathbf{P}_k^1$. (Here we take $s = t/x$, where x is a local parameter for X at ξ. Thus the locus of $(s = \infty)$ is the proper transform of $\bar{\xi}$.)

Let $\{e_1, \ldots, e_r\} \subset E$ be a supplementary generating set for Γ with respect to G, and let $\sigma_1, \ldots, \sigma_r$ be distinct points of $\mathbf{P}_k^1$ other than $s = 0, \infty$. By hypothesis we may choose a point $\eta \in Y$ over $\xi \in X$ for which the inertia group $I \subset G$ normalizes E. Let $g \in G$ be the canonical generator of the inertia group I. Thus the subgroup $E_0 \subset \Gamma$ generated by E and g is of the form $E_0 = E \rtimes I$, and hence its order is not divisible by p. Let $E_1 \subset E_0$ be the subgroup generated by $e_1, \ldots, e_r, g$, and let $h = (e_1 \cdots e_r)^{-1} g$. Thus p also does not divide the order of E_1, and $g^{-1} e_1 \cdots e_r h = 1$. As discussed in §2 above (and cf. [Gr, XIII, Cor. 2.12]), there exists a smooth connected E_1-Galois cover $M \to \mathbf{P}_k^1$ branched at $0, \sigma_1, \ldots, \sigma_r, \infty$ with description $(g^{-1}, e_1, \ldots, e_r, h)$. Let $\mu \in M$ be a point over 0 at which g^{-1} is a canonical generator of inertia. Consider the induced (disconnected) Γ-Galois covers $\mathrm{Ind}_G^\Gamma Y \to X$ and $\mathrm{Ind}_{E_1}^\Gamma M \to \mathbf{P}_k^1$, consisting of disjoint unions of copies of $Y \to X$ and $M \to \mathbf{P}_k^1$, respectively, indexed by the cosets of G and of E_1 in Γ. We may identify Y and M with the identity components of the respective induced covers. Identifying the two points $\gamma(\eta) \in \mathrm{Ind}_G^\Gamma Y$ and $\gamma(\mu) \in \mathrm{Ind}_{E_1}^\Gamma M$ for each $\gamma \in \Gamma$, we obtain a Γ-Galois cover Z_0 of the reducible curve X_0. Moreover Z_0 is admissible over X_0 by construction, and is connected since G and E_1 generate Γ (and cf. also [HS, §4, Pro! p. 2]). By [HS,§2, Cor. to Thm . 2], there is a Γ-Galois cover $Z^* \to X^*$ which is an admissible thickening of $Z_0 \to X_0$ (viz., in the terminology of [HS], the unique solution to the corresponding relative thickening problem).

Let $Z^\circ \to X^\circ$ be the fibre of $Z^* \to X^*$ over the generic point of $\operatorname{Spec} k[[t]]$. Since X^* is the blow-up of $\tilde{X} = X \times_k k[[t]]$ at the closed point of $\bar{\xi}$, there are isomorphisms of K-curves $X^\circ \approx X \times_k K \approx X^* \times_R K$, where $K = k((t))$. Since $Z^* \to X^*$ is a thickening of $Z_0 \to X_0$, the cover in particular restricts to a trivial deformation of the restriction of $\operatorname{Ind}_G^\Gamma Y \to X$ to $X - \{\xi\}$. Hence $Z^\circ \to X^\circ$ is branched at the points of $(B - \{\xi\}) \times_k K$, with the same inertia groups as the corresponding points of $B - \{\xi\}$ for $Y \to X$; and it is branched at no other point of X° except for those whose closure in $\tilde{X}$ passes through the point $(\xi, (t = 0))$. Among points of the latter type, $Z^\circ \to X^\circ$ is branched precisely at $r + 1$ points $\sigma_1^\circ, \ldots, \sigma_r^\circ, \infty^\circ$ whose closures $\sigma_1^*, \ldots, \sigma_r^*, \subset !nfty^*$ in X^* pass through the points $\sigma_1, \ldots, \sigma_r, \infty$. (Note that the singular point of X_0 is an isolated point of the branch locus of $Z^* \to X^*$, as discussed in §2; so it does not contribute to the branch locus of $Z^\circ \to X^\circ$.) Over the point σ_i°, the inertia groups of $Z^\circ \to X^\circ$ are the same as those of $Z^* \to X^*$ over σ_i^*, and one of them has canonical generator e_i. Here the closure of ∞° in $\tilde{X}$ is $\bar{\xi} = \xi \times_k R$. So under the isomorphism $X \times_k K \approx X^\circ$, the branch locus consists of the r points σ_i° and the points of $B \times_k K$ (with $\xi \times_k K$ corresponding to the point ∞° in X°). Also, in the special case $E = 1$, the cover $Z^\circ \to X^\circ$ is just the base change of $Y \to X$ from k to K. Since the above construction commutes with taking quotients, we deduce for arbitrary E that the cover $Z^\circ/N \to X^\circ$ is the base change of $Y/(N \cap G) \to X$ from k to K.

Thus $Z^\circ \to X^\circ \approx X \times_k K$ is a smooth connected Γ-Galois cover whose restriction $W^\circ \to U^\circ := U \times_k K$ has the desired properties for W, but over K instead of over k. Being of finite type, this cover descends to a smooth connected Γ-Galois cover $Z_A \to X_A := X \times_k A$ over some finitely generated k-algebra $A \subset K$, whose restriction to $U_A := U \times_k A$ has the corresponding properties over A. Here $\operatorname{Spec} A$ is an absolutely irreducible variety, since $A \subset K$ and k is algebraically closed. By [Ha2, Prop. 5] (or [FJ, Props. 8.8, 9.29]) we conclude that the specialization $Z_\nu \to X$ of $Z_A \to X_A$ at a k-point $\nu \in \operatorname{Spec} A$ restricts to a G-Galois cover $W := Z_\nu \times_X U \to U$ having the desired properties. $\square$

Using the notion of embedding problems (cf. §2), we may rephrase Proposition 3.1 in more group-theoretic terms. In particular, we have the following corollary. In this connection, we recall that an inclusion $U' \hookrightarrow U$ of affine curves induces a surjection $\pi_1(U') \twoheadrightarrow \pi_1(U)$.

Corollary 3.2. *Let X be a smooth connected projective k-curve, let $U \subset X$ be a dense affine open subset, and let $\xi \in X - U$. Let $\Pi = \pi_1(U)$ and let Π^* be the quotient of Π corresponding to covers whose smooth completions are tamely ramified over ξ. Let $\mathcal{E} = (\alpha : \Pi^* \to H, f : \Gamma \to H)$ be a finite embedding problem for Π^*, and let β be a weak solution to $\mathcal{E}$. Suppose that Γ is generated by $G, E \subset \Gamma$, where E is a subgroup of $\ker(f)$ with p not dividing $|E|$. Suppose also that the normalizer of E in Γ contains $\beta(I)$, where $I \subset \Pi^*$ is an inertia group over ξ. Let $r \geq \mathrm{rk}_\Gamma(E, G)$. Then there is an open subset $U' \subset U$ such that $U - U'$ has cardinality r and the induced embedding problem $\mathcal{E}'$ for $\Pi' = \pi_1(U')$ has a proper solution.*

Proof. Let N be the normal closure of E in Γ. Since $E \subset \ker(f)$, it follows that $N \subset \ker(f)$, and H is a quotient of the group $H_1 := \Gamma/N = G/(N \cap G)$. Let $f_1 : \Gamma \twoheadrightarrow H_1$ and $f_0 : G \twoheadrightarrow H_1$ be the natural quotient maps, and let $\alpha_1 = f_0\beta : \Pi^* \to H_1$. Replacing $\mathcal{E} = (\alpha : \Pi^* \to H, f : \Gamma \to H)$ by the embedding problem $\mathcal{E}_1 = (\alpha_1 : \Pi^* \to H_1, f_1 : \Gamma \to H_1)$, we may assume that $H = H_1$, that $f|_G = f_0$, and that $\alpha = f_0\beta$.

Now β is a proper solution to the embedding problem $\mathcal{E}_0 = (\alpha : \Pi^* \to H, f_0 : G \to H)$, where $f_0 = f|_G : G \to H$. Under the Galois correspondence, the surjection $\beta : \Pi^* \twoheadrightarrow G$ corresponds to a connected étale G-Galois cover $V \to U$ whose smooth completion $Y \to X$ is tamely ramified over ξ, and such that some inertia group over ξ normalizes E. By Proposition 3.1, there is a smooth connected Γ-Galois cover $W \to U$ having at most r branch points, such that there is an isomorphism of H-Galois covers $W/N \approx V/(N \cap G)$ of U. (Here, as above, N is the normal closure of E in Γ, and $H = \Gamma/N = G/(N \cap G)$.) So over the complement $U' \subset U$ of the r-point branch locus of $W \to U$, we obtain a Γ-Galois étale cover $W' \to U'$ corresponding to a proper solution to the embedding problem $\mathcal{E}'$. $\square$

Remarks. (a) The above corollary does not rely on the full statement of Proposition 3.1, since neither (ii) nor (iii) there are used. But if Π' is replaced by a suitably refined quotient Π'^* (containing additional information about inertia groups), then a corresponding result can be proven, with the aid of (ii) and (iii) of 3.1, about embedding problems for Π'^*; and this would correspond to the full content of 3.1.

(b) In the other direction, it would be desirable to state a version of Corollary 3.2 just for Π, rather than for Π^*, and without assumptions on normalizers. Correspondingly, it would be desirable to state a version of

Proposition 3.1 without the assumptions on tameness or normalizers. (The proof of 3.1 at least shows that it is possible to weaken the assumption that an inertia group I over ξ normalizes E, by instead assuming that E, I generate a prime-to-p subgroup of Γ.)

(c) The proofs of the above results, and those that follow in this section, require the base field k to be algebraically (or at least separably) closed, because of the use of [Gr, XIII, Cor. 2.12] in the proof of Proposition 3.1. In particular, the condition that k be *large* (cf. [Po2]) does not suffice, at least for the proofs here. See also Remark (b) at the end of Section 4 below.

In particular, in the split embedding problem situation, the above results give rise to the following proposition and corollary:

Proposition 3.3. *Let Γ be a finite group of the form $N \rtimes G$, where p does not divide the order of N, and let $\{n_1, \ldots, n_r\} \subset N$ be a supplementary generating set for Γ with respect to G. Let $V \to U$ be a G-Galois étale cover of smooth connected affine k-curves whose smooth completion $Y \to X$ is tamely ramified over some point ξ of $B = X - U$. Then there is a smooth connected Γ-Galois cover $W \to U$ branched only at r points $\xi_1, \ldots, \xi_r$, with smooth completion $Z \to X$, such that: $W/N \approx V$ as G-Galois covers of U; the element n_i is the canonical generator of an inertia group of $W \to U$ over ξ_i; $Z \to X$ is tamely ramified over ξ; and each inertia group of $Y \to X$ over any point $\chi \in B - \{\xi\}$ is also an inertia group of $Z \to X$ over χ.*

Proof. Since N is normal in Γ, any inertia group of $Y \to X$ over ξ must normalize N. So Proposition 3.1 applies, with $E = N$, and with the H of Proposition 3.1 being the same group as G here. This yields the result (with tameness over ξ following since $Z/N = Y$ and p does not divide the order of N). $\qquad\square$

Remark. In the special case that the cover $Y \to X$ has trivial inertia groups over ξ (so that the given tamely ramified point is not actually a true branch point), the assertion of Proposition 3.3 is closely related to [Ha6, Theorem 3.5] (by taking the point ξ_0 of [Ha6, Theorem 3.5] to be ξ above), and the proofs are also related. But in the general case, the assertion of [Ha6, Theorem 3.5] is weaker than Proposition 3.3 above, since it requires an extra branch point (beyond the r points in Prop. 3.3). The difference is that in the result above, admissible covers can be used to avoid adding the extra branch point, provided that we have a tameness assumption. (The result in [Ha6] also uses a weaker notion of rank.)

Corollary 3.4. *Let X be a smooth connected projective k-curve, let $U \subset X$ be a dense affine open subset, and let $\xi \in X - U$. Let $\Pi = \pi_1(U)$ and let*

Π^* *be the quotient of* Π *corresponding to covers whose smooth completions are tamely ramified over* ξ. *Consider a finite split embedding problem* $\mathcal{E} = (\alpha : \Pi^* \to G, f : \Gamma \to G)$ *for* Π^*, *such that* p *does not divide the order of* $N = \ker(f)$. *Let* $r \geq \mathrm{rk}_\Gamma(N, \iota(G))$, *where* $\iota : G \to \Gamma$ *is a section of* f. *Then there is an open subset* $U' \subset U$ *such that* $U - U'$ *has cardinality* r *and the induced embedding problem* $\mathcal{E}'$ *for* $\Pi' = \pi_1(U')$ *has a proper solution.*

Proof. Identifying G with its image under ι, we may identify Γ with the semidirect product $N \rtimes G$. By the assumption on rank, there is a supplementary generating set $\{n_1, \ldots, n_r\} \subset N$ for Γ with respect to G. Also, the homomorphism $\alpha : \Pi^* \twoheadrightarrow G$ corresponds to a G-Galois connected étale cover of affine k-curves whose smooth completion $Y \to X$ is tamely ramified over ξ. So the hypotheses of Proposition 3.3 are satisfied, yielding a Γ-Galois cover $W \to U$ that is étale over some $U' \subset U$ with $U - U'$ of cardinality r. This cover corresponds to a homomorphism $\Pi' \twoheadrightarrow \Gamma$ that is a solution to the induced embedding problem $\mathcal{E}'$. $\square$

Remarks. (a) The proper solution to $\mathcal{E}'$ in 3.4 is automatically a proper solution to the induced embedding problem for Π'^*, the quotient of Π' corresponding to covers of U' whose smooth completions are tamely ramified over ξ. As in 3.3, this is because p does not divide the order of N.

(b) Corollary 3.4 can also be deduced directly from Corollary 3.2, by taking $H = G$, $N = E$, and $\beta = \iota\alpha$.

As discussed in Section 2, results about split embedding problems for a group Π can sometimes be extended to results about arbitrary embedding problems for Π, e.g. in situations in which the group Π is projective. By [Se2, Proposition 1], the fundamental group of an affine k-curve has cohomological dimension ≤ 1; and hence it is a projective group [Se1, I.5.9, Prop. 45]. Using this projectivity, we obtain the following variant of Proposition 3.3 that applies even in the non-split case. It does, however, provide a bit less control on the number of punctures needed (and cf. Remark (c) after the proof of Corollary 3.6 below).

Proposition 3.5. *Let* Γ *be a finite group, let* N *be a normal subgroup of order prime to* p, *and let* $G = \Gamma/N$. *Let* $\{n_1, \ldots, n_r\} \subset N$ *be a relative generating set for* N *in* Γ. *Let* $V \to U$ *be a* G-*Galois étale cover of smooth connected affine* k-curves whose smooth completion $Y \to X$ *is tamely ramified over some point* ξ *of* $B = X - U$. *Then there is a smooth connected* Γ-*Galois cover* $W \to U$ *branched only at* r *points* $\xi_1, \ldots, \xi_r$, *such that* $W/N \approx V$ *as* G-*Galois covers of* U; n_i *is the canonical generator of* an inertia group over ξ_i; and the smooth completion of $W \to U$ is tamely ramified over ξ.

Proof. The fundamental group $\Pi := \pi_1(U)$ is a projective group, since $\mathrm{cd}(\Pi) \leq 1$ [Se2, Prop. 1]). So the surjective homomorphism $\Pi \twoheadrightarrow G$ corresponding to $V \to U$ lifts to a homomorphism $\Pi \to \Gamma$, say with image G_0. Let $V_0 \to U$ be the G_0-Galois cover corresponding to this lift. Thus we have an unramified $N \cap G_0$-Galois cover $V_0 \to V$. Let Y_0 be the smooth completion of V_0. Then p does not divide the degree of $Y_0 \to Y$, since that degree divides $|N|$. Since $Y \to X$ is tame over ξ, it follows that so is $Y_0 \to X$. Moreover any inertia group of $Y_0 \to X$ over ξ must normalize N, since N is normal in Γ.

Since $G_0 \twoheadrightarrow G = \Gamma/N$, the group Γ is generated by N and G_0. Hence Γ is generated by $n_1, \ldots, n_r$ and G_0; i.e. $\{n_1, \ldots, n_r\} \subset N$ is a supplementary generating set for Γ with respect to G_0. So by Proposition 3.1, there is a Γ-Galois cover $W \to U$ having at most r branch points $\xi_1, \ldots, \xi_r$ such that the G-Galois cover $W/N \to U$ agrees with $V_0/(N \cap G_0) \to U$, and such that n_i is the canonical generator of an inertia group of $W \to U$ over ξ_i. Since $V_0/(N \cap G_0)$ is isomorphic to V as a G-Galois cover of U, and since p does not divide the order of N, it follows that $W \to U$ is as desired. $\square$

Corollary 3.6. *The assertion of Corollary 3.4 carries over to finite embedding problems that are not necessarily split, provided that one instead takes $r \geq \mathrm{rk}_\Gamma(N)$.*

Proof. Since $r \geq \mathrm{rk}_\Gamma(N)$, there is a relative generating set $\{n_1, \ldots, n_r\} \subset N$ for N in Γ. The proof then proceeds parallel to that of Corollary 3.4, but using Proposition 3.5 instead of Proposition 3.3. $\square$

Remarks. (a) Remark (a) after Corollary 3.4 carries over as well to Corollary 3.6.

(b) Corollary 3.6 can also be proven by applying Proposition 2.1 to Corollary 3.4. This uses that Π is projective; that $\mathrm{rk}_\Gamma(N) \geq \mathrm{rk}_\Gamma(N, G_0)$ (where G_0 is as in the proof of 3.5); and that the G_0-cover $Y_0 \to X$ is tamely ramified over ξ (as in the proof of 3.5).

(c) As mentioned above, 3.3 and 3.4 apply only to split embedding problems, whereas 3.5 and 3.6 apply more generally to embedding problems that need not be split. But in the process of reducing to the split case, we obtain weaker conclusions in 3.5 and 3.6 than in 3.3 and 3.4 (though under more general hypotheses). Specifically, different notions of generators and rank are used in the two pairs of results, and the notion of rank in 3.5 and 3.6 will typically be larger (when both make sense). The need for these variant notions here is due to the fact that one does not in advance know the group $G_0 \subset \Gamma$ that arises in the proof of 3.5 (and indirectly, in 3.6), in the process of reducing to the split case. Another way in which the generalized hypothesis leads to a weaker conclusion here is that one no longer has the same control on the inertia groups over $B - \{\xi\}$ in the situation of 3.5 that one had in 3.3. This is because the map $G_0 \to G$ in 3.5 need ! not be an isomorphism, and because one does not know *a priori* which choice of $G_0 \subset \Gamma$ over G will be needed in the construction.

(d) In the results in this section of the paper, it would be desirable to prove that the positions of the new r branch points can be specified in advance. In a related situation, such a result with control on the additional branch locus appears in Section 4 below. But there, unfortunately, connectivity cannot always be guaranteed.

Section 4. Results via lifting.

In this section another special case of the main theorem in proven, by means of lifting to characteristic 0. As before in Proposition 3.5, we have a finite group Γ and a quotient $G = \Gamma/N$, and a G-Galois étale cover of smooth affine k-curves $V \to U$. And as before, the problem is to show that there is a Γ-Galois cover $W \to U$ dominating $V \to U$, having at most $\mathrm{rk}_\Gamma(N)$ branch points with specified inertia there, under the assumption that N has order prime to p. But unlike the situation of the previous section, we need not make any tameness assumption here on the smooth completion of $V \to U$. What is shown here (Prop. 4.1) is that if the relative rank $\mathrm{rk}_\Gamma(N)$ is at most 1 (or if $p = 0$), then such a connected W exists, and moreover that the position of the extra branch point can be given in advance. (On the other hand if $\mathrm{rk}_\Gamma(N) > 1$ and $p > 0$, then we still obtain a W with specified branch locus and inertia, but conceivably it might not be connected.) Thus if $\mathrm{rk}_\Gamma(N) \le 1$ or $p = 0$ then we can obtain a proper solution to the corresponding embedding problem (Cor. 4.2).

The method of lifting and specializing to characteristic 0, in order to study fundamental groups in characteristic p, was used by Grothendieck (cf. [Gr], [GM]) in the situation of the *tame* fundamental group — with the strongest conclusions obtained on the maximal prime-to-p quotient of π_1. The idea is to work with a mixed characteristic complete discrete valuation ring R, whose residue field is the given algebraically closed field k of characteristic p. By using the knowledge of π_1 in characteristic 0, one can construct a cover over the general fibre; close this up over R; and then specialize to the closed fibre to obtain a cover over k. The main difficulty in extending this method to more general covers is that the restriction to the closed fibre may be inseparable over the generic point or it may have wild ramification there. Nevertheless, in [Ra, §6], Raynaud was able to use this method, in conjunction with a careful analysis of r! amification along the closed fibre of a semistable model, in order to construct covers of the affine line in characteristic p in a key case (and thereby complete the proof of Abhyankar's Conjecture for $\mathbf{A}^1$).

In addition to the problem of specializing characteristic 0 covers to characteristic p, there is also the problem of lifting a given characteristic p cover to characteristic 0. Again, this was done by Grothendieck in the tame case ([Gr], [GM]). In the wild case, this is not in general possible, since some characteristic p curves violate the Hurwitz bound on the number of automorphisms that a curve of genus g can have. But M. Garuti has proven a modified lifting result, which will be sufficient for our purposes. Namely, he has shown [Ga, Thm. 2] that if we are given a G-Galois cover $Y \to X$ over k, and if a lift X^* of X to R as above is given, then (possibly after enlarging R) there is a normal G-Galois cover $Y^* \to X^*$ over R whose closed fiber $Y_k^* \to X$ is closely related to $Y \to X$. Specifically, Y_k^* is an irreducible curve whose only singularities are cusps over wildly ramified branch points of $Y \to X$, and Y is the nor! malization of Y_k^*.

Using Garuti's result to lift, followed by a construction in characteristic 0 and then descent to characteristic p, we obtain the following version (Proposition 4.1) of the main theorem of the paper. Note that in the proof, after constructing a Γ-Galois cover W^* in characteristic 0, we do not in general know that its closed fibre W_k^* is irreducible. So instead we will choose a suitable irreducible component W of W_k^*, which will be Γ'-Galois for some $\Gamma' \subset \Gamma$. But if $\mathrm{rk}_\Gamma(N) \leq 1$ then W_k^* will in fact be irreducible, and so we will have $\Gamma' = \Gamma$ in this special case.

We state the result in a slightly more general form, in which we specify in advance the extra inertial elements $n_1, \ldots, n_r$, but do not require them to constitute a relative generating set for N. In this generality we still obtain a Γ-Galois cover with the desired properties except for connectivity (and so the Galois group of a connected component will be a subgroup of Γ). But when the n_i form a relative generating set, and $r \leq 1$ or $p = 0$, then we do obtain connectivity (cf. part (c) below).

Proposition 4.1. *Let Γ be a finite group, let N be a normal subgroup of order not divisible by p, and let $G = \Gamma/N$. Let $S = \{n_1, \ldots, n_r\}$ be a finite subset of N, with $r \geq 0$. Let $V \to U$ be a G-Galois étale cover of smooth connected affine k-curves, and let $\xi_1, \ldots, \xi_r \in U$ be distinct points.*

a) Then there is a subgroup $\Gamma' \subset \Gamma$ and a smooth connected Γ'-Galois cover

$W \to U$ branched only at $\xi_1, \ldots, \xi_r$, such that $\Gamma = N\Gamma'$ and $n_1 \in \Gamma'$ (if $r \geq 1$); $W/(N \cap \Gamma')$ is isomorphic to V as a G-Galois cover of U; and the canonical generator of each inertia group over ξ_i is conjugate to n_i in Γ, with n_i being equal to the canonical generator of some inertia group over ξ_i if $r = 1$ or $p = 0$.

b) If $r \leq 1$ or $p = 0$ then we may also require that $S \subset \Gamma'$.

c) If S is a relative generating set for N in Γ, and if either $r \leq 1$ or $p = 0$, then we may take $\Gamma' = \Gamma$.

Proof. Let $Y \to X$ be the smooth completion of $V \to U$. Let $B = X - U$, which is a non-empty finite set of $|B|$ points. Let $B' = \{\xi_1, \ldots, \xi_r\}$, which is a subset of U; and let $U' = U - B'$. There are two cases to consider:

Case A: $p = 0$.

(a), (b): Let g be the genus of X and let $n = |B|$. Thus $n \geq 1$. By [Gr, XIII, Cor. 2.12], the fundamental group $\pi_1(U)$ is generated by elements $a_1, \ldots, a_g, b_1, \ldots, b_g, c_1, \ldots, c_n$ subject to the presentation $(*)$ of §2 above. Similarly, $\pi_1(U')$ is generated by elements $\tilde{a}_1, \ldots, \tilde{a}_g, \tilde{b}_1, \ldots, \tilde{b}_g, \tilde{c}_1, \ldots, \tilde{c}_{n+r}$ subject to the analogous presentation $(*)'$. The natural map $\pi_1(U') \twoheadrightarrow \pi_1(U)$ takes $\tilde{a}_i \mapsto a_i$, $\tilde{b}_i \mapsto b_i$, $\tilde{c}_j \mapsto c_j$ for $1 \leq j \leq n$, and $\tilde{c}_j \mapsto 1$ for $n < j \leq n+r$.

The G-Galois étale cover $V \to U$ corresponds to a surjective group homomorphism $\phi : \pi_1(U) \to G$. Let $\alpha_i = \phi(a_i)$ and $\beta_i = \phi(b_i)$ for $1 \leq i \leq g$, and let $\gamma_j = \phi(c_j)$ for $1 < j \leq n$. For each such i and j choose elements $\tilde{\alpha}_i, \tilde{\beta}_i, \tilde{\gamma}_j \in \Gamma$ over $\alpha_i, \beta_i, \gamma_j \in G$. Also, let $\tilde{\gamma}_{n+j} = n_j \in N$ for $1 \leq j \leq r$. Then there is a unique homomorphism $\tilde{\phi} : \pi_1(U') \to \Gamma$ given by $\tilde{\phi}(\tilde{a}_i) = \tilde{\alpha}_i$, $\tilde{\phi}(\tilde{b}_i) = \tilde{\beta}_i$, and $\tilde{\phi}(\tilde{c}_j) = \tilde{\gamma}_j$ for $1 < j \leq n+r$. (The image of $\tilde{c}_1$ is uniquely determined by the single relation $(*)'$ for $\pi_1(U')$ and by the requirement that $\tilde{\phi}$ be a homomorphism.) Let Γ' be the image of $\tilde{\phi}$. Thus $S \subset \Gamma'$. Also, $\tilde{\phi}$ lifts the surjection ϕ, and so $N\Gamma' = \Gamma$. Thus $\tilde{\phi}$ corresponds to a connected Γ'-Galois étale cover $W \to U$ that dominates $V \to U$. The smooth completion $Z \to X$ of $W \to U$ has description $(\tilde{\phi}(\tilde{c}_1), \ldots, \tilde{\phi}(\tilde{c}_{n+r}))$ (cf. §2 above). In particular, $n_j = \tilde{\phi}(\tilde{c}_{n+j}) \in N$ is a canonical generator of inertia above ξ_j for $1 \leq j \leq r$. This cover is thus as desired, for parts (a) and (b).

(c): By part (b), Γ' contains S. But S is assumed to be a relative generating set for N. Since $N\Gamma' = \Gamma$ it follows that $\Gamma' = \Gamma$.

Case B: $p > 0$.

(a): Let R be a complete discrete valuation ring of mixed characteristic and residue field k (e.g. the Witt ring over k). By [Gr, III, Cor. 7.4], there is a smooth complete R-curve X^* whose closed fibre is X. By [Ga, Thm. 2], it follows that after replacing R by a finite extension (and X^* by its pullback to that extension), there is an irreducible G-Galois cover $Y^* \to X^*$ of proper R-curves with Y^* normal, such that Y is the normalization of the closed fibre Y_k^* of Y^*, and such that Y_k^* is everywhere unibranched, with the morphism $Y \to Y_k^*$ being an isomorphism away from the wildly ramified points of Y. (Here the residue field of the enlarged R is still k, since k is algebraically closed.)

Suppose that $R \subset T$ is a finite extension of complete discrete valuation rings, and let $Y_T^* \to X_T^*$ be the normalized pullback of $Y^* \to X^*$ from R to T. Thus Y_T^* is the normalization of $Y^* \times_{X^*} X_T^*$. Now the closed fibre of $Y^* \to X^*$ is generically étale, but the closed fibre of $X_T^* \to X^*$ is totally ramified. So the irreducible schemes Y^* and X_T^* dominate no common non-trivial covers of X^*. But Y^* and X_T^* are normal; so the intersection of their function fields is that of X^*. Also, since Y^* is Galois over X^*, the corresponding extension of function fields is also Galois. So by [FJ, p. 110], the function fields of Y and X_T^* are linearly disjoint over that of X^*. Thus $Y^* \times_{X^*} X_T^*$ is irreducible and hence so is Y_T^*. The conclusion is that the generic fibre Y° of Y^* is geometrically irreducible.

Since X^* is regular and Y^* is normal, Purity of Branch Locus applies to $Y^* \to X^*$ [Na, 41.1]; and since the closed fibre is generically étale, it follows that the branch locus B^* defines a cover of Spec R. Also, since $X^* \to \text{Spec } R$ is smooth, it follows from [Gr, III, Thm. 3.1] that there are R-points ξ_i^* of X^* that lift the k-points ξ_i of $X = X_k^*$. The ξ_i^* have pairwise disjoint support, since the points ξ_i are distinct and since R is a complete local ring. Let B'^* be the union of the loci of the ξ_i^*, and write $U^* = X^* - B^*$, $U'^* = U^* - B'^*$, $V^* = Y^* \times_{X^*} U^*$, and $V'^* = V^* \times_{U^*} U'^*$.

Let K be the fraction field of R, and let $\bar{K}$ be the algebraic closure of K. Let $\bar{X}^\circ = X^* \times_R \bar{K}$; $\bar{B}^\circ = B^* \times_R \bar{K}$; $\bar{U}^\circ = \bar{X}^\circ - \bar{B}^\circ$; and $\bar{V}^\circ = V^* \times_{U^*} \bar{U}^\circ$. Similarly, write $\bar{B}'^\circ = B'^* \times_R \bar{K}$; $\bar{U}'^\circ = \bar{U}^\circ - \bar{B}'^\circ$; and $\bar{V}'^\circ = \bar{V}^\circ \times_{\bar{U}^\circ} \bar{U}'^\circ$. Thus $\bar{V}^\circ \to \bar{U}^\circ$ and $\bar{V}'^\circ \to \bar{U}'^\circ$ are G-Galois étale covers which are connected (since Y° is geometrically irreducible). Also, $\bar{B}'^\circ$ consists of r distinct $\bar{K}$-points $\bar{\xi}_1^\circ, \ldots, \bar{\xi}_r^\circ$ specializing respectively to the k-points $\xi_1, \ldots, \xi_r$ — viz. $\bar{\xi}_i^\circ = \xi_i^* \times_R \bar{K}$. By Hensel's Lemma, the compatible system of roots of unity $\{\zeta_n\}$ in k lifts uniquely to such a system in K, and so in $\bar{K}$. With respect to this system, we may consider the canonical generators of inertia of covers defined over K or $\bar{K}$.

By Case A over the characteristic 0 field $\bar{K}$, there is a subgroup $\Gamma_0 \subset \Gamma$ containing $n_1, \ldots, n_r$ such that $N\Gamma_0 = \Gamma$, together with a smooth connected Γ_0-Galois cover $\bar{W}^\circ \to \bar{U}^\circ$ of $\bar{K}$-curves that dominates $\bar{V}^\circ \to \bar{U}^\circ$ and whose canonical generators of inertia at points $\bar{\omega}_i^{\circ}$ over $\bar{\xi}_i^{\circ}$ are equal to n_i. Let $N_0 = N \cap \Gamma_0$, so that $\Gamma_0/N_0 \approx G$. For some subfield $\tilde{K} \subset \bar{K}$ that is finite over K, this cover descends to a connected Γ_0-Galois cover of $\tilde{K}$-curves with the corresponding properties (and so in particular it dominates the induced G-Galois cover $\tilde{V}^{\circ} \to \tilde{U}^{\circ}$ of $\tilde{K}$-curves). Replacing K by $\tilde{K}$, and R by its integral closure in $\tilde{K}$, we may assume that there is a smooth connected Γ_0-Galois cover $W^{\circ} \to U^{\circ}$ that dominates $V^{\circ} \to U^{\circ}$ and is étale away from the general fibre B'° of B'^{*}; and that there is a K-point ω_i° on W° over ξ_i° at which the canonical generator of inertia is n_i.

Let W^{*} be the normalization of U^{*} in W°, and let $W'^{*} = W^{*} \times_{U^{*}} U'^{*}$. Thus $W^{*} \to U^{*}$ is Γ_0-Galois and $W^{*}/N_0 \approx V^{*}$ as G-Galois covers of U^{*}, and similarly for $W'^{*} \to U'^{*}$. Moreover $W'^{*} \to U'^{*}$ is étale except possibly on the closed fibre. Since p does not divide the order of $N_0 \subset N$, any ramification of $W'^{*} \to V'^{*}$ along the closed fibre is tame. So applying Abhyankar's Lemma, and after replacing K by a finite separable extension (and R by its normalization in this extension), we may assume that $W'^{*} \to V'^{*}$ is étale along the general point of the closed fibre. Since $V'^{*} \to U'^{*}$ is étale, it follows that the Γ_0-Galois cover $W'^{*} \to U'^{*}$ is étale along the general point of the closed fibre, as well as away from the closed fibre. But U'^{*} is regular and W'^{*} is normal. So Purity of Branch Locus implies that $W'^{*} \to U'^{*}$ is étale. Hence the closed fibre $W_k^{*} \to U_k^{*}$ of $W^{*} \to U^{*}$ is étale over U'.

Let ω_i^{*} be the closure of ω_i° in W^{*} and let ω_i be the closed point of ω_i^{*}. Thus ω_i^{*} is an R-point of W^{*} over ξ_i^{*}, and ω_i is a k-point of W_k^{*} over ξ.

Claim: The k-curve W_k^{*} is smooth at the point ω_i, and the inertia group I_i there is cyclic with canonical generator n_i.

To prove the claim, we first apply [GM, Corollary 2.3.6] over the complete local ring of X^{*} at $\xi_i \in X \subset X^{*}$. In our situation, that result says that the restriction of the tame I_i-Galois cover $\mathrm{Spec}\,\hat{\mathcal{O}}_{W^{*},\omega_i} \to \mathrm{Spec}\,\hat{\mathcal{O}}_{U^{*},\xi_i}$ over its closed fibre is also tame (in the sense of [GM, Def. 2.2.2]). In particular, this restriction $\mathrm{Spec}\,\hat{\mathcal{O}}_{W_k^{*},\omega_i} \to \mathrm{Spec}\,\hat{\mathcal{O}}_{U,\xi_i}$ is normal, and the inertia group I_i is a cyclic group. Thus the k-curve W_k^{*} is regular at ω_i, and hence smooth there (since k is perfect). Moreover I_i is abelian, and $n_i \in I_i$ (since n_i is in the inertia group of W° at ω_i°), so $\langle n_i \rangle$ is a normal subgroup of I_i. The intermediate cover $\mathrm{Spec}\,\hat{\mathcal{O}}_{W^{*},\omega_i}/\langle n_i \rangle \to \mathrm{Spec}\,\hat{\mathcal{O}}_{U^{*},\xi_i}$ is then a normal $I_i/\langle n_i \rangle$-Galois cover which is unramified except at the closed point (where it is totally ramified). By Purity of Branch Locus it follows that this intermediate cover is trivial, and so $I_i = \langle n_i \rangle$. That is, the inertia group of $W^{*} \to U^{*}$ at ω_i is generated by n_i. (This conclusion can also be reached by reasoning as in the proof of [Fu, Theorem 3.3, Case 1], instead of using [GM].) So $\hat{\mathcal{O}}_{W^*,\omega_i}$ is of the form $\hat{\mathcal{O}}_{U,\xi_i}[f_i^{1/m_i}]$, where f_i is a local uniformizer along ξ_i^*, where m_i is the order of n_i. Since n_i is the canonical generator of W° at ω_i°, it follows that the generator $n_i \in I_i$ acts on the overring by $f_i^{1/m_i} \mapsto \zeta_{m_i} f_i^{1/m_i}$. Restricting to the closed fibre, we find that n_i is the canonical generator of $W_k^* \to U$ at ω_i. This proves the claim.

Now $W_k^* \to U$ is Galois, étale over $U' = U - \{\xi_1, \ldots, \xi_n\}$, and smooth at the point ω_i over ξ_i. So W_k^* is smooth. Hence there is a unique irreducible component W of W_k^* that contains ω_1. Let $\Gamma' \subset \Gamma_0$ be the decomposition group of the generic point of W. Thus $W \to U$ is a Γ'-Galois smooth connected cover that is branched only at $\{\xi_1, \ldots, \xi_n\}$. Also, by the claim, the canonical generator of inertia at any point of W_k^* over ξ_i is conjugate to n_i in Γ_0; and in particular this is true for the points of W over ξ_i. Moreover the inertia group at $\omega_1 \in W$ has canonical generator n_1, and so we have that $n_1 \in \Gamma'$. And since V is irreducible, the composition $W \hookrightarrow W_k^* \to W_k^*/N_0 \approx V$ (which is a morphism of covers of U) is surjective on points. Thus $W/(N_0 \cap \Gamma') \approx V$ as G-Galois covers of U. This implies that $N_0\Gamma'/N_0 \approx \Gamma'/(N_0\cap\Gamma') \approx G \approx \Gamma_0/N_0$; but $N_0, \Gamma' \subset \Gamma_0$. So $N_0\Gamma' = \Gamma_0$ and thus $N\Gamma' = N\Gamma_0 = \Gamma$. Thus W has the desired properties.

(b): This is automatic: If $r = 0$ then S is empty; and if $r = 1$ then $S = \{n_1\}$, and $n_1 \in \Gamma'$ by (a).

(c): The same proof works as in Case A(c). □

Reinterpreting the above result in terms of embedding problems, we obtain:

Corollary 4.2. *Let X be a smooth connected projective k-curve, let $U \subset X$ be a dense affine open subset, and let $\Pi = \pi_1(U)$. Let $\mathcal{E} = (\alpha : \Pi \to G,$ $f : \Gamma \to G)$ be a finite embedding problem for Π, such that p does not divide the order of $N = \ker(f)$. Let $U' \subset U$ such that $U - U'$ has cardinality $r \geq \mathrm{rk}_\Gamma(N)$. If $p = 0$ or $r \leq 1$, then the induced embedding problem $\mathcal{E}'$ for $\Pi' = \pi_1(U')$ has a proper solution.*

Proof. The surjection $\alpha : \Pi \twoheadrightarrow G$ corresponds to a connected G-Galois étale cover $V \to U$. By Proposition 4.1, there is a smooth connected Γ-Galois cover $W \to U$ branched only at the r points of $B' = U - U'$, and which dominates $V \to U$. Its restriction $W' \to U'$ is étale, and dominates the restriction $V' \to U'$. Hence it corresponds to a proper solution to the induced embedding problem $\mathcal{E}'$ for $\Pi' = \pi_1(U')$. □

Remarks. (a) It would be desirable to extend Proposition 4.1(c) to the case that $r > 1$ with $p > 0$. While the main theorem of the paper (Theorem 5.4) does give a connected Γ-Galois cover $W \to U$ that dominates a given G-Galois étale cover $V \to U$ and has r additional branch points (even if $r > 1$), it does not allow control over the positions of those branch points (except for the first).

The difficulty in extending the proof of 4.1(c) to $r > 1$, $p > 0$ is this: In the proof of Case B of 4.1, if S is a relative generating set for N in Γ, then the cover $W^* \to U^*$ is irreducible and Γ-Galois (using (c) in the characteristic 0 version); but it is unclear whether its closed fibre W_k^* is also irreducible, if $r > 1$. It would suffice to show that $Z_k^* \to X$ is unbranched at its wildly ramified points, where Z_k^* is the closed fibre of the normalization Z^* of X^* in W^*. One approach to this would be to use Purity of Branch Locus over Y^*; but for this, one wants Y^* to be regular. This raises the question of whether Garuti's result [Ga, Thm. 2] can be strengthened to show that Y^* can always be chosen to be regular.

(b) The proofs of the above results require the base field k to be algebraically closed, because of the use of [Ga, Thm. 2] in Proposition 4.1.

Section 5. The main result.

This section contains the main result of this paper, that a given Γ/N-Galois étale cover $V \to U$ of an affine k-curve is dominated by a Γ-Galois cover of U that is branched at $\mathrm{rk}_{\bar{\Gamma}}(\bar{N})$ points of U, where $\bar{\Gamma}, \bar{N}$ are the reductions of Γ, N modulo $p(N)$. In the case that the order of N is prime to p (so that $\bar{\Gamma} = \Gamma$ and $\bar{N} = N$), we prove this essentially by combining the special cases in which it has already been shown: the case that the smooth completion of $V \to U$ has a tamely ramified branch point (cf. §3), and the case where $\mathrm{rk}_\Gamma(N) \leq 1$ (cf. §4). This is done in Proposition 5.1. Afterwards, this result is combined with a result of F. Pop (cf. Thm. 5.2 and Cor. 5.3 below) to prove the general case (Theorem 5.4). This is then interpreted in terms of embedding problems (Corollary 5.5).

Proposition 5.1. *Let Γ be a finite group, let N be a normal subgroup whose order is not divisible by p, and let $G = \Gamma/N$. Let $\{n_1, \ldots, n_r\} \subset N$ be a relative generating set for N in Γ, with $r \geq 0$. Let $V \to U$ be a G-Galois étale cover of smooth connected affine k-curves. Then there is a smooth connected Γ-Galois cover $W \to U$ branched only at r distinct points $\xi_1, \ldots, \xi_r \in U$, such that W/N is isomorphic to V as a G-Galois cover of U, and n_i is a canonical generator of inertia over ξ_i for $1 < i \leq r$. Moreover we may specify the position of ξ_1 in advance, if $r > 0$.*

Proof. If $r \leq 1$ then this assertion is contained in the statement of Proposition 4.1. So we may assume $r > 1$.

Let $\xi_1 \in U$ be any point. We may apply Proposition 4.1 with $S = \{n_1\}$. Doing so yields a subgroup $\tilde{G} \subset \Gamma$ and a smooth connected $\tilde{G}$-Galois cover $\tilde{V} \to U$ branched only at ξ_1, such that $\Gamma = N\tilde{G}$ and $n_1 \in \tilde{G}$; $\tilde{V}/(N \cap \tilde{G})$ is isomorphic to V as a G-Galois cover of U; and n_1 is the canonical generator of an inertia group over ξ_1.

Let X be the smooth completion of U, let $U' = U - \{\xi_1\}$, let $\tilde{V}'$ be the restriction of $\tilde{V}$ to U', and let $\tilde{Y}$ be the normalization of X in $\tilde{V}'$ (or equivalently, in $\tilde{V}$). Thus $\tilde{Y} \to X$ is a $\tilde{G}$-Galois cover that is tamely ramified over the point $\xi_1 \in B' := X - U'$. Moreover n_1 is the canonical generator of some inertia group over ξ_1, and this inertia group normalizes N since N is normal. In addition, $\Gamma = N\tilde{G}$, and so the relative generating set $\{n_1, \ldots, n_r\}$ for N is in particular a supplementary generating set for Γ with respect to $\tilde{G}$. Since $n_1 \in \tilde{G}$, it follows that $\{n_2, \ldots, n_r\}$ is a supplementary generating set for Γ with respect to $\tilde{G}$.

So we may apply Proposition 3.1 (with $\tilde{G}, N, r-1, \tilde{V}' \to U', n_2, \ldots, n_r$ playing the roles of $G, E, r, V \to U, e_1, \ldots, e_r$ there). As a result, we obtain a smooth connected Γ-Galois cover $W' \to U'$ having at most $r - 1$ branch points $\xi_2, \ldots, \xi_r \in U'$, satisfying the three conditions (i)-(iii) there. That is, the G-Galois cover $W'/N \to U'$ agrees with $\tilde{V}'/(N \cap \tilde{G}) \to U'$; there are inertia groups of $W' \to U'$ over the branch points $\xi_2, \ldots, \xi_r$ having canonical generators $n_2, \ldots, n_r$ respectively; and the inertia groups of $\tilde{Y} \to X$, over each point of $B' - \{\xi_1\} = X - U$, are also inertia groups of $Z \to X$ over that point, where Z is the smooth completion of W'. Let W be the normalization of U in W'. Thus $W \to U$ is a smooth connected Γ-Galois cover branched only at $\xi_1, \ldots, \xi_r$, with n_i a canonical generator of inertia over ξ_i for each $i > 1$. As a G-Galois cover of U, the intermediate cover W/N is isomorphic to $\tilde{V}/(N \cap \tilde{G})$ and hence to V. $\square$

Remarks. (a) Unlike Proposition 4.1, the above result yields a connected Γ-Galois cover over the given G-Galois cover even if there are two or more elements in the relative generating set, in characteristic $p > 0$. But on the other hand, in 5.1 we lose control of the positions of the branch points other than ξ_1, and of the inertia group over ξ_1. Similarly, unlike the results of §3, the above result does not require that the smooth completion $Y \to X$ of $V \to U$ have a tamely ramified branch point. But on the other hand, in 5.1 we have less control over inertia groups, and do not have a version that is analogous to 3.1 using the smaller number $\mathrm{rk}(E, G)$ of new branch points.

(b) In the proof of the above result, it is tempting to try to invoke

Proposition 4.1 repeatedly on successive n_i's, rather than to use Proposition 3.1 (and thus to control the positions of all of the branch points, though losing control over the inertia groups). The strategy would be to take the subgroup $\tilde{G} \subset \Gamma$ containing n_1 and surjecting onto G, given by the use of Prop. 4.1 as above; to take the minimal $\tilde{G}$-invariant subgroup $N_1 \subset N$ that contains n_2; and then to form the semi-direct product $\Gamma' = N_1 \rtimes \tilde{G}$ with respect to conjugation action of $\tilde{G}$ on N_1 in Γ. Applying 4.1 again would yield a subgroup $\tilde{G}' \subset \Gamma'$ that contains $(n_2, 1)$ and surjects onto $\tilde{G}$ under the second projection; and then one could take the image of $\tilde{G}'$ in Γ under the multiplication homomorphism $\mu : \Gamma' \to \Gamma$ given by $(n, g) \mapsto ng$. This image $\mu(\tilde{G}') \subset \Gamma$ would then contain n_2 and surject onto G, and one might hope to repeat the process. But unfortunately, $\mu(\tilde{G}')$ need no longer contain n_1. For example, suppose that we are given $p > 3$; $N = S_3$; $G = \langle g \rangle$, cyclic of order p; and $\Gamma = N \times G$; with $n_1 = (12)$ and $n_2 = (13)$ being (relative) generators of N. Then we could have $\tilde{G} = \langle (12), g \rangle \subset \Gamma$ (viewing N, G as subgroups of Γ); $N_1 = N$; $\Gamma' = N \rtimes \tilde{G}$; $\tilde{G}' = \langle ((13), 1), ((12), (12)), (1, g) \rangle \subset N \rtimes \tilde{G} = \Gamma'$; and $\mu(\tilde{G}') = \langle (13), g \rangle \subset \Gamma$, which is strictly smaller than $\Gamma = N \times G$ and does not contain the first generator $n_1 = (12)$ of N.

Finally, we combine the above proposition with a result of F. Pop, to obtain our main theorem (Theorem 5.4 below). In this theorem, unlike the previous results in this paper, we permit the order of the kernel N to be divisible by p. The following is a rephrasing of Pop's result [Po1, Theorem B]:

Theorem 5.2. [Po1, Thm. B] *Let $\Gamma = Q \rtimes G$ where Q is a quasi-p group; let $Y \to X$ be a smooth connected G-Galois cover of k-curves; and let $\xi \in X$ (possibly a branch point of $Y \to X$). Then there is a smooth connected Γ-Galois cover $Z \to X$ dominating $Y \to X$, such that $Z \to Y$ is branched only at points of Y over ξ, and the inertia groups of $Z \to Y$ over those points are the Sylow p-subgroups of Q.*

Although this result is stated just for the split case, it implies a result in the more general case of group extensions by a quasi-p kernel. Namely, as the following corollary states, if U is a smooth connected affine k-curve and $\mathcal{E} = (\alpha, f)$ is a finite embedding problem for $\pi_1(U)$ with $\ker(f)$ a quasi-p group, then $\mathcal{E}$ has a proper solution.

Corollary 5.3. *Let Q be a normal quasi-p subgroup of a finite group Γ, and let $G = \Gamma/Q$. Let $V \to U$ be a connected G-Galois étale cover of smooth affine k-curves. Then there is a connected Γ-Galois étale cover $W \to U$ dominating $V \to U$.*

Proof. By [Se2, Prop. 1], the group $\Pi = \pi_1(U)$ has cohomological dimension 1 and hence is a projective group. The G-Galois étale cover $V \to U$ corresponds to a surjection $\alpha : \Pi \to G$, and there is a natural quotient map $f : \Gamma \to G$. Thus $\mathcal{E} = (f : \Gamma \to G, \alpha : \Pi \to G)$ is an embedding problem for the projective group Π, and so it has a weak solution $\alpha_0 : \Pi \to \Gamma$, say with image G_0. Let $\Gamma_0 = Q \rtimes G_0$, where the semi-direct product is taken with respect to the conjugation action of G_0 on $Q \triangleleft \Gamma$, and let $f_0 : \Gamma_0 \to G_0$ be the natural quotient map.

The surjection $\alpha_0 : \Pi \to G_0$ corresponds to a connected G_0-Galois étale cover $V_0 \to U$, say with smooth completion $Y_0 \to X$. Theorem 5.2 now applies to the group Γ_0, the G_0-Galois cover $Y_0 \to X$, and a point $\xi \in B := X - U$. So there is a smooth connected Γ_0-Galois cover $Z_0 \to X$ dominating $Y_0 \to X$ such that $Z_0 \to Y_0$ is branched only at points over ξ. Thus $Z_0 \to X$ is étale over U, and corresponds to a surjection $\beta_0 : \Pi \to \Gamma_0$.

Now $f\beta_0 = \alpha_0$ since $Z_0 \to X$ dominates $Y_0 \to X$. So β_0 is a proper solution to the split embedding problem $\mathcal{E}_0 = (\alpha_0 : \Pi \to G_0, f_0 : \Gamma_0 \to G_0)$. Thus by Proposition 2.1, the original embedding problem $\mathcal{E} = (f : \Gamma \to G, \alpha : \Pi \to G)$ has a proper solution $\beta : \Pi \to \Gamma$. Here $f\beta = \alpha$. The map β corresponds to a connected Γ-Galois étale cover $W \to U$, and this cover dominates $V \to U$ because $f\beta = \alpha$. $\square$

Combining the above with Proposition 5.1 yields the main theorem of the paper:

Theorem 5.4. *Let N be a normal subgroup of a finite group Γ, and let $G = \Gamma/N$. Let $\bar{N} = N/p(N)$ and $\bar{\Gamma} = \Gamma/p(N)$, and let $r = \mathrm{rk}_{\bar{\Gamma}}(\bar{N})$. Let $V \to U$ be a G-Galois étale cover of smooth connected affine k-curves. Then there is a smooth connected Γ-Galois cover $W \to U$ branched only at r distinct points $\xi_1, \ldots, \xi_r \in U$, such that W/N is isomorphic to V as a G-Galois cover of U. Moreover we may specify the position of ξ_1 in advance, if $r > 0$.*

Proof. First observe here that $p(N)$ is a characteristic subgroup of N, and so is a normal subgroup of Γ. Hence $\bar{\Gamma} = \Gamma/p(N)$ is well defined.

Pick $\xi_1 \in U$. Since $r = \mathrm{rk}_{\bar{\Gamma}}(\bar{N})$, there is a relative generating set $\{\bar{n}_1, \ldots, \bar{n}_r\} \subset \bar{N}$ for $\bar{N}$ in $\bar{\Gamma}$. Also, $\bar{\Gamma}/\bar{N} = (\Gamma/p(N))/(N/p(N)) \approx \Gamma/N = G$. So by Proposition 5.1, there are distinct points $\xi_2, \ldots, \xi_r \in U - \{\xi_1\}$ together with a smooth connected $\bar{\Gamma}$-Galois cover $\bar{W} \to U$ branched only at $S = \{\xi_1, \ldots, \xi_r\}$ such that $\bar{W}/\bar{N}$ is isomorphic to V as a G-Galois cover of U. Let $U' = U - S$ and let $\bar{W}'$ be the inverse image of U' in $\bar{W}$. We may now apply Corollary 5.3 with $Q = p(N)$, and with $\bar{W}' \to U'$ playing the role of $V \to U$ there. As a result, we obtain a connected Γ-Galois étale cover $W' \to U'$ dominating $\bar{W}' \to U'$. The normalization $W \to U$ of U in $W' \to U'$ is then as desired. □

Remark. In the situation of Theorem 5.4 above, if the short exact sequence $1 \to p(N) \to \Gamma \to \bar{\Gamma} \to 1$ is split, then in the proof we can apply Pop's Theorem (5.2 above) rather than Corollary 5.3. Doing so gives more information about inertia. In particular, suppose we are given an integer $r \geq \mathrm{rk}_{\bar{\Gamma}}(\bar{N})$ and a relative generating set $\{n_1, \ldots, n_r\} \subset N$ for N in Γ. Then in the split situation we may choose the Γ-Galois cover $W \to U$ so that the branching over $\xi_2, \ldots, \xi_r$ is tame, and so that n_i is the canonical generator of inertia at some point over ξ_i for each $i > 1$.

Reinterpreting the above result in terms of embedding problems, we immediately obtain:

Corollary 5.5. *Let U be a smooth connected affine k-curve, let $\Pi = \pi_1(U)$, and let $\mathcal{E} = (\alpha : \Pi \to G, f : \Gamma \to G)$ be a finite embedding problem for Π. Let $N = \ker(f)$ and let $r = \mathrm{rk}_{\bar{\Gamma}}(\bar{N})$. Then for some set $S \subset U$ of cardinality r, the induced embedding problem $\mathcal{E}'$ for $\Pi' = \pi_1(U - S)$ has a proper solution. Moreover, if $r > 0$, then one of the points of S can be chosen in advance.*

References

[FJ]   M. Fried, M. Jarden. "Field Arithmetic." Ergeb. Math. Grenzgeb., Vol. 11, Springer-Verlag, Berlin/New York, 1986.

[Fu]   W. Fulton, Hurwitz schemes and irreducibility of moduli of algebraic curves. Ann. Math. **90** (1969), 542-575.

[Ga]   M. Garuti. Prolongement de revêtements galoisiens en géométrie rigide. Compositio Math. **104** (1996), 305-331.

[Gr]   A. Grothendieck. "Revêtements étales et groupe fondamental" (SGA 1). Lecture Notes in Mathematics, Vol. 224, Springer-Verlag, Berlin-Heidelberg-New York, 1971.

[GM]   A. Grothendieck, J. Murre. "The tame fundamental group of a formal neighborhood of a divisor with normal crossings on a scheme." Lecture Notes in Mathematics, Vol. 208, Springer-Verlag, Berlin-Heidelberg-New York, 1970.

[Ha1]   D. Harbater. Galois coverings of the arithmetic line, in "Number Theory: New York, 1984-85." Lecture Notes in Mathematics, Vol. 1240, Springer-Verlag, Berlin-Heidelberg-New York, 1987, pp. 165-195.

[Ha2]   D. Harbater. Formal patching and adding branch points. Amer. J. Math. **115** (1993), 487-508.

[Ha3]   D. Harbater. Abhyankar's conjecture on Galois groups over curves. Inventiones Math. **117** (1994), 1-25.

[Ha4]   D. Harbater. Galois groups with prescribed ramification. In: "Proceedings of a Conference on Arithmetic Geometry (Arizona State Univ., 1993)," (N. Childress and J. Jones, eds.); AMS Contemporary Math. Series, vol. 174, 1994, pp. 35-60.

[Ha5]   D. Harbater. Fundamental groups of curves in characteristic p. In: "Proceedings of the International Congress of Mathematicians, Zürich 1994." Birkhäuser, Basel, 1995, pp. 656-666.

[Ha6]   D. Harbater. Fundamental groups and embedding problems in characteristic p. In "Recent Developments in the Inverse Galois Problem: Proceedings of a conference at Univ. of Washington, July 17-23, 1993" (M. Fried ed.); AMS Contemporary Math. Series, vol. 186, 1995, pp. 353-370.

[HS]   D. Harbater, K. Stevenson. Patching and thickening problems. 1997 manuscript.

[Na]   M. Nagata, "Local Rings." Intersc. Tracts in Pure and Appl. Math. **13**, New York, 1962.

[Po1]   F. Pop. Étale Galois covers over smooth affine curves. Invent. Math. **120** (1995), 555-578.

[Po2]   F. Pop. Embedding problems over large fields. Annals of Math. **144** (1996), 1-34.

[Ra]   M. Raynaud. Revêtements de la droite affine en caractéristique $p > 0$ et conjecture d'Abhyankar. Invent. Math. **116** (1994), 425-462.

[Sa]   M. Saïdi. Revêtements étales et groupe fondamental de graphes de groupes, C. R. Acad. Sci. Paris **318** Série I (1994), 1115-1118.

[Se1]   J.-P. Serre, "Cohomologie Galoisienne." Lecture Notes in Mathematics, Vol. 5, Springer-Verlag, Berlin-Heidelberg-New York, 1964.

[Se2]   J.-P. Serre. Construction de revêtements étales de la droite affine en caractéristique p. Comptes Rendus **311** (1990), 341-346.

[St]   K. F. Stevenson. Galois groups of unramified covers of projective curves in characteristic p. J. of Alg.**182** (1996) 770-804.

[Ta]   A. Tamagawa. On the fundamental groups of curves over algebraically closed fields of characteristic > 0. 1997 manuscript.

Department of Mathematics, University of Pennsylvania, Philadelphia, PA 19104-6395, USA.

E-mail address: harbater@math.upenn.edu

On beta and gamma functions associated with the Grothendieck-Teichmüller groups

Yasutaka Ihara[*]

February 15, 1999

Introduction

The absolute Galois group $G_{\mathbb{Q}}$ over the rational number field $\mathbb{Q}$ can be considered as a subgroup of the Grothendieck-Teichmüller group GT which acts faithfully on the free profinite group $\hat{F}_2$ of rank 2 (the algebraic fundamental group of the projective line minus 3 points). A key property characterizing GT as a subgroup of the automorphism group Aut $\hat{F}_2$ of $\hat{F}_2$ is the so-called 5-cycle relation. It is not known whether $G_{\mathbb{Q}} = \text{GT}$.

In this article, we shall study the action of GT on the double commutator quotient $\hat{F}_2/\hat{F}_2''$ of $\hat{F}_2$. In particular, we shall show that the "beta function" B_{σ} ($\sigma \in \text{GT}$), which together with the "cyclotomic character" λ_{σ} describes this action completely, enjoys the "Γ_{σ}-factorization property" as in the case of $\sigma \in G_{\mathbb{Q}}$ proved earlier by G. W. Anderson. The key to our proof is the 5-cycle relation for GT. We shall then discuss which of the main known properties of Γ_{σ} for $\sigma \in G_{\mathbb{Q}}$ can possibly be generalized to the case of $\sigma \in \text{GT}$. As basic tools, we shall use profinite free differential calculus and profinite Blanchfield-Lyndon theorem, stated with proofs in the Appendix, which may be of independent interest.

Although the first main property of $B_{\sigma}(\sigma \in G_{\mathbb{Q}})$ is thus shared by $B_{\sigma}(\sigma \in \text{GT})$, the author suspects that $G_{\mathbb{Q}} \subsetneq \text{GT}$, and also that Anderson's multiplication formula for $\Gamma_{\sigma}(\sigma \in G_{\mathbb{Q}})$, which is arithmetic in nature, does not hold if σ is an arbitrary element of GT (whose definition is basically geometric and the key "5-cycle relation" seems to be "used up" to prove the Γ_{σ}-factorizability of B_{σ}).

The definitions and main results are stated in §1. They are "generalizations" of Anderson's, [A$_2$], from the case of $G_{\mathbb{Q}}$ to GT. First, we recall the definition of the group GT and define for each $\sigma \in \text{GT}$ the associated adelic beta function B_{σ}, which is a function on $(\mathbb{Q}/\mathbb{Z})^2$ with values in

---

[*]Research Institute for Mathematical Sciences, Kyoto University, 606-8502 Kyoto, Japan. *E-mail address*: ihara@kurims.kyoto-u.ac.jp

$(\hat{\mathbb{Z}} \otimes \mathbb{Q}_{ab})^{\times}$ ($\mathbb{Q}_{ab}$ is the maximal abelian extension of $\mathbb{Q}$). Then we state some main properties of B_{σ} (§1.6 (Theorem 1, etc.)). As a consequence, B_{σ} factors as

$$B_{\sigma}(s_1, s_2) = \frac{\Gamma_{\sigma}(s_1)\Gamma_{\sigma}(s_2)}{\Gamma_{\sigma}(s_1 + s_2)} \quad (s_1, s_2 \in \mathbb{Q}/\mathbb{Z}), \qquad (1)$$

with some $(\mathbb{W} \otimes \mathbb{Q}_{ab})^{\times}$-valued function $\Gamma_{\sigma}(s)$ on $\mathbb{Q}/\mathbb{Z}$, where $\mathbb{W} = \Pi_p W(\bar{\mathbb{F}}_p)$ (the product of the ring of Witt vectors). This defines the hyperadelic gamma function Γ_{σ} ($\sigma \in GT$), modulo some elementary factors. We shall then examine which of the main known properties of $\Gamma_{\sigma}(s)$ for $\sigma \in G_{\mathbb{Q}}$ still holds when $\sigma \in GT$. One of them is Anderson's formula [A₂](i) Th.5, connecting Dlog Γ_{σ} ($\sigma \in G_{\mathbb{Q}}$) with the Kummer 1-cocycles associated with certain cyclotomic elements. We shall show that this remains valid for $\sigma \in GT$ in an appropriate sense (Theorem 2, §1.8). Another is his multiplication formula [A₂](i)Cor. 8.6.3 for $\Gamma_{\sigma}(\sigma \in G_{\mathbb{Q}})$:

$$(\prod_{nb=a} \Gamma_{\sigma}(b))(\prod_{nc=0} \Gamma_{\sigma}(c))^{-1}\Gamma_{\sigma}(a)^{-1} = 1 \otimes (n^{-(a)})^{\sigma-1} \qquad (2)$$

($a \in \mathbb{Q}/\mathbb{Z}, n \in \mathbb{N}$), which has its origin in the Gauss multiplication formula for the classical gamma function. This appears to the author rather as a key property which possibly distinguishes Γ_{σ} for $\sigma \in G_{\mathbb{Q}}$ from those for $\sigma \in GT$. (See §1.8 for the definition of "$(n^{-(a)})^{\sigma-1}$" for $\sigma \in GT$.) We note that the above formula (2), related to the $G_{\mathbb{Q}}$-action on the double commutator quotient of the fundamental group of $\mathbb{P}^1 - \{0, 1, \infty\}$, is *invisible* from the $G_{\mathbb{Q}}$-action on its quotient modulo any fixed member of the lower central series (see §1.9).

In §2, we shall give proofs of statements given in §1. As for the proof of Theorem 1, instead of using the 5-cycle relation in its bare form, we shall use the essentially equivalent "active" form, the extendability of the action of GT on the fundamental group of $\mathbb{P}^1 - \{0, 1, \infty\}$ to that on the sphere braid group on 5 strings.

The final section §3 is for the Appendix.

1 Statement of main results and discussions

1.1

First, we shall fix notations related to free groups and braid groups.

F_r: the (abstract) free group of rank r ($r \geq 1$),

B_n: the Artin braid group on n strings ($n \geq 2$)

$= \langle b_1, \cdots, b_{n-1} \rangle$: the standard generators,

$$b_i: \quad \begin{array}{ccccc} 1 & & i & i+1 & n \\ \times & \cdots & \times & \times & \cdots \times \end{array}$$

P_n: the pure Artin braid group on n strings $(P_n \lhd B_n)$
$= \langle p_{ij} \rangle_{1 \leq i < j \leq n}, \ p_{ij} = (b_{j-1} \cdots b_{i+1}) b_i^2 (b_{j-1} \cdots b_{i+1})^{-1}$

$$\begin{array}{cccc} 1 & i & j & n \end{array}$$

$p_{ij}:$ $\times \cdots \fbox{$\times$} \cdots \times \cdots \times$ $p_{ji} := p_{ij}.$

The composition of braids is from the right $\quad b'b = b' \circ b \ (b, b' \in B_n)$. Define the elements $p_i \ (2 \leq i \leq n)$ of P_n by

$$\begin{array}{ccc} 1 & i & n \end{array}$$

$\times \cdots \times \cdots \times$ $p_i = p_{1i} p_{2i} \cdots p_{i-1,i}.$

Note that the p_i's are mutually commutative. The (2π)-rotation $w_n = p_2 \cdots p_n$ generates the center $(\simeq \mathbb{Z})$ of B_n. Consider the quotient groups B_n^*, P_n^* of B_n, P_n defined by the extra relations "$p_n = w_n = 1$", and for each $b \in B_n$, denote its image on B_n^* by b^*. The relation "$p_n = 1$" corresponds to adding the point at infinity, and "$w_n = 1$" corresponds to dividing by the (2π)-rotation. The group P_n^* is the fundamental group of

$$((\mathbb{P}_{\mathbb{C}}^1)^n - \Delta)/PGL_2(\mathbb{C}),$$

where $\mathbb{P}_{\mathbb{C}}^1$ is the complex projective line and Δ is the hyperdiagonal

$$\Delta = \{(x_1, \cdots, x_n) \in (\mathbb{P}_{\mathbb{C}}^1)^n; x_i = x_j \quad \text{for some} \quad i \neq j\}.$$

1.2

Now let us recall the definition of the Grothendieck-Teichmüller group GT. In what follows, for any discrete group Γ, $\hat{\Gamma}$ will denote its profinite completion. The groups Γ, for which we shall consider $\hat{\Gamma}$ in this article, are always residually finite, i.e., the canonical homomorphism $\Gamma \to \hat{\Gamma}$ is injective. We shall use the same symbol for an element of Γ and its image in $\hat{\Gamma}$. Let G be any topological group. Then its abelianized quotient (resp. its commutator subgroup) will be denoted by G^{ab} (resp. G'), so that $G^{ab} = G/G'$. The double commutator subgroup of G is $G'' = (G')'$. By Aut G, we shall mean the full automorphism group of the topological group G.

Now let F_2 be a free group of rank 2, with a given set of free generators x, y. Then, by definition ([D], [I_3]; see also [I-M] A.2),

$$\text{GT} = \left\{ \sigma \in \text{Aut } \hat{F}_2 \ \middle| \ \begin{array}{l} \exists \lambda \in \hat{\mathbb{Z}}^\times, \exists f \in \hat{F}_2', \text{s.t. } \sigma(x) = x^\lambda, \ \sigma(y) = f^{-1} y^\lambda f, \\ \text{and} \quad \lambda, f \quad \text{satisfy (2)(3)(4) below} \end{array} \right\}. \tag{1}$$

$$f(x, y)f(y, x) = 1 \qquad \text{(2-cycle relation)}, \tag{2}$$

$$f(z,x)z^m f(y,z)y^m f(x,y)x^m = 1, \quad \text{if} \quad xyz = 1, m = \tfrac{1}{2}(\lambda - 1),$$
$$\text{(3-cycle relation)}, \qquad (3)$$

$$f(x_{12}, x_{23})f(x_{34}, x_{45})f(x_{51}, x_{12})f(x_{23}, x_{34})f(x_{45}, x_{51}) = 1$$
$$\text{in} \quad \hat{P}_5^*, \quad \text{where} \quad x_{ij} = p_{ij}^* \quad \text{(5-cycle relation)}. \qquad (4)$$

Here, for a profinite group G and $\alpha, \beta \in G$, $f(\alpha, \beta)$ will denote the image of $f \in \hat{F}_2$ under the unique homomorphism $\hat{F}_2 \to G$ that maps x to α and y to β.

It is easy to see that λ, f are uniquely determined by σ. They will sometimes be denoted by λ_σ, f_σ, respectively. It is known that GT forms a *subgroup* of Aut $\hat{F}_2$, and that the absolute Galois group $G_{\mathbb{Q}} = \text{Gal}(\bar{\mathbb{Q}}/\mathbb{Q})$ of $\mathbb{Q}$ ($\mathbb{Q}$: the rational number field, $\bar{\mathbb{Q}}$: its algebraic closure in $\mathbb{C}$) can be embedded into GT through its action on the algebraic fundamental group

$$\hat{F}_2 = \pi_1(\mathbb{P}_{\bar{\mathbb{Q}}}^1 - \{0, 1, \infty\}, \vec{01}) = \langle x, y \rangle, \qquad (5)$$

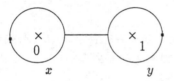

(cf [D][I$_4$]). The homomorphism GT $\to \hat{\mathbb{Z}}^\times$ defined by $\sigma \to \lambda_\sigma$ coincides with the cyclotomic character χ on $G_{\mathbb{Q}}$. We also recall that each $\sigma \in$ GT maps $z = (xy)^{-1}$ (the loop around ∞) to $g_\sigma^{-1} z^\lambda g_\sigma$, where $g_\sigma = f_\sigma(x, z)x^{\frac{1}{2}(1-\lambda_\sigma)}$ (cf [I$_3$]).

1.3

In general, for any profinite group G, $\hat{\mathbb{Z}}[[G]]$ will denote the completed group algebra of G. It is a profinite ring obtained as the projective limit of all finite group algebras $(\mathbb{Z}/n)[G/N]$, where n (resp. N) runs over the positive integers (resp. open normal subgroups of G). When G is free profinite, or free abelian profinite, we write:

$$\Lambda_r = \hat{\mathbb{Z}}[[\hat{F}_r]], \qquad (1)$$
$$A_r = \hat{\mathbb{Z}}[[\hat{\mathbb{Z}}^r]] \qquad (r \geq 1). \qquad (2)$$

In practice, $\hat{F}_r$ will always be given together with its free generators $x_1, \cdots, x_r$, and $\hat{\mathbb{Z}}^r$ will appear as its abelianization $\hat{F}_r^{ab}$. The image of x_i on $\hat{F}_r^{ab}$ will usually be denoted by the boldface $\mathbf{x}_i$ ($1 \leq i \leq r$), and $\hat{\mathbb{Z}}^r$ (though written additively) will be considered as a multiplicative group, which could better be expressed as

$$\mathbf{x}_1^{\hat{z}} \mathbf{x}_2^{\hat{z}} \cdots \mathbf{x}_r^{\hat{z}}. \tag{3}$$

The image of $\lambda \in \Lambda_r$ under the projection

$$\pi_r : \Lambda_r \longrightarrow A_r \tag{4}$$

is usually denoted by λ^{ab} (except that x_i^{ab} are abbreviated as $\mathbf{x}_i$). In this article, the two cases $r = 2, 3$ appear frequently.

1.4

Now the group GT acts on $\hat{F}_2 = \langle x, y \rangle$ (by its very definition), and hence also on its characteristic subgroups $\hat{F}_2', \hat{F}_2''$ and their quotients $(\hat{F}_2')^{ab} = \hat{F}_2'/\hat{F}_2''$, etc. Our main aim is to study its action on $(\hat{F}_2')^{ab}$.

First, since $\hat{F}_2^{ab} = \hat{F}_2/\hat{F}_2'$ is a free abelian profinite group of rank 2 generated by the images $\mathbf{x}, \mathbf{y}$ of x, y, we shall identify $\hat{F}_2^{ab}$ with $\hat{\mathbb{Z}}^2$ via its basis $\mathbf{x}, \mathbf{y}$, and write $A_2 = \hat{\mathbb{Z}}[[\hat{F}_2^{ab}]]$. The action of $\hat{F}_2^{ab}$ on $(\hat{F}_2')^{ab}$ induced from the conjugation $(n \rightarrow fnf^{-1} \ (n \in \hat{F}_2', f \in \hat{F}_2))$ equips $(\hat{F}_2')^{ab}$ with a structure of A_2-module.

Proposition 1.4.1 $(\hat{F}_2')^{ab}$ *is a free A_2-module of rank 1 generated by $\overline{(x, y)}$, the class mod $\hat{F}_2''$ of $(x, y) = xyx^{-1}y^{-1}$.*

The pro-ℓ version of this statement was proved in [I_1], and the profinite version stated without proof in [I_3]. We shall provide its proof in §2.1.

Now let us study the GT-actions on $\hat{F}_2^{ab}$ and $(\hat{F}_2')^{ab}$. First, $\sigma \in$ GT acts on the abelian group $\hat{F}_2^{ab}$ as the λ_σ-th power automorphism, because it maps the generators $\mathbf{x}, \mathbf{y}$ to their λ_σ-th power. Extend this to a continuous ring automorphism action of σ on $A_2 = \hat{\mathbb{Z}}[[\hat{F}_2^{ab}]]$ in the unique way. The natural action of σ on the A_2-module $(\hat{F}_2')^{ab}$ is semi-linear, and is expressed by a single element B_σ' of $A_2^\times$ defined by

$$\sigma\overline{(x, y)} = B_\sigma' \cdot \overline{(x, y)} \tag{1}$$

(cf. Proposition 1.4.1). Define $B_\sigma \in A_2^\times$, for each $\sigma \in$ GT, by the equality

$$B_\sigma' = \frac{\mathbf{x}^{\lambda_\sigma} - 1}{\mathbf{x} - 1} \cdot \frac{\mathbf{y}^{\lambda_\sigma} - 1}{\mathbf{y} - 1} \cdot B_\sigma. \tag{2}$$

This makes sense because $\mathbf{x} - 1, \mathbf{y} - 1$ are not zero-divisors of A_2, and the two factors in front of B_σ are units of A_2 (Proposition 2.1.1 below). We shall see later (§2.2) that B_σ can be expressed as the abelianization of a certain *derivative* of f_σ.

1.5

G. W. Anderson has shown [A$_2$] that it is theoretically useful to embed the completed group algebra $A_r = \hat{\mathbb{Z}}[[\hat{\mathbb{Z}}^r]]$ of $\hat{\mathbb{Z}}^r$ ($r \geq 1$) into the ring

$$\mathrm{Fct}((\mathbb{Q}/\mathbb{Z})^r, \hat{\mathbb{Z}} \otimes \mathbb{Q}_{ab}) \tag{1}$$

of $(\hat{\mathbb{Z}} \otimes \mathbb{Q}_{ab})$-valued functions on $(\mathbb{Q}/\mathbb{Z})^r$, where $\mathbb{Q}_{ab}$ denotes the maximal abelian extension field over $\mathbb{Q}$ in $\mathbb{C}$. His embedding

$$
\begin{array}{ccc}
A_r & \hookrightarrow & \mathrm{Fct}((\mathbb{Q}/\mathbb{Z})^r, \hat{\mathbb{Z}} \otimes \mathbb{Q}_{ab}) \\
\cup\!\!\!| & & \cup\!\!\!| \\
b & \rightarrow & B
\end{array}
\tag{2}
$$

is defined as follows. Take any $(s_1, \cdots, s_r) \in (\mathbb{Q}/\mathbb{Z})^r$, and choose any positive integer N such that $Ns_i = 0$ for all i. Express the image of b on $\hat{\mathbb{Z}}[(\mathbb{Z}/N)^r]$ as

$$b_N = \sum_a c_a \cdot a, \tag{3}$$

where $a = (a_1, \cdots, a_r)$ runs over $(\mathbb{Z}/N)^r$, and $c_a \in \hat{\mathbb{Z}}$. Define the function B by the formula

$$B(s_1, \cdots, s_r) = \sum_a c_a \otimes \exp(-2\pi\sqrt{-1}(a_1 s_1 + \cdots + a_r s_r)). \tag{4}$$

Then $b \to B$ is an injective ring homomorphism.

For example, the element $(0, \cdots, 0, \overset{j}{1}, 0, \cdots, 0) \in \hat{\mathbb{Z}}^r$ corresponds to the function $B(s_1, \cdots, s_r) = \exp(-2\pi\sqrt{-1}s_j)$.

By means of the embedding (2) for $r = 2$, we shall regard B_σ ($\sigma \in \mathrm{GT}$) as a $(\hat{\mathbb{Z}} \otimes \mathbb{Q}_{ab})$-valued function on $(\mathbb{Q}/\mathbb{Z})^2$, and call it *the adelic beta function* (*or simply, the beta function*) *associated with* σ. Since B_σ is invertible in A_2, it is $(\hat{\mathbb{Z}} \otimes \mathbb{Q}_{ab})^\times$-valued. Moreover, it follows directly from the definitions that

$$B_{\sigma\tau}(s_1, s_2) = B_\sigma(s_1, s_2) B_\tau(\lambda_\sigma s_1, \lambda_\sigma s_2) \tag{5}$$

($\sigma, \tau \in \mathrm{GT}$). Moreover, when $\sigma \in G_{\mathbb{Q}}$, $B_\sigma(s_1, s_2)$ coincides with the *adelic beta function* defined and studied by Anderson [A$_2$] (which generalizes the ℓ-adic power series studied in [I$_1$]) (see [A$_2$] (ii) §5.6). We shall study some basic properties of $B_\sigma(s_1, s_2)$ for $\sigma \in GT$.

Remark The minus sign in the exponential in (4) is for adjusting our definition of B_σ ($\sigma \in G_{\mathbb{Q}}$) with Anderson's. His generator of $\hat{\mathbb{Z}}(1)^2$ arising from that of $\hat{\mathbb{Z}}(1)$ called ζ_∞ corresponds to our generator $\mathbf{x}^{-1}, \mathbf{y}^{-1}$ of $\hat{F}_2^{ab}$. His association in [A$_2$](i)§13.1, $z \to (1+u)^{r(z)}(1+v)^{s(z)}$ should be replaced by $z \to (1+u)^{-r(z)}(1+v)^{-s(z)}$.

1.6

First, we shall list some properties of $B_\sigma (\sigma \in \mathrm{GT})$ that can be proved without using the 5-cycle relation, in the following

Proposition 1.6.1 *Let $\sigma \in \mathrm{GT}$. Then*

(i) $B_\sigma(s_1, 0) = B_\sigma(0, s_2) = 1 \otimes 1,$

(ii) $B_\sigma(s, -s) = \lambda_\sigma \otimes \dfrac{e^{i\pi s} - e^{-i\pi s}}{e^{i\pi s \lambda_\sigma} - e^{-i\pi s \lambda_\sigma}}$ $(i = \sqrt{-1}),$

(iii) $B_\sigma(s_1, s_2) = B_\sigma(s_2, s_1),$

(iv) $B_\sigma(s_1, s_2) B_\sigma(s_3, -s_3)$ *is S_3-symmetric on $s_1 + s_2 + s_3 = 0$.*

A deeper result which uses the 5-cycle relation is:

Theorem 1 *If $\sigma \in \mathrm{GT}$, then*

$$B_\sigma(s_1, s_2) B_\sigma(s_1 + s_2, s_3) \tag{1}$$

is S_3-symmetric in s_1, s_2, s_3. Under (iv) of Proposition 1.6.1, this is also equivalent to saying that

$$B_\sigma(s_1, s_2) B_\sigma(s_3, s_4) B_\sigma(s_1 + s_2, s_3 + s_4) \tag{2}$$

is S_4-symmetric on $s_1 + s_2 + s_3 + s_4 = 0$.

The proofs will be given in §2.

1.7

In [A₂], G. Anderson has shown that his adelic beta function $B_\sigma (\sigma \in G_\mathbb{Q})$ has a factorization, which is analogous to the factorization of the classical beta function in terms of the classical gamma function, and studied some fundamental properties of its "factor", the hyperadelic gamma function Γ_σ. We are now going to examine which of these properties can be generalized to the case where $\sigma \in GT$.

Let $\mathbb{W} = \prod_p \mathbb{W}_p$ be the product over all primes p of the ring $\mathbb{W}_p = W(\bar{\mathbb{F}}_p)$ of Witt vectors over $\bar{\mathbb{F}}_p$ (an algebraic closure of $\mathbb{F}_p = \mathbb{Z}/p$). It is linearly topologized by taking $\{n\mathbb{W}\}_{n \in \mathbb{N}}$ as the fundamental system of neighborhoods of 0. For each $r \geq 1$, we consider the completed group algebra

$$\tilde{A}_r = \mathbb{W}[[\hat{\mathbb{Z}}^r]] = \varprojlim_{n, N}((\mathbb{W}/n\mathbb{W})[(\mathbb{Z}/N)^r]), \tag{1}$$

and embed this into the ring

$$\mathrm{Fct}((\mathbb{Q}/\mathbb{Z})^r, \mathbb{W} \otimes \mathbb{Q}_{ab})$$

of $(\mathbb{W} \otimes \mathbb{Q}_{ab})$-valued functions on $(\mathbb{Q}/\mathbb{Z})^r$, exactly in the same way as in the case of A_r described in §1.5 (cf. [A$_2$] (i) for more details). The following key observation is due to Anderson ([A$_2$] (ii) §2.3; Theorem 6). (It is a (mild) generalization of the well-known fact that, over $\mathbb{W}_p$, there exist no non-trivial extensions of the formal completion of the multiplicative group $\mathbb{G}_m$ by itself.)

Lemma (G. Anderson) *Let B be a unit of the ring $\tilde{A}_2 = \mathbb{W}[[\hat{\mathbb{Z}}^2]]$ satisfying*
  (i) $B(s_1, s_2) = B(s_2, s_1)$ $\qquad\qquad\qquad\qquad\qquad (s_1, s_2 \in \mathbb{Q}/\mathbb{Z})$,
  (ii) $B(s_1, s_2)B(s_1 + s_2, s_3) = B(s_1, s_2 + s_3)B(s_2, s_3)$ $\quad (s_1, s_2, s_3 \in \mathbb{Q}/\mathbb{Z})$.
Then there exists a unit Γ of the ring $\tilde{A}_1 = \mathbb{W}[[\hat{\mathbb{Z}}]]$ such that

$$B(s_1, s_2) = \frac{\Gamma(s_1)\Gamma(s_2)}{\Gamma(s_1 + s_2)} \qquad\qquad (2)$$

holds for all $s_1, s_2 \in \mathbb{Q}/\mathbb{Z}$.

For each B, the element Γ is determined only modulo multiples of elements of the group

$$\mathcal{E} = \{E \in \tilde{A}_1^{\times}; \quad E(s_1 + s_2) = E(s_1)E(s_2) \quad \forall s_1, s_2 \in \mathbb{Q}/\mathbb{Z}\}. \qquad (3)$$

Anderson has also shown ([A$_2$] (i) Proposition 7.3.1) that $\mathcal{E}$ consists of exactly those elements of $A_1^{\times}$ (a priori $\tilde{A}_1^{\times}$, but in fact $A_1^{\times}$) of the form $\sum_p e_p \mathbf{x}_1^{c_p}$, where $\mathbf{x}_1$ is as in §1.3, p runs over all primes, $e_p \in \hat{\mathbb{Z}}$ (the coefficient ring) is the idempotent associated with the projection $\hat{\mathbb{Z}} \longrightarrow \mathbb{Z}_p$, and $(c_p)_p$ is an arbitrary element of $\prod_p \hat{\mathbb{Z}}$. (Note that this series converges in $A_1^{\times}$.)

Now let σ be any element of GT, and $B_\sigma \in A_2^{\times}$ be the associated beta function. Then $B = B_\sigma$ satisfies (i)(ii) of the above lemma, by Proposition 1.6.1(iii) and Theorem 1. Therefore, there exists $\Gamma_\sigma \in \tilde{A}_1^{\times}$ satisfying

$$B_\sigma(s_1, s_2) = \frac{\Gamma_\sigma(s_1)\Gamma_\sigma(s_2)}{\Gamma_\sigma(s_1 + s_2)} \quad (s_1, s_2 \in \mathbb{Q}/\mathbb{Z}), \qquad\qquad (2)_\sigma$$

and Γ_σ is unique modulo $\mathcal{E}$ (multiplicatively). Each Γ_σ satisfying $(2)_\sigma$ will be called *a branch of the hyperadelic gamma function (or simply, the gamma function) associated with σ.* When $\sigma \in G_{\mathbb{Q}}$, this corresponds to Anderson's "branch of the hyperadelic gamma function". It is easy to see that if $\sigma, \tau \in GT$, and $\Gamma_\sigma, \Gamma_\tau$ are some branches associated with σ, τ, then $\Gamma_{\sigma\tau}(s) = \Gamma_\sigma(s)\Gamma_\tau(\lambda_\sigma s)$ is a branch associated with $\sigma\tau$. Since the elements of $\mathcal{E}$ are group homomorphisms from $\mathbb{Q}/\mathbb{Z}$ into $(\hat{\mathbb{Z}} \otimes \mathbb{Q}_{ab})^{\times}$, it is clear that for any $s_1, \cdots, s_k \in \mathbb{Q}/\mathbb{Z}$ and $n_1 \cdots, n_k \in \mathbb{Z}$ satisfying

$$\sum_{i=1}^{k} n_i s_i = 0 \qquad (in \quad \mathbb{Q}/\mathbb{Z}), \tag{4}$$

the value of the product

$$\prod_{i=1}^{k} \Gamma_{\sigma}(s_i)^{n_i} \in (\mathbb{W} \otimes \mathbb{Q}_{ab})^{\times} \tag{5}_\sigma$$

is independent of the choice of a branch Γ_σ. *For which $s_1, \cdots, s_k$ and $n_1, \cdots, n_k$ can we compute the value of $(5)_\sigma$?* The first obvious examples are given in

Proposition 1.7.1 *Let $\sigma \in GT$, and Γ_σ be a branch of the gamma function associated with σ. Then*
  (i) $\Gamma_\sigma(0) = 1 \otimes 1$,
  (ii) $\Gamma_\sigma(s)\Gamma_\sigma(-s) = \lambda_\sigma \otimes \dfrac{e^{i\pi s} - e^{-i\pi s}}{e^{i\pi s \lambda_\sigma} - e^{-i\pi s \lambda_\sigma}} \qquad (s \in \mathbb{Q}/\mathbb{Z}, i = \sqrt{-1}).$

Proof. These follow immediately from Proposition 1.6.1 (i)(ii) (respectively).

$\square$

For each $s \in \mathbb{Q}/\mathbb{Z}$, $\langle s \rangle$ will denote its representative in $\mathbb{Q}$ in the interval $[0,1)$. When $s_1 \cdots, s_k \in \mathbb{Q}/\mathbb{Z}$ and $n_1, \cdots, n_k \in \mathbb{Z}$ are such that $s_1, \cdots, s_k \neq 0$, and moreover that the sum

$$w = \sum_{i=1}^{k} n_i \langle \lambda s_i \rangle \qquad (\lambda \in \hat{\mathbb{Z}}^{\times}) \tag{6}$$

is *independent of λ* and belongs to $\mathbb{Z}$, Anderson has connected the value of $(5)_\sigma$ with that of

$$(2\pi\sqrt{-1})^{-w} \prod_{i=1}^{k} \Gamma(\langle s_i \rangle)^{n_i}, \tag{5}_\mathbb{C}$$

where Γ is the classical gamma function. In fact, then the value of $(5)_\mathbb{C}$ is known to be algebraic (N. Koblitz and A. Ogus), and by using Deligne's theory of absolute Hodge cycles, he proved a beautiful formula connecting $(5)_\sigma$ with "$(5)_\mathbb{C}^{\sigma-1}$", *i.e.*,

Theorem (Anderson [A$_2$](i) Theorem 2) *Under the above assumptions on $s_1, \cdots, s_k \in \mathbb{Q}/\mathbb{Z}$ and $n_1, \cdots, n_k$, one has*

$$\prod_{i=1}^{k} \Gamma_{\sigma}(s_i)^{n_i} = \chi(\sigma)^w \otimes \{(2\pi\sqrt{-1})^{-w} \prod_{i=1}^{k} \Gamma(\langle s_i \rangle)^{n_i}\}^{\sigma-1}, \tag{7}$$

for any $\sigma \in G_{\mathbb{Q}}$ and any branch Γ_σ of the gamma function associated with σ. Here, $\chi(\sigma)$ is the cyclotomic character.

As a direct application, using the Gauss multiplication formula

$$(\prod_{j=0}^{n-1} \Gamma(\frac{s+j}{n}))(\prod_{j=1}^{n-1} \Gamma(\frac{j}{n}))^{-1}\Gamma(s)^{-1} = n^{1-s} \qquad (8)_{\mathrm{C}}$$

$(n \in \mathbb{N}, s \in \mathbb{C})$, he proved

Corollary   ([A$_2$](i) Corollary 8.6.3) *Let $\sigma \in G_{\mathbb{Q}}$ and Γ_σ be any branch of the gamma function associated with σ. Let n be any natural number, and $a \in \mathbb{Q}/\mathbb{Z}$. Then*

$$(\prod_{nb=a} \Gamma_\sigma(b))(\prod_{nc=0} \Gamma_\sigma(c))^{-1}\Gamma_\sigma(a)^{-1} = 1 \otimes (n^{-(a)})^{\sigma-1}. \qquad (8)_\sigma$$

Remark   As for the value of $\prod_{nc=0} \Gamma_\sigma(c)$, it can be computed either from Proposition 1.7.1, or by combining Anderson's connection formula (7) with the classical formula

$$\prod_{j=1}^{n-1} \Gamma(\frac{j}{n}) = (2\pi)^{\frac{n-1}{2}} n^{-\frac{1}{2}}. \qquad (9)_{\mathrm{C}}$$

Note that the "sign" of $\Gamma_\sigma(1/2)$ depends on the choice of a branch Γ_σ. The result reads as follows.

$$\begin{cases} \prod_{\substack{nc=0 \\ c \neq 1/2}} \Gamma_\sigma(c) &= \chi(\sigma)^{\frac{n-1}{2}} \otimes ((\sqrt{-1})^{\frac{1-n}{2}}\sqrt{n})^{\sigma-1} \cdots n \text{ :odd,} \\[2mm] &= \chi(\sigma)^{\frac{n}{2}-1} \otimes ((\sqrt{-1})^{1-\frac{n}{2}}\sqrt{2n})^{\sigma-1} \cdots n \text{ :even,} \qquad (9)_\sigma \\[2mm] \Gamma_\sigma(\frac{1}{2})^2 &= \chi(\sigma) \otimes (\sqrt{-1})^{\sigma-1}. \end{cases}$$

Suppose now that σ is an arbitrary element of GT. Then since we still have Γ_σ for such σ, the left hand side of $(8)_\sigma$ is defined, which is an element of $(W \otimes \mathbb{Q}_{ab})^\times$ depending only on σ. And as we shall see in §1.8, it is easy to *define* an object corresponding to $(n^{-(a)})^{\sigma-1}$ for any $\sigma \in GT$, which is again a root of unity (whose order dividing the denominator of a). A basic question is whether $(8)_\sigma$ remains to hold for all $\sigma \in GT$. The author suspects that $(8)_\sigma$ could be a key property that distinguishes $G_{\mathbb{Q}}$ from GT. We shall come back to this subject in §1.9.

1.8

In this section, we shall define, for each $n \in \mathbb{N}$, $i \in \mathbb{Z}/n$ and $r \in \mathbb{Q}/\mathbb{Z}$, a certain 1-cocycle $\kappa(\sigma) = \kappa_{n,r}^{(i)}(\sigma)(\sigma \in GT)$. When $\sigma \in G_{\mathbb{Q}}$ and $i = 0$ (resp.

$i \neq 0$), this is a Kummer 1-cocycle associated with n^{-r} (resp. $(1 - \zeta_n^{-i})^r$), where $\zeta_n = \exp(2\pi\sqrt{-1}/n)$. We shall also show that a basic result of Anderson (Theorem 5 in [A$_2$](i) §11.4), which connects $D \log \Gamma_\sigma$ with the system of $\kappa_{n,r}^{(i)}(\sigma)$ when $\sigma \in G_{\mathbb{Q}}$, remains valid on GT. This will be stated as Theorem 2, whose proof will be given in §2.6. (This gives an alternative proof for the case $\sigma \in G_{\mathbb{Q}}$.)

We shall give two equivalent definitions of $\kappa_{n,r}^{(i)}(\sigma)$, of which the first is geometric and the second, group theoretic. Let $\sigma \in GT$. The first definition is based on the interpretation of $f_\sigma \in \hat{F}_2'$ given in [I$_4$](§1). Let M be the maximal Galois extension of $\bar{\mathbb{Q}}(t)$ unramified outside $t = 0, 1, \infty$, embedded in the field $\bar{\mathbb{Q}}\{\{t\}\}$ of all formal Puiseux series in t. We shall identify $\hat{F}_2 = \langle x, y \rangle$ with the Galois group $\mathrm{Gal}(M/\bar{\mathbb{Q}}(t))$ in the way described there. Since $t^{1/n} \in M$ is abelian over $\bar{\mathbb{Q}}(t)$, and since $f_\sigma \in \hat{F}_2'$, f_σ stabilizes each element of M of the form $1 - \zeta_n^{-i}t^{1/n}$. Moreover, all roots of $1 - \zeta_n^{-i}t^{1/n}$ are contained in M. Therefore, for each $r \in \mathbb{Q}/\mathbb{Z}$, there is a unique root of unity $\kappa(\sigma) = \kappa_{n,r}^{(i)}(\sigma)$ such that

$$f_\sigma((1 - \zeta_n^{-i}t^{1/n})^{\langle r \rangle}) = \kappa_{n,r}^{(i)}(\sigma) \cdot (1 - \zeta_n^{-i}t^{1/n})^{\langle r \rangle}, \tag{1}$$

where the $\langle r \rangle$-th power of $1 - \zeta_n^{-i}t^{1/n}$ is chosen to be the one with value 1 at $t = 0$.

Proposition 1.8.1 *If $\sigma \in G_{\mathbb{Q}}$, then*

$$\begin{aligned}
\kappa_{n,r}^{(i)}(\sigma) &= \sigma(n^{-\langle r \rangle})/n^{-\langle r \rangle} & \cdots i = 0, \\
&= \sigma((1 - \zeta_n^{-\chi(\sigma)^{-1}i})^{\langle r \rangle})/(1 - \zeta_n^{-i})^{\langle r \rangle} \cdots i \neq 0. \tag{2}
\end{aligned}$$

Here, for any root of unity $\zeta \neq 1$, the branch of $\log(1-\zeta)$ which determines our choice of $(1 - \zeta)^{\langle r \rangle}$ is the one whose imaginary part has absolute value $< \pi/2$.

Proof. This follows easily by using the description of the action of f_σ on M given in [I$_4$]§1 (i.e., first, let σ^{-1} act on the Puiseux coefficients at $t = 0$, then make an analytic continuation along $(0,1)$, then let σ act on the Puiseux coefficients at $t = 1$ in $1 - t$, then continue analytically back to $t = 0$ on $(0,1)$). □

Now, to give an equivalent group theoretic definition of $\kappa_{n,r}^{(i)}(\sigma)$, consider the kernel H_n of the homomorphism $\hat{F}_2 \to \mathbb{Z}/n$ defined by $x \to 1, y \to 0$. Then H_n is normally generated by y and x^n, and is free (profinite) of rank $n + 1$ with free generators $x^{-i}yx^i (0 \leq i < n)$ and x^n. As $\hat{F}_2' \subset H_n, f_\sigma$ belongs to H_n.

Proposition 1.8.2 *For any $\sigma \in GT$ and $i \in \mathbb{Z}/n$, $\kappa_{n,r}^{(i)}(\sigma)$ is the image of f_σ under the unique homomorphism $H_n \to \mu_\infty$ (the group of roots of unity in $\mathbb{C}$) defined by*

$$x^n \to 1, \quad x^{-i}yx^i \to \exp(2\pi\sqrt{-1}r), \quad x^{-j}yx^j \to 1(j \neq i). \tag{3}$$

Proof. First, note that $H_n = \mathrm{Gal}(M/\bar{\mathbb{Q}}(t^{1/n}))$, and that the inertia groups above the places $t^{1/n} = 0, \zeta_n^j(0 \leq j < n)$ are generated by some H_n-conjugate of $x^n, x^{-j}yx^j$, respectively. Moreover,

$$(x^{-i}yx^i)(1 - \zeta_n^{-i}t^{1/n})^{\langle r \rangle} = \exp(2\pi\sqrt{-1}r)(1 - \zeta_n^{-i}t^{1/n})^{\langle r \rangle}. \tag{4}$$

Our assertion is now obvious. □

Proposition 1.8.3 *For each $\sigma \in GT$, there exists a unique element $\Psi_\sigma \in A_1$ such that if we write*

$$\Psi_\sigma \equiv \sum_{i(\mathrm{mod}\ n)} \Psi_n^{(i)}(\sigma)\mathbf{x}^{-i} \ \mathrm{mod}\ (\mathbf{x}^n - 1) \tag{5}$$

for each $n \geq 1$, then each coefficient $\Psi_n^{(i)}(\sigma) \in \hat{\mathbb{Z}}$ satisfies

$$\kappa_{n,1/m}^{(i)}(\sigma) = \zeta_m^{\Psi_n^{(i)}(\sigma)} \tag{6}$$

for all $m = 1, 2, \cdots$.

Proof. Since $r \to \kappa_{n,r}^{(i)}(\sigma)$, for fixed σ, n, i, transforms addition to multiplication, the formula (6) for all m determines $\Psi_n^{(i)}(\sigma) \in \hat{\mathbb{Z}}$. Since the product of $\kappa_{nd,r}^{(j)}(\sigma)$, for all j (mod nd) which project to a fixed i (mod n) is equal to $\kappa_{n,r}^{(i)}(\sigma)$, as can be checked directly from the definition of $\kappa_{n,r}^{(i)}(\sigma)$, the simultaneous congruences (5) have a unique solution $\Psi_\sigma \in A_1$. □

Now let $D : \tilde{A}_1 \to \tilde{A}_1$ be the unique continuous W-linear derivation such that $D(\mathbf{x}^{-1}) = \mathbf{x}^{-1}$, or equivalently, $D(\mathbf{x}) = -\mathbf{x}$ (cf. [A$_2$](i)§11.1). This can also be defined as follows. Let $\mathbf{x}, \mathbf{y}$ be free generators of $\hat{\mathbb{Z}}^2$, let $\tilde{A}_2 = \mathrm{W}[[\hat{\mathbb{Z}}^2]]$, and for each $\lambda \in \tilde{A}_1$, denote by $\lambda\{\mathbf{x}\}$ resp. $\lambda\{\mathbf{xy}\}$ its image in $\tilde{A}_2$ under the continuous W-algebra embeddings $\tilde{A}_1 \to \tilde{A}_2$ defined by $\mathbf{x} \to \mathbf{x}$ resp. $\mathbf{x} \to \mathbf{xy}$. Then $D(\lambda)$ is the image of

$$-\frac{\lambda\{\mathbf{xy}\} - \lambda\{\mathbf{x}\}}{\mathbf{y} - 1} \tag{7}$$

under the projection $\tilde{A}_2 \to \tilde{A}_1$ defined by $\mathbf{y} \to 1$. As in [A$_2$], for each $\lambda \in \tilde{A}_1^\times$, we define

$$D \log \lambda := \lambda^{-1}D(\lambda). \tag{8}$$

The following theorem is an enlargement, from the case $\sigma \in G_\mathbb{Q}$ to $\sigma \in GT$, of Anderson's Theorem 5 of [A$_2$](i) (with different methods of proof).

Theorem 2 *For any $\sigma \in GT$,*

$$D \log \Gamma_\sigma - (D \log \Gamma_\sigma)(0) = \Psi_\sigma \qquad (9)$$

holds for any branch Γ_σ of the gamma function associated with σ.

The proof will be postponed until §2.6, as some results of §2.2 will be used. At any rate, the proof is quite elementary. We only note here that for any $E \in \mathcal{E}, D \log E \in \hat{\mathbb{Z}}$ (a constant) (cf. [A$_2$](i) §11.1); hence the left hand side of (9) is a priori independent of the choice of a branch of Γ_σ.

1.9

Anderson's multiplication formula $(8)_\sigma$ reflects the existence of a family of $G_{\mathbb{Q}_{ab}}$-stable abelian subquotients of $\hat{F}_2$ whose members are mutually independent as subquotients but are closely related to each other as $G_{\mathbb{Q}_{ab}}$-modules. To explain this in a simplified situation, let us fix an odd prime number l, and let $s_1, s_2 \in \mathbb{Q}/\mathbb{Z}$ be such that $s_1, s_2, s_1 + s_2$ have the same exact denominator l^n for some $n \geq 1$. Let $C = C(s_1, s_2)$ be the curve over $\mathbb{Q}(\mu_{l^\infty})$ with function field

$$K(s_1, s_2) = \mathbb{Q}(\mu_{l^\infty})(t, \ t^{(s_1)}(1-t)^{(s_2)}),$$

and $J = J(s_1, s_2)$ be its Jacobian. Let $\theta^* = \theta^*(s_1, s_2)$ be the automorphism of $K(s_1, s_2)$ over $\mathbb{Q}(\mu_{l^\infty})$ that leaves t invariant and multiplies $t^{(s_1)}(1-t)^{(s_2)}$ by $\zeta_{l^n} = \exp(2\pi\sqrt{-1}/l^n)$. Let $\theta = \theta(s_1, s_2)$ be the automorphism of C such that $(\theta^* f)(\theta P) = f(P)$ holds for all geometric points P of C and $f \in K(s_1, s_2)$; the induced endomorphism of J will also be denoted by θ. Denote by $A = A(s_1, s_2)$ the quotient of J modulo the image of

$$\sum_{i=0}^{l-1} \theta^{i l^{n-1}}.$$

Then A is an abelian variety with complex multiplication by the ring of integers of $\mathbb{Q}(\zeta_{l^n})$ (identified with $\mathbb{Q}(\theta)$ via $\zeta_{l^n} \leftrightarrow \theta$), and the Tate module $T(s_1, s_2) := T_l(A)$ is a free $\mathbb{Z}_l \otimes \mathbb{Z}[\zeta_{l^n}]$-module of rank 1. We note that $A(s_1, s_2)$ is the *primitive* part of $J(s_1, s_2)$, and so, as abelian varieties, $A(s_1, s_2)$ and $A(ls_1, ls_2)$ are (at least in general) *unrelated* to each other. Still, as $G_{\mathbb{Q}(\mu_{l^\infty})}$-modules, $T(s_1, s_2)$ and $T(ls_1, ls_2)$ are related to each other as follows. Let $\sigma \in G_{\mathbb{Q}(\mu_{l^\infty})}$. Then σ acts on $T(s_1, s_2)$ as a multiplication of some $\beta_\sigma = \beta_\sigma(s_1, s_2) \in \mathbb{Z}_l \otimes \mathbb{Z}[\zeta_{l^n}]$. The result of [I$_1$] §II, Cor. of Theorem 4, restated in terms of B_σ, reads as

Proposition 1.9.1 *The multiplier $\beta_\sigma(s_1, s_2)$ is equal to the projection of $B_\sigma(-s_1, -s_2)$ on $\mathbb{Z}_l \otimes \mathbb{Q}_{ab}$.*

Since the conjugates of $\beta_\sigma(s_1, s_2)$ over $\mathbb{Q}_l(\zeta_{l^{n-1}})$ are $\beta_\sigma(\lambda s_1, \lambda s_2)$ with $\lambda \equiv 1 \pmod{l^{n-1}}$ counted mod l^n, Anderson's multiplication formula §1.7 $(8)_\sigma$ (for $n = l, a = s_1, s_2$) gives directly the following

Corollary 1.9.2

$$N_{\mathbb{Q}_l(\zeta_{l^n})/\mathbb{Q}_l(\zeta_{l^{n-1}})}(\beta_\sigma(s_1, s_2)) = \beta_\sigma(ls_1, ls_2) \qquad (n \geq 2).$$

Let me stress again that this is not a mere compatibility, but is something more surprising.

The equality of this type is quite analogous to Coleman's norm relation, and indeed, Coleman himself made significant contributions to the discovery and the use of the above relation (cf. [C][I-K]).

Now let $F_2^{(l)}$ denote the maximal pro-l quotient of $\hat{F}_2$. Then, as described in [I$_1$]§II, each $T(s_1, s_2)$ is canonically isomorphic, as $G_{\mathbb{Q}(\mu_{l^\infty})}$-module, to a certain quotient of $(F_2^{(l)})'/(F_2^{(l)})''$, and as its quotient, $T(s_1, s_2)$ and $T(ls_1, ls_2)$ are *unrelated*. If we truncate $F_2^{(l)}$ by a member of its lower central series, then, only some finite (over $\mathbb{F}_l$) quotient of each $T(s_1, s_2)$ remains. So, the phenomena as shown in Corollary 1.9.2 cannot be observed if we only consider the Galois actions on the Lie algebra over $\mathbb{Q}_l$ of $F_2^{(l)}$ modulo a fixed member of its lower central series. It is easy to see that each quotient $T(s_1, s_2)$ of $(F^{(l)})'/(F_2^{(l)})''$ is stable under the action of GT. But it is hard to believe that one can prove such a relation as Corollary 1.9.2 (and hence such a relation as Anderson's multiplication formula $(8)_\sigma$) for any element $\sigma \in GT$. A key property defining GT as a subgroup of $\mathrm{Aut}\hat{F}_2$ is the 5-cycle relation which seems to be already "used up" in proving the decomposability of B_σ in terms of Γ_σ. As a closely related matter, it seems to be an interesting problem to determine the image of the 1-cocycle

$$G_{\mathbb{Q}} \ni \sigma \rightarrow \Gamma_\sigma \in \tilde{A}_1^\times/\varepsilon. \qquad (1)$$

In the pro-l case, this study has already been done by Coleman, Ichimura and Kaneko (cf. [I-K]).

1.10

Some additional remarks. In §1.8, we defined for each $n \in \mathbb{N}$, $i \in \mathbb{Z}/n$, $r \in \mathbb{Q}/\mathbb{Z}$ and $\sigma \in GT$ a certain root of unity $\kappa_{n,r}^{(i)}(\sigma)$. They satisfy

$$\kappa_{n,r+r'}^{(i)}(\sigma) = \kappa_{n,r}^{(i)}(\sigma)\kappa_{n,r'}^{(i)}(\sigma) \qquad (r, r' \in \mathbb{Q}/\mathbb{Z}), \qquad (1)$$

$$\prod_{\substack{j(\bmod\ nd) \\ j \to i}} \kappa_{nd,r}^{(j)}(\sigma) = \kappa_{n,r}^{(i)}(\sigma) \qquad (n, d \in \mathbb{N}), \qquad (2)$$

(as already noted in the proof of Proposition 1.8.3), and moreover,

$$\kappa_{n,r}^{(i)}(\sigma\tau) = \kappa_{n,r}^{(i)}(\sigma)(\kappa_{n,r}^{(\lambda_\sigma^{-1}i)}(\tau))^{\lambda_\sigma} \qquad (\sigma,\tau \in \mathrm{GT}). \qquad (3)$$

The last equality follows easily from Proposition 1.8.2, using the obvious identity $f_{\sigma\tau} = f_\sigma \cdot \sigma(f_\tau)$. In particular, $\kappa_{n,r}^{(0)}(\sigma)$ is a 1-cocycle on GT. When $\sigma \in G_{\mathbb{Q}}$, this was the Kummer 1-cocycle w.r.t. $n^{-(r)}$ (Proposition 1.8.1). Thus, in view of Anderson's multiplication formula (§1.7 $(8)_\sigma$), we are lead to define the following subgroup of GT containing $G_{\mathbb{Q}}$;

GTA *is the collection of all elements* σ *of* GT *satisfying*

$$(\prod_{nb=a} \Gamma_\sigma(b))(\prod_{nc=0} \Gamma_\sigma(c))^{-1}\Gamma_\sigma(a)^{-1} = 1\otimes\kappa_{n,a}^{(0)}(\sigma) \quad (n \in \mathbb{N},\, a \in \mathbb{Q}/\mathbb{Z}). \quad (4)_\sigma$$

It forms a subgroup because $\kappa_{n,r}^{(0)}(\sigma)$ is a 1-cocycle. A natural question is how close is GTA to $G_{\mathbb{Q}}$ (or to GT). One related remark. When $\sigma \in G_{\mathbb{Q}}$, $\kappa_{n,r}^{(0)}(\sigma)$ (for each fixed r) is obviously *multiplicative in* n, but the author does not know whether this remains valid for any $\sigma \in \mathrm{GT}$. However, if $\sigma \in \mathrm{GTA}$, then $\kappa_{n,r}^{(0)}(\sigma)$ is necessarily multiplicative in n. This can be deduced easily by rewriting the left hand side of $(4)_\sigma$ for $n = n_1 n_2$ by performing the division in two steps, and by using $(4)_\sigma$ for $n = n_1$ and $n = n_2$.

Another related remark. There is another important property of $\Gamma_\sigma(\sigma \in G_{\mathbb{Q}})$ proved in [A$_2$]; namely, Theorem 3 in [A$_2$](ii). But this is a formal consequence of his multiplication formula; hence its generalization to $\sigma \in$ GT would simply be a formal consequence of $(4)_\sigma$.

2   Proofs of the main results

In §2.1, after some preliminary remarks on the completed group algebras A_r, we shall provide a proof of Proposition 1.4.1. Then, in §2.2, we shall write down some basic formulae relating B_σ with f_σ, and in §2.3, using these and the 2-cycle and the 3-cycle relations for f_σ, we shall settle the proof of Proposition 1.6.1. The proof of Theorem 1 will be given in §§2.4,2.5, and that of Theorem 2, in §2.6. The readers are advised to look at §3.1 (Theorem A-1, the basic rules (i)-(v) for free differential calculus, and Theorem A-2) beforehand.

2.1

We begin with the following elementary

Proposition 2.1.1 *Let* $A_r = \hat{\mathbb{Z}}[[\hat{\mathbb{Z}}^r]]$ *be the completed group algebra of a free abelian profinite group of rank* r *with generators* $\mathbf{x}_1,\cdots,\mathbf{x}_r$. *Then, in the ring* A_r, *for each* i ($1 \leq i \leq r$),

(i) $\mathbf{x}_i - 1$ *is not a zero-divisor,*

(ii) *for any* $\alpha \in \hat{\mathbb{Z}}$, $\mathbf{x}_i^\alpha - 1$ *is divisible by* $\mathbf{x}_i - 1$,

(iii) *for any* $\alpha \in \hat{\mathbb{Z}}^\times$, $(\mathbf{x}_i^\alpha - 1)/(\mathbf{x}_i - 1)$ *is invertible.*

Proof. (i) It suffices to show that $\mathbf{x}_1 - 1$ is not a zero divisor. Suppose $\mathbf{y} \in A_r$ satisfies $\mathbf{y}(\mathbf{x}_1 - 1) = 0$. The goal is to show that $\mathbf{y} = 0$, or equivalently, that, for any integers $m, n \geq 1$, the projection of $\mathbf{y}$ on $(\mathbb{Z}/m)[(\mathbb{Z}/n)^r]$ must vanish. Consider the projection η of $\mathbf{y}$ on a "finer" quotient

$$(\mathbb{Z}/m)[(\mathbb{Z}/mn) \times (\mathbb{Z}/n)^{r-1}],$$

and call ξ_i the image of $\mathbf{x}_i$ on $(\mathbb{Z}/mn) \times (\mathbb{Z}/n)^{r-1} (1 \leq i \leq r)$. Thus,

$$(\mathbb{Z}/mn) \times (\mathbb{Z}/n)^{r-1} = \{\xi_1^{a_1} \cdots \xi_r^{a_r}; a_1 \in \mathbb{Z}/mn, a_i \in \mathbb{Z}/n \ (2 \leq i \leq r)\}.$$

Write $\eta = \sum_I a_I(\xi_1)\xi_2^{i_2} \cdots \xi_r^{i_r}$, where $I = (i_2, \cdots, i_r)$ runs over $(\mathbb{Z}/n)^{r-1}$, and $a_I(\xi_1) \in (\mathbb{Z}/m)[\xi_1]$ is of degree $< mn$. Now, since $\eta(\xi_1 - 1) = 0$, we have $a_I(\xi_1)(\xi_1 - 1) \equiv 0 \mod (\xi_1^{mn} - 1)$ for each I. Therefore, each $a_I(\xi_1)$ must be a constant multiple of $\sum_{i=0}^{mn-1} \xi_1^i$. But then, the image of $a_I(\xi_1)$ on $(\mathbb{Z}/m)[(\mathbb{Z}/n)]$ must vanish (being killed by the multiplicity m). Therefore, the projection of $\mathbf{y}$ (which is the same as that of η) on $(\mathbb{Z}/m)[(\mathbb{Z}/n)^r]$ must be 0, as desired.

(ii) The closed ideal $A_r(\mathbf{x}_i - 1)$ contains $\mathbf{x}_i^n - 1$ for all $n \in \mathbb{Z}$. But $\mathbf{x}_i^\alpha - 1 (\alpha \in \hat{\mathbb{Z}})$ can be approximated by such elements; hence $\mathbf{x}_i^\alpha - 1 \in A_r(\mathbf{x}_i - 1)$ for any $\alpha \in \hat{\mathbb{Z}}$.

(iii) Apply (ii) to $\mathbf{x}_i^\alpha$ instead of $\mathbf{x}_i$. $\qquad\square$

Proof of Proposition 1.4.1 We shall apply Theorem A-2 (§3) for the case where $\mathcal{F} = \hat{F}_2, x_1 = x, x_2 = y$ and $\mathcal{N} = \hat{F}_2'$. It asserts that there is an A_2-isomorphism from $(\hat{F}_2')^{ab}$ onto

$$\mathcal{I}_2 = \{(a, a') \in (A_2)^{\oplus 2}; a(\mathbf{x} - 1) + a'(\mathbf{y} - 1) = 0\}, \tag{1}$$

which is induced from the mapping

$$\hat{F}_2' \ni n \rightarrow ((\frac{\partial n}{\partial x})^{ab}, (\frac{\partial n}{\partial y})^{ab}) \in (A_2)^{\oplus 2}, \tag{2}$$

where ab denotes the abelianization $\Lambda_2 = \hat{\mathbb{Z}}[[\hat{F}_2]] \rightarrow A_2$. First, let us check that $\mathcal{I}_2$ is free of rank 1 generated by $(1 - \mathbf{y}, \mathbf{x} - 1)$. (The only "arithmetic" used in §2!) For this purpose, consider the projection $\hat{F}_2^{ab} = \langle \mathbf{x}, \mathbf{y} \rangle \rightarrow \hat{F}_1^{ab} = \langle \mathbf{x} \rangle$ defined by $\mathbf{y} \rightarrow 1$. The associated group algebra homomorphism $A_2 \rightarrow A_1$ has the kernel $(\mathbf{y} - 1)A_2$. Therefore, if $(a, a') \in \mathcal{I}_2$, then the image of $a(\mathbf{x} - 1)$ on A_1 vanishes. By Proposition 2.1.1 (i) for $r = 1$, the image of a on A_1 must vanish; hence a must be divisible by $\mathbf{y} - 1$. Similarly, a' is divisible by $\mathbf{x} - 1$. Since $(\mathbf{x} - 1)(\mathbf{y} - 1)$ is also a non zero-divisor

(Proposition 2.1.1 (i)), this proves that $\mathcal{I}_2$ is free of rank 1 generated by $(1 - \mathbf{y}, \mathbf{x} - 1)$.

On the other hand, if $n = (x, y) = xyx^{-1}y^{-1} \in \hat{F}_2'$, then

$$\frac{\partial n}{\partial x} = 1 - xyx^{-1} \xrightarrow{ab} 1 - \mathbf{y} \tag{3}$$

$$\frac{\partial n}{\partial y} = x - xyx^{-1}y^{-1} \xrightarrow{ab} \mathbf{x} - 1; \tag{4}$$

hence $n \pmod{\hat{F}_2''}$ corresponds with $(1 - \mathbf{y}, \mathbf{x} - 1)$. Therefore, $\overline{(x,y)}$ gives a free generator of the A_2-module $(\hat{F}_2')^{ab}$. □

2.2

Now we shall present some basic formulae connecting f_σ (§1.2) and B_σ (§1.4) for any $\sigma \in GT$. Roughly speaking, B_σ is the *abelianization* of some *derivative* of f_σ. By the profinite Fox theorem applied to $\hat{F}_2 = \langle x, y \rangle$, every element λ of $\Lambda_2 = \hat{\mathbb{Z}}[[\hat{F}_2]]$ can be expressed uniquely as

$$\lambda = s(\lambda) + \lambda_1 \cdot (x - 1) + \lambda_2 \cdot (y - 1) \quad (\lambda, \lambda_2 \in \Lambda_2), \tag{1}$$

and $\lambda_1 = \frac{\partial \lambda}{\partial x}, \lambda_2 = \frac{\partial \lambda}{\partial y}$ (see §3 (Theorem A-1,etc.)) In particular, for each $\sigma \in GT$,

$$f_\sigma^{-1} = 1 + \frac{\partial(f_\sigma^{-1})}{\partial x}(x - 1) + \frac{\partial(f_\sigma^{-1})}{\partial y}(y - 1). \tag{2}$$

Define an element ψ_σ of Λ_2 by

$$\psi_\sigma = 1 + \frac{\partial(f_\sigma^{-1})}{\partial x}(x - 1) = f_\sigma^{-1} - \frac{\partial(f_\sigma^{-1})}{\partial y}(y - 1)$$

$$= 1 - f_\sigma^{-1}\frac{\partial f_\sigma}{\partial x}(x - 1) = f_\sigma^{-1} + f_\sigma^{-1}\frac{\partial f_\sigma}{\partial y}(y - 1). \tag{3}$$

This is a profinite analogue of the power series defined and studied in [I₂] (see also [I₃] §6.5 Remark).

Proposition 2.2.1

$$f_\sigma(x, y)\psi_\sigma(x, y) = \psi_\sigma(y, x) \quad (\sigma \in GT).$$

Proof. The 2-cycle relation gives

$$f_\sigma(y, x)f_\sigma(x, y) = 1. \tag{4}$$

By operating the free differentiation $\frac{\partial}{\partial x}$ (w.r.t. x, y) on both sides of (4), using the formulae (i)-(v) of §3.2, we obtain

$$\frac{\partial f_\sigma}{\partial y}(y, x) + f_\sigma(y, x)\frac{\partial f_\sigma}{\partial x}(x, y) = 0. \tag{5}$$

On the other hand, by (3) (the second row equalities), we deduce both

$$f_\sigma(x, y)\psi_\sigma(x, y) = f_\sigma(x, y) - \frac{\partial f_\sigma}{\partial x}(x, y)(x - 1), \tag{6}$$

and

$$\psi_\sigma(y, x) = f_\sigma(y, x)^{-1} + f_\sigma(y, x)^{-1}(\frac{\partial f_\sigma}{\partial y})(y, x)(x - 1)$$

$$= f_\sigma(x, y) + f_\sigma(x, y)\frac{\partial f_\sigma}{\partial y}(y, x)(x - 1). \tag{7}$$

We obtain the desired formula by comparison of (6) and (7) using (5). □

The image of each element λ of Λ_2 under the abelianization map $\Lambda_2 \rightarrow A_2$ is denoted by λ^{ab}. Since $f_\sigma \in \hat{F}'_2$, we have

$$f_\sigma^{ab} = 1 \qquad (\sigma \in GT). \tag{8}$$

This is an obvious but crucial remark. Note also that (3) and (8) give

$$\psi_\sigma^{ab} = 1 - (\frac{\partial f_\sigma}{\partial x})^{ab}(\mathbf{x} - 1) = 1 + (\frac{\partial f_\sigma}{\partial y})^{ab}(\mathbf{y} - 1). \tag{9}$$

Proposition 2.2.2 $B_\sigma = \psi_\sigma^{ab} \qquad (\sigma \in GT).$

Proof. Put $n = (x, y) = xyx^{-1}y^{-1}$, and let $\sigma = (\lambda, f) \in GT$. Then $\sigma(n) = (x^\lambda, f^{-1}y^\lambda f)$. Let us compute $(\frac{\partial \sigma(n)}{\partial x})^{ab}$, which is simpler than $\frac{\partial(\sigma(n))}{\partial x}$ itself, as we may replace any element of $\hat{F}'_2$ (e.g. f_σ) by 1 in the abelianization process. A straightforward computation gives

$$\frac{\partial(\sigma(n))}{\partial x} \equiv \frac{x^\lambda - 1}{x - 1} - x^\lambda\frac{\partial f}{\partial x} + x^\lambda y^\lambda\frac{\partial f}{\partial x} + x^\lambda y^\lambda\frac{x^{-\lambda} - 1}{x - 1}$$

$$-y^\lambda\frac{\partial f}{\partial x} + \frac{\partial f}{\partial x},$$

where $\equiv$ is the congruence modulo the kernel of abelianization. Therefore,

$$(\frac{\partial(\sigma(n))}{\partial x})^{ab} = (1 - \mathbf{x}^\lambda)(1 - \mathbf{y}^\lambda)\{(1 - \mathbf{x})^{-1} + (\frac{\partial f}{\partial x})^{ab}\}.$$

But since $(\frac{\partial n}{\partial x})^{ab} = 1 - \mathbf{y}$ (§2.1(3)), this gives

$$\sigma(\overline{n}) \equiv \frac{(1-\mathbf{x}^\lambda)(1-\mathbf{y}^\lambda)}{(1-\mathbf{x})(1-\mathbf{y})}\{1 - (\mathbf{x}-1)(\frac{\partial f_\sigma}{\partial x})^{ab}\} \cdot \overline{n}$$

in $(\hat{F}'_2)^{ab}$, where $\overline{n} = n$ (mod $\hat{F}''_2$). From the definition of B_σ and (9), it follows that

$$B_\sigma = 1 - (\mathbf{x}-1)(\frac{\partial f_\sigma}{\partial x})^{ab} = \psi_\sigma^{ab}.$$

□

Proposition 2.2.3 *Express the image of f_σ on $(\hat{F}'_2)^{ab}$ as an A_2-multiple of the image of (x,y) on $(\hat{F}'_2)^{ab}$, and call this element $C_\sigma \in A_2$. Then*

$$\psi_\sigma^{ab} = 1 + (\mathbf{x}-1)(\mathbf{y}-1)C_\sigma \qquad (\sigma \in GT).$$

Proof. The first coordinates in $\mathcal{I}_2$ of the images of f_σ and (x,y) on $(\hat{F}'_2)^{ab}$ are $(\frac{\partial f_\sigma}{\partial x})^{ab}$, resp. $1 - \mathbf{y}$. Therefore,

$$(\frac{\partial f_\sigma}{\partial x})^{ab} = C_\sigma(1-\mathbf{y});$$

hence we obtain the desired formula by using (9). □

Corollary 2.2.4 $B_\sigma \equiv 1 \mod (\mathbf{x}-1)(\mathbf{y}-1)$ $(\sigma \in GT)$.

Remark From Proposition 2.2.3, it also follows that if $\sigma \in GT$ acts trivially on $\hat{F}'/\hat{F}''$, so that $B_\sigma = 1$, then $C_\sigma = 0$ which implies $f_\sigma \in \hat{F}''_2$. Therefore, if σ acts also trivially on $\hat{F}/\hat{F}'$, so that $\lambda_\sigma = 1$, then σ must act trivially on $\hat{F}/\hat{F}''$.

2.3

Now we are going to prove Proposition 1.6.1(i)-(iv). Let $\sigma \in GT$, and put $m = m_\sigma = \frac{1}{2}(\lambda_\sigma - 1), B = B_\sigma$. Let τ be the automorphism of A_2 defined by $\mathbf{x} \to \mathbf{y}, \mathbf{y} \to \mathbf{z}, \mathbf{z} \to \mathbf{x}$, where $\mathbf{z} = (\mathbf{xy})^{-1}$, and write $B = B\{\mathbf{x},\mathbf{y}\}, B^\tau = B\{\mathbf{y},\mathbf{z}\}, B^{\tau^2} = B\{\mathbf{z},\mathbf{x}\}$. Then one checks easily that (i) $\sim$ (iv) are respectively equivalent with the following

[i] $B\{\mathbf{x},\mathbf{y}\} \equiv 1 \mod (\mathbf{x}-1)(\mathbf{y}-1)$
[ii] $B\{\mathbf{x},\mathbf{x}^{-1}\} = \lambda \mathbf{x}^m \frac{1-\mathbf{x}}{1-\mathbf{x}^\lambda}$
[iii] $B\{\mathbf{x},\mathbf{y}\} = B\{\mathbf{y},\mathbf{x}\}$
[iv] $B\{\mathbf{x},\mathbf{y}\}B\{\mathbf{z},\mathbf{z}^{-1}\}$ is symmetric in $\mathbf{x},\mathbf{y},\mathbf{z}$.

Proofs of [i] ~ [iv].

   [i] This is Corollary 2.2.4 just proved.

   [iii] By Proposition 2.2.1, $f_\sigma(x,y)\psi_\sigma(x,y) = \psi_\sigma(y,x)$. But $f_\sigma^{ab} = 1$; hence ψ_σ^{ab} is symmetric in x, y. By Proposition 2.2.2, this gives [iii].

   To prove [ii],[iv], we need further the following

Lemma 2.3.1 *The notations being as above, we have*

$$\mathbf{y}^m(1 - \mathbf{xy})B\{\mathbf{x}, \mathbf{y}\} + (\mathbf{x} - 1)B\{\mathbf{y}, \mathbf{z}\} + \mathbf{x}^{m+1}\mathbf{y}^m(\mathbf{y} - 1)B\{\mathbf{z}, \mathbf{x}\} = 0. \quad (1)$$

Proof of Lemma 2.3.1. This follows from the 3-cycle relation for $f = f_\sigma$:

$$f(z, x)z^m f(y, z)y^m f(x, y)x^m = 1, \quad (2)$$

by taking $\frac{\partial}{\partial x}$ of both sides (the free differentiation w.r.t. x, y), then passing to the abelian quotient, and then by using the following equalities (cf. Proposition 2.2.2 and §2.2 (9)):

$$B\{\mathbf{x}, \mathbf{y}\}(= \psi^{ab}) = 1 - (f_x)^{ab} \cdot (\mathbf{x} - 1) = 1 + (f_y)^{ab} \cdot (\mathbf{y} - 1) \quad (3)$$

$(f_x := \frac{\partial f}{\partial x}, f_y := \frac{\partial f}{\partial y})$. In fact, by letting $\frac{\partial}{\partial x}$ operate on both sides of (2), using the identities $\frac{\partial z}{\partial x} = -z$, $\frac{\partial f(z,x)}{\partial x} = -f_x(z,x)z + f_y(z,x)$, and by passing to the abelianization, we obtain

$$(-f_x^{ab}(\mathbf{z}, \mathbf{x})\mathbf{z} + f_y^{ab}(\mathbf{z}, \mathbf{x})) - \frac{z^m - 1}{z - 1}z - z^{m+1}f_y^{ab}(\mathbf{y}, \mathbf{z})$$

$$+\mathbf{x}^{-m} f_x^{ab}(\mathbf{x}, \mathbf{y}) + \mathbf{x}^{-m}\frac{\mathbf{x}^m - 1}{\mathbf{x} - 1} = 0. \quad (4)$$

By multiplying $(\mathbf{x} - 1)(\mathbf{z} - 1)$ on both sides of (4) and by using (3) (to express f_x^{ab}, f_y^{ab} in terms of B), we obtain (term by term)

$$(B\{\mathbf{z}, \mathbf{x}\} - 1)(\mathbf{xz} - 1) - \mathbf{z}(\mathbf{z}^m - 1)(\mathbf{x} - 1) - \mathbf{z}^{m+1}(\mathbf{x} - 1)(B\{\mathbf{y}, \mathbf{z}\} - 1)$$

$$+\mathbf{x}^{-m}(\mathbf{z} - 1)(1 - B\{\mathbf{x}, \mathbf{y}\}) + (\mathbf{z} - 1)(1 - \mathbf{x}^{-m}) = 0. \quad (5)$$

We see that the sum of terms not involving $B\{,\}$ vanishes, and by multiplication of $-\mathbf{x}^{m+1}\mathbf{y}^{m+1}$ on both sides of (5), we obtain the desired formula (1). □

Remark   We only obtain the same formula if we use $\frac{\partial}{\partial y}$ instead.

   Proof of Proposition 1.6.1 (ii), (iv). Now by applying the involutive automorphism $\mathbf{x} \to \mathbf{y}, \mathbf{y} \to \mathbf{x}$ of A_2 to the formula (1) of Lemma 2.3.1, and by using the already established equality $B\{\mathbf{y}, \mathbf{x}\} = B\{\mathbf{x}, \mathbf{y}\}$, we obtain

$$\mathbf{x}^m(1 - \mathbf{xy})B\{\mathbf{x}, \mathbf{y}\} + (\mathbf{y} - 1)B\{\mathbf{z}, \mathbf{x}\} + \mathbf{x}^m\mathbf{y}^{m+1}(\mathbf{x} - 1)B\{\mathbf{y}, \mathbf{z}\} = 0. \quad (1)'$$

Multiplication of $x^{m+1}y^m$ on both sides of (1)$'$ and subtraction from (1) give

$$y^m(1 - x^\lambda)(1 - xy)B\{x, y\} + (x - 1)(1 - x^\lambda y^\lambda)B\{y, z\} = 0.$$

Therefore,

$$B\{y, z\} = \frac{(1 - x^\lambda)(1 - xy)}{(1 - x)(1 - x^\lambda y^\lambda)}y^m \cdot B\{x, y\}. \tag{6}$$

By letting $z \to 1$, we obtain

$$B\{x, x^{-1}\} = \lambda x^m \frac{1 - x}{1 - x^\lambda}, \tag{7}$$

which is equivalent with [ii]. Moreover, (6) and (7) give

$$B\{y, z\}B\{x, x^{-1}\} = B\{x, y\}B\{z, z^{-1}\}, \tag{8}$$

which implies that (8) is symmetric in x, y, z, whence [iv]. $\square$

2.4

Before going into the proof of Theorem 1, we shall first show why the two statements in Theorem 1 are easy formal consequences of each other, under Proposition 1.6.1 (iv).

Let $\sigma \in GT$, and write $B(s_1, s_2) = B_\sigma(s_1, s_2)$. First, assume that

$$B(s_1, s_2)B(s_1 + s_2, s_3) \tag{1}$$

is S_3-symmetric, and put $s_4 = -(s_1 + s_2 + s_3)$. Then

$$
\begin{aligned}
& B(s_1, s_2)B(s_3, s_4)B(s_1 + s_2, s_3 + s_4) && (2) \\
= \ & B(s_1, s_2)B(s_1 + s_2, s_4)B(s_3, -s_3) && \text{(by (iv))} \\
= \ & B(s_4, s_2)B(s_4 + s_2, s_1)B(s_3, -s_3) && (S_3\text{-symmetry of (1))} \\
= \ & B(s_4, s_2)B(s_3, s_1)B(s_4 + s_2, s_3 + s_1) && \text{(by (iv))}.
\end{aligned}
$$

Therefore, (2) is invariant under the transposition (1,4). Since (2) is obviously invariant under (3,4) and (1,2), it follows that (2) is S_4-invariant. Secondly, assume that (2) is S_4-symmetric. Then

$$
\begin{aligned}
& B(s_1, s_2)B(s_1 + s_2, s_3)B(s_4, -s_4) && (3) \\
= \ & B(s_1, s_2)B(s_3, s_4)B(s_1 + s_2, s_3 + s_4) && \text{(by (iv))} \\
= \ & B(s_3, s_2)B(s_1, s_4)B(s_2 + s_3, s_1 + s_4) && (S_4\text{-symmetry of (2))} \\
= \ & B(s_3, s_2)B(s_2 + s_3, s_1)B(s_4 - s_4) && \text{(by (iv))}.
\end{aligned}
$$

Therefore, (1) is S_3-symmetric. $\square$

2.5

Now we are going to prove Theorem 1 (the S_4-symmetry of (2)). Our proof will use a GT-action on a suitable subquotient of $\hat{B}_5^*$. First recall ([D][I-M]) that GT acts on the profinite completion $\hat{B}_n$ of the Artin braid group B_n for any $n \geq 4$, in the following manner;

$$\sigma(b_1) = b_1^\lambda, \quad \sigma(b_i) = f(p_i, b_i^2)^{-1} b_i^\lambda f(p_i, b_i^2) \ (2 \leq i \leq n-1). \tag{1}$$

Here, $\sigma = (\lambda, f) \in GT$, and $b_i, p_i \in \hat{B}_n$ are as in §1.1. This GT action on $\hat{B}_n$ and the trivial GT action on the symmetric group S_n are compatible with the projection $\hat{B}_n \to S_n$, and hence GT acts on its kernel which is $\hat{P}_n$. Each $\sigma = (\lambda, f) \in GT$ maps each of the standard generators p_{ij} of $\hat{P}_n$ to some $\hat{P}_n$-conjugate of p_{ij}^λ, because p_{ij} is $\hat{B}_n$-conjugate with $p_{12} = b_1^2$ and σ acts trivially on $\hat{B}_n/\hat{P}_n$. Moreover, we have $\sigma(p_i) = p_i^\lambda$ ($2 \leq i \leq n$) and $\sigma(w_n) = w_n^\lambda$ (cf. [I-M] A.3); hence the above GT actions on $\hat{B}_n, \hat{P}_n$ induce those on $\hat{B}_n^*, \hat{P}_n^*$. Its action on $\hat{P}_4^*$ exactly corresponds to the original GT action on $\hat{F}_2$, via the isomorphism

$$\hat{P}_4^* \xrightarrow{\sim} \hat{F}_2; \qquad p_{12}^* = p_{34}^* \longmapsto x, \tag{2}$$
$$p_{23}^* = p_{14}^* \longmapsto y,$$
$$p_{13}^* = p_{24}^* \longrightarrow z = (xy)^{-1}.$$

Now let N be the kernel of the canonical homomorphism $P_5^* \to P_4^*$ defined by $p_{i5}^* \to 1 (1 \leq i \leq 4), p_{kl}^* \to p_{kl}^* (1 \leq k < l \leq 4)$. It is the homomorphism defined by removing the 5-th string. Then N is generated by $t_i = p_{i5}^* (1 \leq i \leq 4)$ (not just normally but also) in the usual sense, because, as is well-known and easy to see, every P_5^*-conjugate of t_i is an N-conjugate of t_i. Moreover, N is free of rank 3 on t_1, t_2, t_3, and we have $t_1 t_2 t_3 t_4 = 1$. Since N is free, so that $\hat{N}$ is center-free, the induced homomorphism $\hat{N} \to \hat{P}_5^*$ is still injective ([A*] Proposition 3), and so we shall consider $\hat{N}$ as a (normal) subgroup of $\hat{P}_5^*$. We have an exact sequence

$$\begin{array}{ccccccccc} 1 & \to & \hat{N} & \to & \hat{P}_5^* & \to & \hat{P}_4^* & \to & 1 \\ & & \| \wr & & & & \| & & \\ & & \hat{F}_3 = \langle t_1, t_2, t_3 \rangle & & & & \hat{F}_2 & & \end{array} \tag{3}$$

of profinite groups, on which GT acts compatibly, because $\hat{N}$ is GT-stable and the standard embedding $\hat{P}_4^* \to \hat{P}_5^*$ which splits (3) is compatible with the GT-action by the formulae (1). Now define an open subgroup $H \subset \hat{B}_5^*$ as the unique intermediate group $\hat{P}_5^* \subset H \subset \hat{B}_5^*$ that corresponds to the stabilizer of the letter 5 in $S_5 = \hat{B}_5^*/\hat{P}_5^*$. Then $\hat{N}$ is normal in H, and hence H also acts on $\hat{N}$ by conjugation $n \to hnh^{-1} (h \in H, n \in \hat{N})$. We denote by $s : H \to S_4$ (on $\{1, 2, 3, 4\}$) the canonical projection. The key point of the proof is the study of the actions of GT and H on $\hat{N}'/\hat{N}''$, and to see how close to being "mutually commutative" these two actions are. (The quotient $\hat{P}_5^{*'}/\hat{P}_5^{*''}$ is useless for this purpose.) But first, we need:

Proposition 2.5.1 $\hat{N}^{ab} = \hat{N}/\hat{N}'$ *is a free abelian profinite group of rank 3 generated by the images* $\mathbf{t}_i$ *of* $t_i (1 \leq i \leq 4)$ *which satisfy a single relation* $\mathbf{t}_1 \mathbf{t}_2 \mathbf{t}_3 \mathbf{t}_4 = 1$. *The groups GT (resp. H) act on* $\hat{N}^{ab}$ *by* $\sigma(\mathbf{t}_i) = \mathbf{t}_i^{\lambda_\sigma}$ *(resp.* $h(\mathbf{t}_i) = \mathbf{t}_{s(h)i}$ *)* $(\sigma \in GT, h \in H, 1 \leq i \leq 4)$. *In particular, the normal subgroups* $GT_1 = \{\sigma \in GT; \lambda_\sigma = 1\}$ *(resp.* $\hat{P}_5^*$ *) of GT (resp. H) act trivially on* $\hat{N}^{ab}$.

Proof. The first statement is obvious. As for the second, since $\sigma(t_i) = g_i t_i^{\lambda_\sigma} g_i^{-1}$ with some $g_i \in \hat{P}_5^*$ for each $i (1 \leq i \leq 4)$, and $g_i t_i g_i^{-1}$ is $\hat{N}$-conjugate with t_i (as noted above), $\sigma(t_i)$ is an $\hat{N}$-conjugate of $t_i^{\lambda_\sigma}$; hence σ acts on $\hat{N}^{ab}$ as the λ_σ-th power map. On the other hand, $ht_i h^{-1}(h \in H)$ is an $\hat{N}$-conjugate of $t_{s(h)i}$. The rest is also obvious. $\square$

Now let us introduce the completed group algebra $\hat{\mathbb{Z}}[[\hat{N}^{ab}]]$ of $\hat{N}^{ab}$. Since $\hat{N}^{ab}$ is a free abelian profinite group of rank 3 generated by $\mathbf{t}_i (1 \leq i \leq 3)$, where $\mathbf{t}_i (1 \leq i \leq 4)$ denotes the image of $t_i \in \hat{N}$ on $\hat{N}^{ab}$, we may identify $\hat{\mathbb{Z}}[[\hat{N}^{ab}]]$ with $A_3 = \hat{\mathbb{Z}}[[\hat{\mathbb{Z}}^3]]$, *via* the basis $\mathbf{t}_1, \mathbf{t}_2, \mathbf{t}_3$ of $\hat{N}^{ab}$, and call it A_3. Thus, $A_3 = \hat{\mathbb{Z}}[[\hat{N}^{ab}]]$ and $\mathbf{t}_1 \mathbf{t}_2 \mathbf{t}_3 \mathbf{t}_4 = 1$. Using $\mathbf{t}_i (1 \leq i \leq 4)$, we can embed A_3 into the ring

$$\text{Fct}(\{(s_1, s_2, s_3, s_4) \in (\mathbb{Q}/\mathbb{Z})^4; \sum_{i=1}^{4} s_i = 0\}, \hat{\mathbb{Z}} \otimes \mathbb{Q}_{ab}) \qquad (4)$$

of functions as described in §1.5. Let A_3^0 be the total ring of fractions of A_3, obtained from A_3 by allowing all non zero-divisors as denominators. By Proposition 2.1.1, $1 - \mathbf{t}(\mathbf{t} \in \hat{N}^{ab})$ is a non zero-divisor of A_3 if $\mathbf{t}$ can serve as one of the free generators of $\hat{N}^{ab}$; for example, $\mathbf{t} = \mathbf{t}_i, \mathbf{t}_i \mathbf{t}_j (i \neq j)$, etc. The actions of $\sigma = (\lambda, f) \in GT$ and of $h \in H$ on $\hat{N}^{ab}$ extend naturally to continuous ring automorphisms of A_3 and A_3^0. The σ-action on A_3^0 is $\mathbf{t}_i \rightarrow \mathbf{t}_i^{\lambda} (1 \leq i \leq 4)$, which will be denoted by (λ). The h-action on A_3^0 is the substitution $\mathbf{t}_i \rightarrow \mathbf{t}_{s(h)i}$, denoted also by $s(h)$. In terms of functions on $(\mathbb{Q}/\mathbb{Z})^3$, the first action is the multiplication of λ on $(s_1, \cdots, s_4)$, and the second is the corresponding substitution of indices of $s_1, \cdots, s_4$.

Now let us turn our attention to the next commutator quotient $\hat{N}'^{ab} = \hat{N}'/\hat{N}''$. By the conjugation $x \rightarrow fxf^{-1}(x \in \hat{N}', f \in \hat{N}), \hat{N}'^{ab}$ is considered as an $\hat{N}^{ab}$-module, and hence also as a module over $A_3 = \hat{\mathbb{Z}}[[\hat{N}^{ab}]]$ in the natural way. By the profinite Blanchfield-Lyndon theorem (Th A-2), we can (and shall) identify $\hat{N}'^{ab}$ with the A_3-module

$$\mathcal{I}_3 = \{(\alpha_1, \alpha_2, \alpha_3) \in A_3^{\oplus 3}; \sum_{i=1}^{3} \alpha_i(\mathbf{t}_i - 1) = 0\}, \qquad (5)$$

via

$$\hat{N}' \ni n \longrightarrow ((\frac{\partial n}{\partial t_1})^{ab}, (\frac{\partial n}{\partial t_2})^{ab}, (\frac{\partial n}{\partial t_3})^{ab})) \in \mathcal{I}_3, \tag{6}$$

where

$$\Lambda_3 = \hat{\mathbb{Z}}[[\hat{N}]] \overset{ab}{\to} \hat{\mathbb{Z}}[[\hat{N}^{ab}]] = A_3. \tag{7}$$

is the projection induced from the abelianization of $\hat{N}$.

Put

$$V = \hat{N}'^{ab} \otimes_{A_3} A_3^0 = \mathcal{I}_3 \otimes_{A_3} A_3^0.$$

Proposition 2.5.2 *V is a free A_3^0-module of rank 2 with basis*

$$\mathbf{e} = (t_4, t_3) \pmod{\hat{N}''}, \quad \mathbf{e}' = (t_2, t_1) \pmod{\hat{N}''}.$$

Proof. We shall compute the "$\mathcal{I}_3$-coordinates" of $\mathbf{e}, \mathbf{e}'$. First, $(t_4, t_3) = ((t_1 t_2 t_3)^{-1}, t_3) = t_3^{-1} t_2^{-1} t_1^{-1} t_3 t_1 t_2$; hence

$$\begin{cases} \frac{\partial(t_4, t_3)}{\partial t_1} = -t_3^{-1} t_2^{-1} t_1^{-1} + t_3^{-1} t_2^{-1} t_1^{-1} t_3 \overset{ab}{\to} -t_4 + t_3 t_4, \\ \frac{\partial(t_4, t_3)}{\partial t_2} = -t_3^{-1} t_2^{-1} + t_3^{-1} t_2^{-1} t_1^{-1} t_3 t_1 \overset{ab}{\to} -t_1 t_4 + t_2^{-1}, \\ \frac{\partial(t_4, t_3)}{\partial t_3} = -t_3^{-1} + t_3^{-1} t_2^{-1} t_1^{-1} \overset{ab}{\to} -t_3^{-1} + t_4; \end{cases} \tag{8}$$

hence $\mathbf{e}$ has the coordinates

$$\mathbf{e} = \mathbf{t_4}(\mathbf{t_3} - 1, \mathbf{t_1}(\mathbf{t_3} - 1), 1 - \mathbf{t_1}\mathbf{t_2}). \tag{9}$$

As for $\mathbf{e}'$, the computation is even easier;

$$\mathbf{e}' = (\mathbf{t_2} - 1, 1 - \mathbf{t_1}, 0). \tag{10}$$

Since $1 - \mathbf{t_1}\mathbf{t_2}, 1 - \mathbf{t_1}$ are non zero divisors of A_3 and hence are units of A_3^0, it is now clear that $\{\mathbf{e}, \mathbf{e}'\}$ is a free A_3^0-basis of V. $\quad\square$

Now consider the special elements $b_i^*(1 \leq i \leq 3)$ of B_5^* (§1.1). Since $s(b_i^*)$ is the transposition $(i, i+1)$ and $i \leq 3$, each such b_i^* belongs to H.

Proposition 2.5.3 (i) *The groups GT and H act on the A_3^0-module V semi-linearly, i.e., $\sigma(av) = a^{(\lambda_\sigma)}\sigma(v), h(av) = a^{(s(h))}h(v)(\sigma \in GT, h \in H, a \in A_3^0, v \in V)$. In particular, GT_1 and $\hat{P}_5^*$ act on V A_3^0-linearly.* (ii) *For any $\sigma \in GT$ and $1 \leq i \leq 3$, there exists some $\delta \in \hat{P}_5^{*'}$ such that*

$$\sigma \circ b_i^* = \delta \circ (b_i^*)^{\lambda_\sigma} \circ \sigma \tag{11}$$

on V.

Proof. (i) is obvious, and so is (ii) if we put $\delta = \sigma(b_i^*)(b_i^*)^{-\lambda_\sigma}$. Note that $\sigma(b_i^*)$ is a $(\hat{P}_5^*)'$-conjugate of $(b_i^*)^{\lambda_\sigma}$ (as $f_\sigma \in \hat{F}_2'$) and hence that $\delta \in (\hat{P}_5^*)'$. $\quad\square$

Now we shall compute the actions of σ and of b_2^* on the basis $\mathbf{e}, \mathbf{e}'$ of V. First, we need

Proposition 2.5.4 *For each $\sigma = (\lambda, f) \in GT$ and $1 \le i \le 4$, we have*

$$\sigma(t_i) = n_i t_i^\lambda n_i^{-1}, \tag{12}$$

with $n_i \in \hat{N}(1 \le i \le 4)$ given by

$$n_4 = 1, \qquad n_3 = t_4^m f(t_4, t_3)^{-1}, \qquad n_2 = f(t_1 t_2, t_4)(t_1 t_2)^{-m} f(t_2, t_1),$$
$$n_1 = n_2 \cdot t_2^m f(t_2, t_1)^{-1}, \tag{13}$$

where $m = \frac{1}{2}(\lambda - 1)$.

Proof. By almost straightforward computations using the following basic formulae, where $\beta_i = b_i^*(1 \le i \le 4), \pi_i = p_i^*(2 \le i \le 5)$.

$$\sigma(\beta_1) = \beta_1^\lambda, \qquad \sigma(\beta_i) = f(\pi_i, \beta_i^2)^{-1} \beta_i^\lambda f(\pi_i, \beta_i^2) \quad (2 \le i \le 4), \tag{14}$$
$$t_4 = \beta_4^2, \qquad t_i = (\beta_4 \cdots \beta_{i+1}) \beta_i^2 (\beta_4 \cdots \beta_{i+1})^{-1} \quad (1 \le i \le 3), \tag{15}$$

and the obvious sphere braid commutation relations:

$$\begin{cases} \pi_4 = t_4^{-1}, \\ \beta_4 \pi_3 \beta_4^{-1} = \pi_3, \ (\beta_4 \beta_3) \pi_3 (\beta_4 \beta_3)^{-1} = t_1 t_2, \\ (\beta_4 \beta_3) \pi_2 (\beta_4 \beta_3)^{-1} = \pi_2, \ (\beta_4 \beta_3 \beta_2) \pi_2 (\beta_4 \beta_3 \beta_2)^{-1} = t_1. \end{cases} \tag{16}$$

The following remarks are also used. Since $f \in \hat{F}_2'$, we have $f(\xi, \eta) = 1$ if ξ commutes with η, and also $f(\xi\zeta, \eta) = f(\xi, \zeta\eta) = f(\xi, \eta)$ holds if ζ commutes with both ξ and η (see [I-M] Appendix A2). For example, $f(\pi_3, t_3) = f(p_{12}^* \pi_3, t_3) = f(t_4, t_3)$ (note that $p_{12}^* \pi_3 = p_{12}^* p_3^* = p_{45}^* = t_4$, because $w_5^* = p_5^* = 1$ on B_5^*). Finally, the 3-cycle relation for f (applied to $x = t_1 t_2, y = t_3, z = t_4$) is used in order to obtain the formula for n_2.   □

Proposition 2.5.5 *The action of $\sigma = (\lambda, f) \in GT$ on the A_3^0-basis e, e' of V is given by*

$$\sigma(e, e') = (e, e')S, \tag{17}$$

where $S = \begin{pmatrix} \alpha & 0 \\ 0 & \alpha' \end{pmatrix}$ with

$$\begin{cases} \alpha = t_4^m \dfrac{t_4^\lambda - 1}{t_4 - 1} \dfrac{t_3^\lambda - 1}{t_3 - 1} B_\sigma\{t_4, t_3\}, \\ \alpha' = t_1^{-m} \dfrac{t_2^\lambda - 1}{t_2 - 1} \dfrac{t_1^\lambda - 1}{t_1 - 1} B_\sigma\{t_2, t_1\}. \end{cases} \tag{18}$$

Here, $B_\sigma\{t_i, t_j\}$ denotes the image of $B_\sigma \in A_2$ under the continuous ring homomorphism $A_2 \to A_3$ defined by $x \to t_i, y \to t_j$.

Proof. By Proposition 2.5.4, σ leaves the subgroup $\langle t_3, t_4 \rangle$ stable, and maps $\langle t_1, t_2 \rangle$ to $n_2 \langle t_1, t_2 \rangle n_2^{-1}$. Let us compare these σ-actions with the original σ-action on $\hat{F}_2 = \langle x, y \rangle$ defined by $x \to x^\lambda, y \to f^{-1} y^\lambda f$, via the isomorphisms $\langle t_4, t_3 \rangle \xrightarrow{\sim} \hat{F}_2$ ($t_4 \to x, t_3 \to y$) and $\langle t_2, t_1 \rangle \xrightarrow{\sim} \hat{F}_2$ ($t_2 \to x, t_1 \to y$). Then, by the above formulae for the n_i's, we see that the action of σ on $\langle t_4, t_3 \rangle$ is the same as the original one except that here it is followed by t_4^m-conjugation. Also, the action of σ on $\langle t_2, t_1 \rangle$ is the original action followed by $(n_2 t_2^m)$-conjugation. Note that the image of $n_2 t_2^m$ on $\hat{N}^{ab}$ is $\mathbf{t}_1^{-m}$. So, our assertion follows directly from the definition of B_σ. $\qquad \square$

Remark Thus, $\mathbf{e}, \mathbf{e}'$ are eigenvectors of the A_3^0-linear action of GT_1 on V. This depends on our appropriate choice of pairings $\mathbf{e} \equiv (t_4, t_3), \mathbf{e}' \equiv (t_2, t_1) \bmod \hat{F}_2''$ of indices(!).

Proposition 2.5.6 *Let* $\tau = b_2^* \in H$. *Then*

$$\tau(\mathbf{e}, \mathbf{e}') = (\mathbf{e}, \mathbf{e}')T, \tag{19}$$

with

$$T = \mathbf{t}_3^{-1}(1 - \mathbf{t}_1 \mathbf{t}_2)^{-1} \cdot \begin{pmatrix} \mathbf{t}_2 - 1, & \mathbf{t}_1 \mathbf{t}_2 \mathbf{t}_3^2 (1 - \mathbf{t}_1) \\ \mathbf{t}_4 - 1, & \mathbf{t}_1 \mathbf{t}_3 (1 - \mathbf{t}_3) \end{pmatrix}. \tag{20}$$

Proof. Since $\tau(t_4) = t_4$, $\tau(t_3) = t_3^{-1} t_2 t_3$, $\tau(t_2) = t_3$ and $\tau(t_1) = t_1$, we have $\tau(t_4, t_3) = (t_4, t_3^{-1} t_2 t_3)$, $\tau(t_2, t_1) = (t_3, t_1)$. We can compute $\tau(\mathbf{e}), \tau(\mathbf{e}')$ by writing down the abelian images of $\frac{\partial}{\partial t_i}$ ($1 \le i \le 3$) of these commutators. First, since

$$\tau(t_4, t_3) = t_3^{-1}(t_3 t_4 t_3^{-1}, t_2)t_3 = t_3^{-1}(t_2^{-1} t_1^{-1} t_3^{-1}, t_2)t_3, \tag{21}$$

$t_3 \cdot \tau(\mathbf{e})$ is the abelian image of $(\frac{\partial}{\partial t_i}(t_2^{-1} t_1^{-1} t_3^{-1} t_2 t_3 t_1))_{1 \le i \le 3}$, which is equal to $(-t_1^{-1} t_2^{-1} + t_1^{-1}, -t_2^{-1} + t_4, -t_4 + t_1^{-1} t_3^{-1})$. Hence $\tau(\mathbf{e})$ is

$$\mathbf{t}_4(\mathbf{t}_2 - 1, -\mathbf{t}_1 + \mathbf{t}_3^{-1}, -\mathbf{t}_3^{-1} + \mathbf{t}_2 \mathbf{t}_3^{-1}). \tag{22}$$

Similarly, $\tau(\mathbf{e}')$ is the abelian image of $(\frac{\partial}{\partial t_1}(t_3, t_1), 0, \frac{\partial}{\partial t_3}(t_3, t_1))$, i.e.,

$$(\mathbf{t}_3 - 1, 0, 1 - \mathbf{t}_1). \tag{23}$$

Our formula follows by comparing these with the coordinates of $\mathbf{e}, \mathbf{e}'$ given by (9), (10). $\qquad \square$

Now, to complete the proof of Theorem 1, apply $\sigma = (\lambda, f) \in GT$ on both sides of (19), which gives

$$(\sigma \tau(\mathbf{e}), \sigma \tau(\mathbf{e}')) = (\sigma(\mathbf{e}), \sigma(\mathbf{e}'))T^{(\lambda)} = (\mathbf{e}, \mathbf{e}')ST^{(\lambda)}. \tag{24}$$

But since $\sigma\tau = \delta\tau^\lambda\sigma$ with some $\delta \in \hat{P}_5^{*'}$ (Proposition 2.5.3), and since $\tau^2 = p_{23}^* \in \hat{P}_5^*$ acts A_3^0-linearly on V as

$$\tau^2(e, e') = (e, e')T \cdot T^{(2,3)},$$

where the superscript $(2,3)$ indicates the action of the transposition $(2,3)$ on the indices of t_j's, we see that (24) is also equal to

$$(\delta\tau^{2m})\tau((e, e')S) = \delta(\tau^{2m}(e, e')T \cdot S^{(2,3)}) = (e, e')D(T \cdot T^{(2,3)})^m T S^{(2,3)},$$
$$(24)'$$

where $D \in M_2(A_3^0)$ is defined by $\delta(e, e') = (e, e')D$. But since $\hat{P}_5^*$ acts A_3^0-linearly on V, and since $\delta \in \hat{P}_5^{*'}$, we have

$$\det(D) = 1. \tag{25}$$

Now by $(24) < (24)'$ (since e, e' form an A_3^0-basis of V), we obtain

$$ST^{(\lambda)} = D(TT^{(2,3)})^m T S^{(2,3)}; \tag{26}$$

hence by $(25)(26)$,

$$\det S = \det(T)\det(T^{-1})^{(\lambda)} \cdot \det(TT^{(2,3)})^m \det(S)^{(2,3)}. \tag{27}$$

But since

$$\det T = -t_3^{-1}(1 - t_1 t_2)^{-1}(1 - t_1 t_3), \tag{28}$$

we obtain

$$\det(TT^{(2,3)}) = (t_2 t_3)^{-1}, \tag{29}$$

and

$$\det(S) = \frac{1 - t_1^\lambda t_2^\lambda}{1 - t_1 t_2} \frac{1 - t_1 t_3}{1 - t_1^\lambda t_3^\lambda} t_2^{-m} t_3^m \det(S)^{(2,3)}. \tag{30}$$

Therefore, by Proposition 2.5.5

$$B_\sigma\{t_2, t_1\}B_\sigma\{t_4, t_3\} = t_2^{-m} t_3^m \frac{1 - t_1^\lambda t_2^\lambda}{1 - t_1 t_2} \frac{1 - t_1 t_3}{1 - t_1^\lambda t_3^\lambda} B_\sigma\{t_3, t_1\}B_\sigma\{t_4, t_2\},$$
$$(31)$$

or equivalently, that

$$\lambda \cdot (t_1 t_2)^m \frac{1 - t_1 t_2}{1 - t_1^\lambda t_2^\lambda} B_\sigma\{t_4, t_3\}B_\sigma\{t_2, t_1\} \tag{32}$$

is invariant under the transposition $(2,3)$. But this implies the S_4-invariance of (32), and hence (in view of the formula §2.3 (7)) the second formulation of Theorem 1. This completes the proof of Theorem 1.

2.6

Proof of Theorem 2. We keep our notation of §1.8. Let σ be any element of GT. By using Proposition 1.8.2 and Theorem A-2, we shall relate $\kappa_{n,r}^{(i)}(\sigma)$ with the derivative of B_σ. First, by Theorem A-2 applied to $\mathcal{N} = H_n$,

$$H_n \ni h \longrightarrow \left(\left(\frac{\partial h}{\partial x} \right)_{y=x^n=1}, \left(\frac{\partial h}{\partial y} \right)_{y=x^n=1} \right) \in (A_1/(\mathbf{x}^n - 1))^{\oplus 2} \quad (1)$$

induces an $(A_1/(\mathbf{x}^n - 1))$-module isomorphism

$$H_n^{ab} \xrightarrow{\sim} \{(\alpha, \beta) \in (A_1/(\mathbf{x}^n - 1))^{\oplus 2}; \alpha(\mathbf{x} - 1) = 0\}. \quad (2)$$

Note that (2) maps the images on H_n^{ab} of $x^n, x^{-j}yx^j$ $(0 \le j < n)$ to

$$\left(\frac{\mathbf{x}^n - 1}{\mathbf{x} - 1}, 0 \right), (0, \mathbf{x}^{-j}),$$

respectively. On the other hand, the image of f_σ is mapped to the specialization of $((f_\sigma)_x^{ab}, (f_\sigma)_y^{ab})$ over $y \to 1, x^n \to 1$. But since

$$B_\sigma = \psi_\sigma^{ab} = 1 - (f_\sigma)_x^{ab}(\mathbf{x} - 1) = 1 + (f_\sigma)_y^{ab}(\mathbf{y} - 1) \quad (3)$$

(§2.2), this specialization has the *second* coordinate equal to

$$\left(\frac{B_\sigma - 1}{\mathbf{y} - 1} \right)_{y=1, \mathbf{x}^n=1}.$$

This means that if we temporarily write

$$\left(\frac{B_\sigma - 1}{\mathbf{y} - 1} \right)_{y=1, \mathbf{x}^n=1} = \sum_{j(\bmod n)} a_j \mathbf{x}^{-j} \quad (a_j \in \hat{\mathbb{Z}}), \quad (4)$$

then the image of f_σ on H_n^{ab} is given by

$$x^{nb} \cdot \prod_{j(\bmod n)} (x^{-j}yx^j)^{a_j}$$

with some $b \in \hat{\mathbb{Z}}$; hence the homomorphism $H_n \to \mu_\infty$ of Proposition 1.8.2 maps f_σ to $\exp(2\pi\sqrt{-1}ra_i)$. Therefore, $\kappa_{n,r}^{(i)}(\sigma) = \exp(2\pi\sqrt{-1}ra_i)$; hence in particular,

$$\kappa_{n,1/m}^{(i)}(\sigma) = \zeta_m^{a_i}. \quad (5)$$

By Proposition 1.8.3, we obtain $a_i = \Psi_n^{(i)}(\sigma)$. Therefore,

$$\left(\frac{B_\sigma - 1}{\mathbf{y} - 1} \right)_{y=1} \equiv \sum_{i(\bmod n)} \Psi_n^{(i)}(\sigma)\mathbf{x}^{-i} \quad \bmod (\mathbf{x}^n - 1) \quad (6)$$

for all $n \in \mathbb{N}$; hence

$$\left(\frac{B_\sigma - 1}{y - 1}\right)_{y=1} = \Psi_\sigma. \tag{7}$$

Now we compute the left hand side of (7). With the notation of §1.8, the equality $B_\sigma(s_1, s_2) = \Gamma_\sigma(s_1)\Gamma_\sigma(s_2)\Gamma_\sigma(s_1 + s_2)^{-1}$ can be expressed as $B_\sigma = \Gamma_\sigma\{\mathbf{x}\}\Gamma_\sigma\{\mathbf{y}\}\Gamma_\sigma\{\mathbf{xy}\}^{-1}$. Therefore,

$$
\begin{aligned}
\left(\frac{B_\sigma - 1}{y - 1}\right)_{y=1} &= \left(\frac{\Gamma_\sigma\{\mathbf{x}\}\Gamma_\sigma\{\mathbf{y}\} - \Gamma_\sigma\{\mathbf{xy}\}}{\Gamma_\sigma\{\mathbf{xy}\}(\mathbf{y} - 1)}\right)_{y=1} \\[2mm]
&= \left(\frac{\Gamma_\sigma\{\mathbf{x}\}(\Gamma_\sigma\{\mathbf{y}\} - 1)}{\Gamma_\sigma\{\mathbf{xy}\}(\mathbf{y} - 1)}\right)_{y=1} \\[2mm]
&\quad - \left(\frac{\Gamma_\sigma\{\mathbf{xy}\} - \Gamma_\sigma\{\mathbf{x}\}}{\Gamma_\sigma\{\mathbf{xy}\}(\mathbf{y} - 1)}\right)_{y=1} \\[2mm]
&= -(D\log\Gamma_\sigma)_{\mathbf{x}=1} + D\log\Gamma_\sigma.
\end{aligned}
\tag{8}
$$

Our theorem follows directly from (7)(8).   □

3 Appendix

Profinite free differential calculus and profinite Blanchfield-Lyndon theorem.

3.1

In the main text, we made use of the profinite versions of two classical theorems on free groups, a theorem of Fox and that of Blanchfield-Lyndon. In this Appendix, we shall provide their general formulations and proofs. Their "almost pro-l" versions have been treated in [I$_2$]. The profinite version of a theorem of Fox (Theorem A-1 below) was, as far as the author knows, first proved by G. W. Anderson (a letter to the author [A$_1$], unpublished) using intersections of cycles on finite branched coverings of the complex projective line. Here, we shall give simple algebraic proofs for this, and also for the profinite Blanchfield-Lyndon theorem (Theorem A-2).

3.2

Let $\mathcal{F} = \mathcal{F}_r$ be a free profinite group of finite rank $r \geq 1$, and $x_1, \cdots, x_r$ be a set of free generators. We denote by $\Lambda = \Lambda_r = \hat{\mathbb{Z}}[[\mathcal{F}]]$ the completed group algebra of $\mathcal{F}$. It is a profinite ring obtained as the projective limit of finite group rings $(\mathbb{Z}/n)[\mathcal{F}/\mathcal{N}]$, where n (resp. $\mathcal{N}$) runs over all positive integers (resp. open normal subgroups of $\mathcal{F}$). Let $s : \Lambda \to \hat{\mathbb{Z}}$ be the augmentation homomorphism which extends the homomorphisms $(\mathbb{Z}/n)[\mathcal{F}/\mathcal{N}] \to \mathbb{Z}/n$ defined by "taking the sum of coefficients".

Theorem A-1 (G. W. Anderson). *Every element λ of Λ can be expressed uniquely in the form*

$$\lambda = s(\lambda) \cdot 1 + \sum_{i=1}^{r} \lambda_i(x_i - 1) \tag{1}$$

with $\lambda_1, \cdots, \lambda_r \in \Lambda$.

This was communicated to the author by G. W. Anderson while we were working on the joint papers [A-I], trying to loosen our pro-l assumption made there. His proof [A$_1$] is closely related to a topological method used there. But here, we shall provide a straightforward algebraic proof.

For each $i(1 \le i \le r)$, assuming Theorem A-1, define the map

$$\frac{\partial}{\partial x_i} : \Lambda \to \Lambda \quad \text{by} \quad \frac{\partial \lambda}{\partial x_i} = \lambda_i. \tag{2}$$

The following properties of $\frac{\partial}{\partial x_i}$ $(1 \le i \le r)$, which follow directly from the definition, are used in the main text.

(i) $\frac{\partial}{\partial x_i}$ *is continuous, $\hat{\mathbb{Z}}$-linear, and* $\frac{\partial(1)}{\partial x_i} = 0$.

(ii) $\frac{\partial x_j}{\partial x_i} = \delta_{ij}$ $(1 \le i, j \le r)$.

(iii) $\frac{\partial(\lambda\lambda')}{\partial x_i} = \frac{\partial \lambda}{\partial x_i} s(\lambda') + \lambda \frac{\partial \lambda'}{\partial x_i}$ $(\lambda, \lambda' \in \Lambda)$; *in particular,*
$\frac{\partial(f^{-1})}{\partial x_i} = -f^{-1}\frac{\partial f}{\partial x_i}$ $(f \in \mathcal{F})$.

(iv) $(f-1)\frac{\partial(f^\alpha)}{\partial x_i} = (f^\alpha - 1)\frac{\partial f}{\partial x_i}$ *for any $f \in \mathcal{F}, \alpha \in \hat{\mathbb{Z}}$.*

(v) *For any $\lambda \in \Lambda$ and $\varphi_1, \cdots, \varphi_r \in \mathcal{F}$, we denote by $\lambda(\varphi_1, \cdots, \varphi_r)$ the image of λ under the endomorphism $\Lambda \to \Lambda$ induced from the homomorphism $\mathcal{F} \to \mathcal{F}$ defined by $x_i \to \varphi_i$ $(1 \le i \le r)$. Then*

$$\frac{\partial}{\partial x_i}(\lambda(\varphi_1, \cdots, \varphi_r)) = \sum_{j=1}^{r}(\frac{\partial \lambda}{\partial x_j})(\varphi_1, \cdots, \varphi_r) \cdot \frac{\partial \varphi_j}{\partial x_i}.$$

Theorem A-2 *Let $\mathcal{N}$ be any closed normal subgroup of the free profinite group $\mathcal{F}$ of rank r on $x_1, \cdots, x_r$. Put $G = \mathcal{F}/\mathcal{N}$, and let $\Lambda = \hat{\mathbb{Z}}[[\mathcal{F}]]$ (resp. $A = \hat{\mathbb{Z}}[[G]]$) be the completed group algebras of $\mathcal{F}$ (resp. G). Denote by $\pi : \Lambda \to A$ the canonical homomorphism. Then the mapping*

$$\mathcal{N} \ni n \to (\pi(\frac{\partial n}{\partial x_1}), \cdots, \pi(\frac{\partial n}{\partial x_r})) \in A^{\oplus r} \tag{3}$$

induces a left A-module isomorphism

$$\mathcal{N}^{ab} \xrightarrow{\sim} \{(a_1, \cdots, a_r) \in A^{\oplus r}; \ \sum_{i=1}^{r} a_i(\pi(x_i) - 1) = 0\}, \qquad (4)$$

where $\mathcal{N}^{ab} = \mathcal{N}/(\mathcal{N}, \mathcal{N})$ is the abelianization of $\mathcal{N}$, whose left A-module structure is the one determined by the G-conjugation on $\mathcal{N}^{ab}$ induced from the $\mathcal{F}$-conjugation

$$n \to fnf^{-1} \quad (n \in \mathcal{N}, f \in \mathcal{F}) \qquad (5)$$

on $\mathcal{N}$.

3.3   Proof of Theorem A-1

Let $F = F_r$ be the free group of rank r on $x_1, \cdots, x_r$ and $\mathcal{F} = \mathcal{F}_r$ be its profinite completion. Let $\mathbb{Z}[F]$ be the group ring of F over $\mathbb{Z}$, and $s : \mathbb{Z}[F] \to \mathbb{Z}$ be the augmentation homomorphism. Then, as is well-known, each element θ of $\mathbb{Z}[F]$ decomposes uniquely as

$$\theta = s(\theta) \cdot 1 + \sum_{i=1}^{r} \theta_i(x_i - 1) \qquad (1)$$

with $\theta_1, \cdots, \theta_r \in \mathbb{Z}[F]$, and the Fox free differential operators $\frac{\partial}{\partial x_i} : \mathbb{Z}[F] \to \mathbb{Z}[F]$ are defined by $\frac{\partial \theta}{\partial x_i} = \theta_i$ $(1 \le i \le r)$. The following properties of $\frac{\partial}{\partial x_i}$ $(1 \le i \le r)$ follow directly from this definition.

   (i) $\frac{\partial}{\partial x_i}$ is $\mathbb{Z}$-linear, and $\frac{\partial(1)}{\partial x_i} = 0$.

   (ii) $\frac{\partial x_j}{\partial x_i} = \delta_{ij}$ $(1 \le j \le r)$.

   (iii) $\frac{\partial(\theta\theta')}{\partial x_i} = \frac{\partial\theta}{\partial x_i} s(\theta') + \theta \frac{\partial\theta'}{\partial x_i}$ $(\theta, \theta' \in \mathbb{Z}[F])$.

   Key lemma   *Fix i $(1 \le i \le r)$, let n be any positive integer, and N be any normal subgroup of F with finite index. Then there is a unique additive homomorphism $\partial_i = \partial_{i,n,N}$ which makes the following diagram commutative. Here, the vertical arrows p_1, p_2 are canonical homomorphisms.*

$$
\begin{array}{ccc}
\mathbb{Z}[F] & \xrightarrow{\frac{\partial}{\partial x_i}} & \mathbb{Z}[F] \\
{\scriptstyle p_1}\downarrow & & {\scriptstyle p_2}\downarrow \\
(\mathbb{Z}/n)[F/(N,N)N^n] & \xrightarrow{\partial_{i,n,N}} & (\mathbb{Z}/n)[F/N].
\end{array}
\qquad (2)
$$

In short, $\frac{\partial}{\partial x_i}$ composed with the projection $\mathbb{Z}[F] \to (\mathbb{Z}/n)[F/N]$ necessarily factors through $(\mathbb{Z}/n)[F/(N,N)N^n]$.

Remark   Note that N is finitely generated; hence $N/(N,N)N^n$ is finite, and hence $F/(N,N)N^n$ is also finite.

Proof. Put $\varphi_i = p_2 \circ \frac{\partial}{\partial x_i}$. Then

$$\varphi_i(\theta\theta') = \varphi_i(\theta)s(\theta') + p_2(\theta)\varphi_i(\theta') \quad (\theta, \theta' \in \mathbb{Z}[F]). \tag{3}$$

Consider the restriction of $\varphi_i : \mathbb{Z}[F] \to (\mathbb{Z}/n)[F/N]$ to the multiplicative group $N \subset \mathbb{Z}[F]$. Then, for any $n, n' \in N$, since $s(n') = p_2(n) = 1$, (3) gives $\varphi_i(nn') = \varphi_i(n) + \varphi_i(n')$; hence $\varphi_i|N$ is a homomorphism from N into the additive group $(\mathbb{Z}/n)[F/N]$ which is abelian and killed by n. Therefore, $\varphi_i|N$ factors through $N/(N,N)N^n$. Now consider the restriction of φ_i to F;

$$\varphi_i|_F : F \to (\mathbb{Z}/n)[F/N], \tag{4}$$

and consider $(\mathbb{Z}/n)[F/N]$ as a left F-module. Then $\varphi_i|_F$ is a 1-cocycle (by (3)), and factors through $F/(N,N)N^n$. Extend this by linearity to

$$\partial_i = \partial_{i,n,N} : (\mathbb{Z}/n)[F/(N,N)N^n] \to (\mathbb{Z}/n)[F/N]. \tag{5}$$

Then, by construction, $p_2 \circ \frac{\partial}{\partial x_i}$ and $\partial_i \circ p_1$ coincide on F, and by linearity, also on $\mathbb{Z}[F]$. The uniqueness statement is obvious by the surjectivity of p_1. $\square$

It is obvious that

$$\partial_i(\theta\theta') = \partial_i(\theta)s(\theta') + p(\theta)\partial_i(\theta') \tag{6}$$

$(\theta, \theta' \in (\mathbb{Z}/n)[F/(N,N)N^n])$, where $p(\theta)$ is the image of θ on $(\mathbb{Z}/n)[F/N]$. By taking the projective limit of $\partial_i = \partial_{i,n,N}$ for all n, N, we obtain a continuous additive map

$$\hat{\partial}_i : \hat{\mathbb{Z}}[[\mathcal{F}]] \to \hat{\mathbb{Z}}[[\mathcal{F}]] \tag{7}$$

satisfying $\hat{\partial}_i(\lambda\lambda') = \hat{\partial}_i(\lambda)s(\lambda') + \lambda\hat{\partial}_i(\lambda')$ $(\lambda, \lambda' \in \Lambda), \hat{\partial}_i(1) = 0, \hat{\partial}_i(x_j) = \delta_{ij}$ $(1 \leq j \leq r)$. Note that $\hat{\partial}_i = \frac{\partial}{\partial x_i}$ on $\mathbb{Z}[F] \subset \hat{\mathbb{Z}}[[\mathcal{F}]]$.

Now, to complete the proof of Theorem A-1, consider the continuous map $\varphi : \hat{\mathbb{Z}}[[\mathcal{F}]] \to \hat{\mathbb{Z}}[[\mathcal{F}]]]$ defined by

$$\varphi(\lambda) = \lambda - (s(\lambda) + \sum_{i=1}^{r} \hat{\partial}_i(\lambda)(x_i - 1)). \tag{8}$$

Then it is continuous and $\varphi(\lambda) = 0$ on the dense subset $\mathbb{Z}[F]$ of $\hat{\mathbb{Z}}[[\mathcal{F}]]$. Therefore, $\varphi = 0$ everywhere. Therefore,

$$\lambda = s(\lambda) + \sum_{i=1}^{r} \lambda_i(x_i - 1) \tag{9}$$

holds for any $\lambda \in \Lambda$ if one puts $\lambda_i = \hat{\partial}_i(\lambda)$. Conversely, if (9) holds for some $\lambda_1, \cdots, \lambda_r \in \Lambda$, then by applying $\hat{\partial}_j$ on both sides of (9) we obtain $\hat{\partial}_j(\lambda) = \lambda_j$ $(1 \leq j \leq r)$. This settles the proof of Theorem A-1.

3.4   Proof of Theorem A-2

First, we shall prove the following

Lemma *Let $I = \mathrm{Ker}(s)$ be the augmentation ideal of $\Lambda = \hat{\mathbb{Z}}[[\mathcal{F}]]$, and f be any element of $\mathcal{F}$. Then $f - 1 \in I^2$ if and only if $f \in \mathcal{F}' = (\mathcal{F}, \mathcal{F})$.*

Proof. The "if" implication is obvious. To prove the "only if" implication, decompose f as $f = x_1^{a_1} \cdots x_r^{a_r} g$, with $a_i \in \hat{\mathbb{Z}}$ $(1 \le i \le r)$ and $g \in \mathcal{F}'$, so that $g - 1 \in I^2$. Since elements of I^2 can be approximated by some linear combinations of elements of the form ii' $(i, i' \in I)$, and since $\frac{\partial(ii')}{\partial x_j} = i \frac{\partial i'}{\partial x_j} \in I$, it follows that $\frac{\partial \lambda}{\partial x_j} \in I$ for any $\lambda \in I^2$. In particular, $\frac{\partial f}{\partial x_j}, \frac{\partial g}{\partial x_j} \in I$. Therefore,

$$0 \equiv \frac{\partial f}{\partial x_j} \;=\; x_1^{a_1} \cdots x_{j-1}^{a_{j-1}} \frac{x_j^{a_j} - 1}{x_j - 1} + x_1^{a_1} \cdots x_r^{a_r} \frac{\partial g}{\partial x_j} \tag{1}$$

$$\equiv\; a_j \bmod I.$$

Therefore, $a_j = 0$ $(1 \le j \le r)$; hence $f = g \in \mathcal{F}'$. $\square$

Now we shall prove Theorem A-2. It is along the same line as in the proof of Theorem 2.2 [I$_2$] (the almost pro-l version). Let K denote the kernel of $\pi : \Lambda \to A$. It is a closed two-sided Λ-ideal contained in I, such that $\Lambda/K \xrightarrow{\sim} A$. Denote by $K \cdot I$ the closed two-sided Λ-ideal generated by elements of the form $k \cdot i$ $(k \in K, i \in I)$. Then the left Λ-module structure of K induces a left A-module structure of K/K^2 and also of its quotient $K/K \cdot I$. Our proof of Theorem A-2 will be divided into two steps. They are to establish two A-isomorphisms

$$K/K \cdot I \xrightarrow{\sim} \{(a_1, \ldots, a_r) \in A^{\oplus r}; \sum_{i=1}^{r} a_i(\pi(x_i) - 1) = 0\}, \tag{1}$$

and

$$\mathcal{N}^{ab} \xrightarrow{\sim} K/K \cdot I, \tag{1}'$$

whose composite gives the desired A-isomorphism. We denote by $\mathcal{I}$ the left A-module on the right hand side of (1).

Proof of $K/K \cdot I \xrightarrow{\sim} \mathcal{I}$.

Consider the map

$$K \ni k \longrightarrow \left(\pi(\frac{\partial k}{\partial x_1}), \ldots, \pi(\frac{\partial k}{\partial x_r}) \right) \in A^{\oplus r}. \tag{2}$$

We shall see that (2) induces $K/K{\cdot}I \xrightarrow{\sim} \mathcal{I}$. First, take any $k \in K$, $i \in I$. Then since $s(i) = \pi(k) = 0$,

$$\frac{\partial(ki)}{\partial x_j} = \frac{\partial k}{\partial x_j} s(i) + k\frac{\partial(i)}{\partial x_j} = k\frac{\partial(i)}{\partial x_j} \tag{3}$$

is mapped to 0 by π. Therefore, (2) factors through $K/K{\cdot}I$. Since

$$k = \sum_j \left(\frac{\partial k}{\partial x_j}\right)(x_j - 1) \tag{4}$$

and $\pi(k) = 0$, the image is contained in $\mathcal{I}$. Therefore, (2) induces a mapping

$$K/K{\cdot}I \to \mathcal{I}, \tag{5}$$

which is an A-module homomorphism, because for any $\lambda \in \Lambda$, $a = \pi(\lambda) \in A$, we have $\pi\left(\frac{\partial(\lambda k)}{\partial x_j}\right) = \pi\left(\lambda\frac{\partial k}{\partial x_j}\right) = a\,\pi\left(\frac{\partial k}{\partial x_j}\right)$ $(1 \le j \le r)$. $\qquad\square$

The injectivity of (5). If k lies in the kernel of (2), then $\frac{\partial k}{\partial x_j} \in K$ for all j; hence $k = \sum_{j=1}^r \frac{\partial k}{\partial x_j}(x_j - 1) \in K{\cdot}I$.

The surjectivity of (5). Take any $(a_j)_{1\le j\le r}$ in $\mathcal{I}$, and for each j, choose $\tilde{a}_j \in \Lambda$ which lifts a_j. Put $k = \sum_{j=1}^r \tilde{a}_j(x_j - 1)$. Then since $\pi(k) = 0$, we have $k \in K$. On the other hand, $\tilde{a}_j = \frac{\partial k}{\partial x_j}$; hence $a_j = \pi\left(\frac{\partial k}{\partial x_j}\right)$. Therefore, (a_j) is the image of $k \mod K{\cdot}I$.

Proof of $\mathcal{N}^{ab} \xrightarrow{\sim} K/K{\cdot}I$.

This will be induced from the map

$$\mathcal{N} \ni n \longrightarrow n - 1 \mod K{\cdot}I \in K/K{\cdot}I. \tag{6}$$

First, since $nn' - 1 = (n-1) + (n'-1) + (n-1)(n'-1) \equiv (n-1) + (n'-1) \mod K{\cdot}I$, (6) is a group homomorphism into the additive group of $K/K{\cdot}I$ which is abelian. Therefore, (6) induces a group homomorphism

$$\mathcal{N}^{ab} \longrightarrow K/K{\cdot}I. \tag{7}$$

Moreover, if $n \in \mathcal{N}$, $f \in \mathcal{F}$, then $fnf^{-1} - 1 = f(n-1) + f(n-1)(f^{-1} - 1)$, and $n - 1 \in K$, $f^{-1} - 1 \in I$; hence $fnf^{-1} - 1 \equiv f(n-1) \mod K{\cdot}I$. Therefore, (7) is an A-module homomorphism.

It is clear that the composite of (7) and (5) gives the desired A-homomorphism in Theorem A-2. So, it remains to prove the bijectivity of (7).

The surjectivity of (7). As additive topological group, K is generated by elements of the form $(n-1)f$ $(n \in \mathcal{N}, f \in \mathcal{F})$. (Note that $f(n-1) = (fnf^{-1} - 1)f$.) But $(n-1)f = (n-1) + (n-1)(f-1) \equiv (n-1) \mod K{\cdot}I$.

Therefore, the image of (7) is dense in K/KI. But since $\mathcal{N}^{ab}$ is compact and (7) is continuous, (7) must be surjective.   □

The injectivity of (7). This is to show that $n \in \mathcal{N}$, $n-1 \in K \cdot I$ implies $n \in (\mathcal{N}, \mathcal{N})$. We shall first prove this when $\mathcal{N}$ is an *open* (normal) subgroup of $\mathcal{F}$. In this case,

$$\mathcal{F} = \bigsqcup_{\nu=1}^{m} \mathcal{N} c_\nu, \quad c_1 = 1, \tag{8}$$

and $\Lambda = \bigoplus_\nu \hat{\mathbb{Z}}[[\mathcal{N}]] c_\nu = \hat{\mathbb{Z}}[[\mathcal{N}]] \oplus \bigoplus_{\nu>1} \hat{\mathbb{Z}}[[\mathcal{N}]](c_\nu - 1)$. Therefore, $I = I_\mathcal{N} \oplus \bigoplus_{\nu>1} \hat{\mathbb{Z}}[[\mathcal{N}]](c_\nu - 1)$, where $I_\mathcal{N}$ is the augmentation ideal of $\hat{\mathbb{Z}}[[\mathcal{N}]]$. Now let $n \in \mathcal{N}$ be such that $n-1 \in K \cdot I$. But since $K = I_\mathcal{N} \cdot \Lambda$, we have $n-1 \in I_\mathcal{N} \cdot I = I_\mathcal{N}^2 \oplus \bigoplus_{\nu>1} I_\mathcal{N}(c_\nu - 1)$. Since $n-1 \in \hat{\mathbb{Z}}[[\mathcal{N}]]$ and the above decomposition of Λ is unique, this implies that $n-1 \in I_\mathcal{N}^2$. Now, since $\mathcal{N}$ is an open subgroup of $\mathcal{F}$, $\mathcal{N}$ is also free of finite rank. Therefore, by the above lemma, we conclude that $n \in (\mathcal{N}, \mathcal{N})$. This settles the case where $\mathcal{N}$ is open.

Finally, in the general case, let $n \in \mathcal{N}$, $n-1 \in K \cdot I$. Let $f : \mathcal{N} \to \Phi$ be any continuous homomorphism onto a finite abelian group. It suffices to prove $f(n) = 1$. Now the kernel of f being open in $\mathcal{N}$, there exists an open normal subgroup $\mathcal{U}$ of $\mathcal{F}$ such that $f|_{\mathcal{N} \cap \mathcal{U}} = (1)$. Put $\mathcal{N}^* = \mathcal{N}\mathcal{U}$ which is an open normal subgroup of $\mathcal{F}$, and let K^* be the Λ-ideal corresponding to $\mathcal{N}^*$. Then since $n-1 \in K \cdot I \subset K^* \cdot I$, we obtain $n \in (\mathcal{N}^*, \mathcal{N}^*)$. This implies that if we extend f to a homomorphism $f^* : \mathcal{N}^* \to \Phi$ by $\mathcal{N}^*/\mathcal{U} \xrightarrow{\sim} \mathcal{N}/\mathcal{N} \cap \mathcal{U}$, then $f^*(n) = 1$. Since $f(n) = f^*(n)$, this proves $f(n) = 1$, and hence that $n \in (\mathcal{N}, \mathcal{N})$.   □

References

[A₁] G. Anderson, Letter to Y. Ihara "Notes for Ihara concerning "profinite ψ""; Feb 23 (1987).

[A₂] ————, (i) The hyperadelic gamma function, Inv. Math. **95** (1989), 63–131.

(ii) The hyperadelic gamma function: A Précis; Advanced Studies in Pure Math. **12** (1987), 1–19.

[A₃] ————, Normalization of the hyperadelic gamma function; In "Galois groups over $\mathbb{Q}$" Publ.MSRI, **16** (1989), 1–31; Springer.

[A-I] G. Anderson and Y. Ihara, Pro-l branched coverings of $\mathbb{P}^1$ and higher circular l-units; (Part 1) Ann. of Math. **128** (1988), 271–293; (Part 2) Int'l J. Math. **1** (1990), 119–148.

[A*] M. P. Anderson, Exactness properties of profinite functors; Topology **13** (1974), 229-239.

[C] R. Coleman, Anderson-Ihara theory: Gauss sums and circular units; Advanced Studies in Pure Math. **17** (1989), 55-72.

[D] V.G. Drinfel'd, On quasitriangular quasi-Hopf algebras and a group closely connected with $Gal(\bar{\mathbb{Q}}/\mathbb{Q})$; Algebra i Analiz **2** (1990), 114-148; English translation. Leningrad Math. J. **2** (1991), 829-860.

[I₁] Y. Ihara, Profinite braid groups, Galois representations, and complex multiplications; Ann of Math. **123** (1986), 43-106.

[I₂] _____, On Galois representations arising from towers of coverings of $\mathbb{P}^1 - \{0, 1, \infty\}$; Inv. Math. **86** (1986), 427-459.

[I₃] _____, Braids, Galois groups, and some arithmetic functions; Proc Int'l Congress of Mathematicians 1990 (I), Math. Soc. Japan, Springer-Verlag. (1991), 99-120.

[I₄] _____, On the embedding of $Gal(\bar{\mathbb{Q}}/\mathbb{Q})$ into GT; in "The Grothendieck Theory of Dessins d'Enfants", London Math. Soc. Lecture Notes Series **200**, Cambridge Univ. Press (1994), 289-305.

[I-K] H. Ichimura and M. Kaneko, On the universal power series for Jacobi sums and the Vandiver conjecture; J. Number Theory **31** (1989), 312-334.

[I-M] Y. Ihara and M. Matsumoto, On Galois actions on profinite completion of braid groups; Contemp. Math. **186** (1995), 173-200.

Arithmetically exceptional functions and elliptic curves

Peter Müller [*]

March 5, 1999

Abstract

Let $f(X) \in \mathbb{Q}(X)$ be a rational function. For almost all primes p we can reduce the coefficients of f and consider $f_p := f \bmod p$ as a function on the projective line $\mathbb{P}^1(\mathbb{F}_p) = \mathbb{F}_p \cup \{\infty\}$. Here we continue the arithmetic aspects of joint work with Guralnick and Saxl, and classify the functions f such that f_p is a bijection for infinitely many primes p. This is the rational function analog of the classical conjecture of Schur (1923), solved by Fried (1970), which considered the case that f is a polynomial.

Thereby we also answer a question of J. G. Thompson about the minimal field of definition of a certain rational function of degree 25.

1 Introduction

A classical problem going back to Schur [Sch23] is the following: Let $f(X) \in \mathbb{Z}[X]$ be a polynomial, which induces a permutation of the residue fields $\mathbb{Z}/p\mathbb{Z}$ for infinitely many primes. Then Schur conjectured (and proved this for prime degree polynomials) that f is a composition of linear polynomials and Dickson polynomials $D_k(a, X)$, which are best defined implicitly by $D_k(a, Z + a/Z) = Z^k + (a/Z)^k$ for $a \in \mathbb{Q}$. Schur's conjecture has been proved by M. Fried in [Fri70], see also [Tur95] and [Mül97]. The obvious generalization of this question to number fields poses no difficulties, result and proof are the same.

In recent joint work [GMS97] with R. Guralnick and J. Saxl we investigated the rational function analog of this question over number fields K. Let $f(X) \in K(X)$ be a rational function, and $\mathcal{O}_K$ be the ring of integers of K. Fix coprime polynomials $r, s \in \mathcal{O}_K[X]$ with $f = r/s$. The coefficients of $f = r/s$ can be reduced modulo all but finitely many prime ideals $\mathfrak{p}$ of $\mathcal{O}_K$ without making s trivial. Such a reduced function induces a map on the

---
[*]Supported by the DFG.

projective line $\mathbb{P}^1(\mathcal{O}_K/\mathfrak{p})$. We say that f is *arithmetically exceptional* if this induced map is bijective for infinitely many prime ideals $\mathfrak{p}$.

It follows from this definition that if an arithmetically exceptional function is a composition $a(b(X))$ of two rational functions $a, b \in K(X)$, then a, b are also arithmetically exceptional. (In contrast to the polynomial case, the converse does not hold even over $\mathbb{Q}$, see [GMS97, Corollary 7.4].) So we can and do restrict to indecomposable functions. Define the degree of $f \in K(X)$ to be the maximum of the degrees of numerator and denominator in a reduced fraction. The degree is the same as the degree of the field extension $K(X)/K(f(X))$.

The aim of this paper is to classify the arithmetically exceptional functions over $\mathbb{Q}$. This also answers a question of J. Thompson [Tho90] raised in a different context.

The classification is in terms of the geometric monodromy group and the branching type. Let f be such a function, then $\mathrm{Gal}(f(X) - t/\mathbb{C}(t))$ is the geometric monodromy group of f, where $\mathrm{Gal}(f(X) - t/\mathbb{C}(t))$ denotes the Galois group of $R(X) - tS(X)$ over $\mathbb{C}(t)$, when $f = R/S$ is a reduced fraction of polynomials. Further, for the finitely many points $b_1, \dots, b_r \in \mathbb{C} \cup \{\infty\}$ with $|f^{-1}(b)| < n = \deg f$ let m_i be the least common multiple of the multiplicities of the points in the fiber $f^{-1}(b_i)$. Then $(m_1, \dots, m_r)$ is the branching type of f. The type, together with the geometric monodromy group, usually gives precise information about the function f. See Section 3 for more details; in particular the numbers m_i will be seen as the orders of the elements of a very specific generating system of $\mathrm{Gal}(f(X) - t/\mathbb{C}(t))$. The result, which completes work from [GMS97], is

Theorem 1.1. *Let $f \in \mathbb{Q}(X)$ be an indecomposable rational function of degree n which is arithmetically exceptional over $\mathbb{Q}$. Set $G := \mathrm{Gal}(f(X) - t/\mathbb{C}(t))$. Then one of the following holds, where p is an odd prime and C_m denotes a cyclic group of order m.*

(a) $n = p$, G is cyclic of type (p, p);

(b) $n = p \geq 5$, G is dihedral of type $(2, 2, p)$;

(c) $n = 4$, $G = C_2 \times C_2$ of type $(2, 2, 2)$;

(d) $n = p \in \{5, 7, 11, 13, 17, 19, 37, 43, 67, 163\}$, G is dihedral of type $(2, 2, 2, 2)$;

(e) $n = p^2$, $G = (C_p \times C_p) \rtimes C_2$ of type $(2, 2, 2, 2)$;

(f) $n = 5^2$, $G = (C_5 \times C_5) \rtimes S_3$ of type $(2, 3, 10)$;

(g) $n = 5^2$, $G = (C_5 \times C_5) \rtimes (C_6 \rtimes C_2)$ of type $(2, 2, 2, 3)$;

(h) $n = 3^2$, $G = (C_3 \times C_3) \rtimes (C_4 \rtimes C_2)$ of types $(2, 2, 2, 4)$ and $(2, 2, 2, 2, 2)$;

(i) $n = 28$, $G = \mathrm{PSL}_2(8)$ *of types* $(2,3,7)$, $(2,3,9)$, *and* $(2,2,2,3)$;

(j) $n = 45$, $G = \mathrm{PSL}_2(9)$ *of type* $(2,4,5)$;

While dealing with the arithmetic of case (g) above, we solve a question raised by John G. Thompson [Tho90] about the minimal field of definition of a certain rational function of degree 25.

Acknowledgment. I thank G. Malle and B. H. Matzat for a careful reading of the manuscript.

2   Arithmetically exceptional rational functions

Let B be a finite permutation group on Ω, and $G \trianglelefteq B$ be a transitive, normal subgroup. We say that the pair (B, G) is exceptional, if none of the orbits $\neq \{\omega\}$ of a point stabilizer G_ω is fixed by B_ω. (This is of course independent of the chosen point $\omega \in \Omega$.) This notion of exceptionality has first appeared in arithmetic questions of finite fields, see [FGS93] and the literature given there.

If the finite group A is acting on Ω, and $G \trianglelefteq A$ is transitive, then we say that (A, G) is *arithmetically exceptional*, if there is a group B with $G \leq B \leq A$, such that (B, G) is exceptional and B/G is cyclic.

Now fix a number field K, and let $f \in K(X)$ be a non–constant rational function. Let t be a transcendental over K, and let L be a splitting field of $f(X) - t$ over the rational function field $K(t)$. Denote by $\hat{K}$ the algebraic closure of K in L. Then $A := \mathrm{Gal}(L/K(t))$ is called the *arithmetic monodromy group* of f, and $G := \mathrm{Gal}(L/\hat{K}(t))$ is called the *geometric monodromy group* of f. Except otherwise said, we regard A and G as permutation groups on the roots of $f(X) - t$. We call $\hat{K}$ the *field of constants* of f. Note that $A/G = \mathrm{Gal}(\hat{K}/K)$.

The following group–theoretic characterization of arithmetically exceptional functions is due to Fried [Fri78], see [GMS97, Theorem 2.1] for a short proof.

Theorem 2.1. *Let f, A, and G be as above. Then f is arithmetically exceptional if and only if the pair (A, G) is arithmetically exceptional.*

Remark 2.2. The proof of this theorem in [GMS97] also characterizes the prime ideals $\mathfrak{p}$ modulo which the function f is bijective if it is arithmetically exceptional. Namely there is a bound C such that if $|\mathcal{O}_K/\mathfrak{p}| > C$, then such an f is bijective modulo $\mathfrak{p}$ if and only if (B, G) is exceptional, where B/G is the decomposition group of a prime of $\hat{K}$ lying above $\mathfrak{p}$.

3   Branch cycle descriptions in geometric monodromy groups

Let $\mathbb{P}^1 = \mathbb{P}^1(\mathbb{C})$ be the Riemann sphere over the complex numbers. We keep the notation from the previous subsection, and regard now f as a covering map from $\mathbb{P}^1$ to $\mathbb{P}^1$. Let n be the degree of f. There is a finite set $\mathcal{B} := \{b_1, b_2, \dots, b_r\} \subset \mathbb{P}^1$ of elements with less than n preimages. We call these elements the *branch points* of f.

Fix a base point $b_0 \in \mathbb{P}^1 \setminus \mathcal{B}$, and denote by π the fundamental group $\pi_1(\mathbb{P}^1 \setminus \mathcal{B}, b_0)$. Then π acts transitively on the points of the fiber $f^{-1}(b_0)$ by lifting of paths. Fix a numbering $1, 2, \dots, n$ of this fiber. Thus we get a homomorphism $\pi \to \mathrm{Sym}_n$. By standard arguments (see [MM] or [Völ96]), the image of π can be identified with the geometric monodromy group G defined above, thus we write G for this group too.

This identification has a combinatorial consequence. Choose a standard homotopy basis of $\mathbb{P}^1 \setminus \mathcal{B}$ as follows. Let γ_i be represented by paths which wind once around b_i clockwise, and around no other branch point, such that $\gamma_1 \gamma_2 \cdots \gamma_r = 1$. Then $\gamma_1, \gamma_2, \dots, \gamma_{r-1}$ freely generate π.

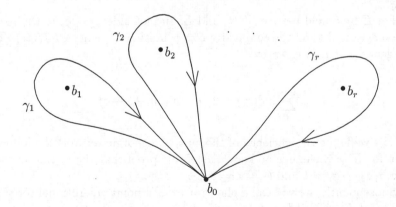

Definition 3.1. Let $\sigma \in \mathrm{Sym}_n$. Then the index $\mathrm{ind}(\sigma)$ is defined to be n minus the number of cycles in σ (where fixed points count as cycles too).

Let σ_i be the image of γ_i in Sym_n. If the points $s_1, \dots, s_m$ in the fiber of b_i have multiplicities $e_1, \dots, e_m$, respectively, then σ_i has cycle lengths $e_1, \dots, e_m$. We say that σ_i has *cycle type* $1^{a_1} 2^{a_2} \cdots$, where a_i is the number of cycle lengths i. Note that $\mathrm{ind}(\sigma_i) = n - |f^{-1}(b_i)| = n - m$.

The Riemann–Hurwitz genus formula (or a more elementary argument using the derivative of f) gives the following basic relation:

$$\sum_i \mathrm{ind}(\sigma_i) = 2(n-1) \tag{1}$$

We call the r–tuple $(\sigma_1, \sigma_2, \ldots, \sigma_r)$ a *branch cycle description* of G, and the unordered tuple $(|\sigma_1|, |\sigma_2|, \ldots, |\sigma_r|)$ the *branching type* of the branch cycle description. Of course the orders of the σ_i do not specify the σ_i, but in most cases where G is fixed the branching types distinguish between the various possibilities for f.

Note that one can arbitrarily order the conjugacy classes of the σ_i using an iteration of the elementary braiding operations Q_i, $i = 1, \ldots, r-1$, which send the tuple $(g_1, g_2, \ldots, g_r)$ to $(g_1, \ldots, g_{i-1}, g_{i+1}, g_{i+1}^{-1} g_i g_{i+1}, g_{i+2}, \ldots, g_r)$.

The elements σ_i can also be seen as inertia group generators in a slightly more general context. Let K be a field of characteristic 0, and $L/K(t)$ be a regular (i.e. K is algebraically closed in L) finite Galois extension with group G. For each ramified place $P_i : t \mapsto b_i$ (or $1/t \mapsto 0$ if $b_i = \infty$) let σ_i be a generator of an inertia group of a place of L lying above P_i. There is a natural choice of σ_i up to conjugacy. Namely set $y := t - b_i$ (or $y := 1/t$ if $b_i = \infty$). There is a minimal integer e such that L embeds into the power series field $\overline{\mathbb{Q}}((y^{1/e}))$. For such an embedding, let σ_i be the restriction to L of the automorphism of $\overline{\mathbb{Q}}((y^{1/e}))$ which is the identity on the coefficients and maps $y^{1/e}$ to $y^{1/e} \exp(2\pi\sqrt{-1}/e)$. The non–uniqueness of the embedding of L accounts for the fact that such σ_i are well–defined only up to conjugacy. We call the conjugacy class of σ_i the *distinguished conjugacy class* associated to b_i.

Let E be a field between $K(t)$ and L, and let $\mathrm{ind}(\sigma_i)$ refer to the permutation action of G on the conjugates of a primitive element of $E/K(t)$. Then the genus g of E is given by

$$\sum_i \mathrm{ind}(\sigma_i) = 2([E : K(t)] - 1 + g). \tag{2}$$

The well–known deficiency of this purely algebraic setup is the following: Even for $K = \mathbb{C}$ there is no known algebraic proof that the σ_i can be chosen with $\sigma_1 \sigma_2 \cdots \sigma_r = 1$ and $G = <\sigma_1, \sigma_2, \ldots, \sigma_r>$.

Subsequently, we will call a place (or branch point) K–rational (or simply rational if $K = \mathbb{Q}$) if $b_i \in K \cup \{\infty\}$.

Sometimes one can read off from the branch cycle description whether the function f is defined over certain fields using the so called branch cycle argument, see [Völ96, 2.8] for a fuller version. Let G be a subgroup of A. We call an element $x \in G$ rational in A, if all powers σ^m with m prime to $|G|$ are conjugate to σ in A. Also, we call a conjugacy class of G rational in A, if it consists of elements rational in A.

Theorem 3.2. *Let K be a field of characteristic 0, and $L/K(t)$ be a finite Galois extension with group A. Set $n = [L\bar{K} : \bar{K}(t)]$. Then $\alpha \in \mathrm{Gal}(L\bar{K}/K(t))$ permutes the branch points of $L\bar{K}/\bar{K}(t)$ among themselves. Let ζ_n be a primitive n–th root of 1, and m be an integer with $\alpha^{-1}(\zeta) = \zeta^m$.*

Let $\mathcal{C}_b$ be the distinguished conjugacy class of inertia generators associated to the place b of $\bar{K}(t)$. Then $\mathcal{C}_{\alpha(b)} = \alpha^*(\mathcal{C}_b)^m$, where α^* is the conjugation map $g \mapsto \alpha g \alpha^{-1}$ on G.

In particular, if b is K–rational, then the class $\mathcal{C}_b$ is rational in A.

A typical application is the following. Suppose $f \in \mathbb{Q}(X)$. Then the absolute Galois group $\mathrm{Gal}(\bar{\mathbb{Q}}/\mathbb{Q})$ permutes the branch points, but also preserves their cycle types. So if there is a branch point whose cycle type appears only once, then it must be rational. If the associated element σ_i is not rational in G, then necessarily $A > G$.

4 Monodromy groups of arithmetically exceptional functions

The main group–theoretic result from [GMS97] is the following:

Theorem 4.1. *Let A be a primitive permutation group of degree n and G be a normal subgroup such that (A, G) is arithmetically exceptional. Further suppose that G has a generating system $\sigma_1, \sigma_2, \ldots, \sigma_r$ with $\sigma_1 \sigma_2 \cdots \sigma_r = 1$ and $\sum \mathrm{ind}(\sigma_i) = 2(n-1)$. Then one of the following holds, where type means the branching type of the generating system.*

(I) *$n = p^e$ for a prime p, $A \le N \rtimes \mathrm{GL}_e(p)$ with $N = \mathbb{F}_p^e$ is an affine group, and one of the following holds:*

    (a) (i) *$n = p \ge 3$, G is cyclic of type (p, p); or*

          (ii) *$n = p \ge 5$, G is dihedral of type $(2, 2, p)$; or*

          (iii) *$n = 4$, $G = C_2 \times C_2$ of type $(2, 2, 2)$, $A = A_4$ or S_4.*

    (b) *$n = p$ or $n = p^2$ for p odd, and*

          (i) *G is of type $(2, 2, 2, 2)$, $G = N \rtimes C_2$, and $n \ge 5$; or*

          (ii) *G is of type $(2, 3, 6)$, $G = N \rtimes C_6$, and $n \equiv 1 \pmod 6$; or*

          (iii) *G is of type $(3, 3, 3)$, $G = N \rtimes C_3$, and $n \equiv 1 \pmod 6$; or*

          (iv) *G is of type $(2, 4, 4)$, $G = N \rtimes C_4$, and $n \equiv 1 \pmod 4$.*

    (c) (i) *$n = 11^2$, G is of type $(2, 3, 8)$, $G/N \cong \mathrm{GL}_2(3)$ and $A = A\Gamma L_1(11^2)$; or*

          (ii) *$n = 5^2$, G is of type $(2, 3, 10)$, $G/N = S_3$ and $A/N \cong S_3 \times C_4$; or*

          (iii) *$n = 5^2$, G is of type $(2, 2, 2, 4)$, G/N is a Sylow 2-subgroup of the subgroup of index 2 in $\mathrm{GL}_2(5)$, and A/G has order 3 or 6; or*

          (iv) *$n = 5^2$, G is of type $(2, 2, 2, 3)$, $G/N = C_6 \rtimes C_2$ and A/G is cyclic of order 2 or 4; or*

(v) $n = 3^2$, G *is of type* $(2,4,6)$, $(2,2,2,4)$, $(2,2,2,6)$, *or*
$(2,2,2,2,2)$, $G/N = C_4 \rtimes C_2$ *and* A/G *has order 2; or*

(vi) $n = 2^4$, G *is of type* $(2,4,5)$ *or* $(2,2,2,4)$, $G/N = C_5 \rtimes C_2$,
and A/G *has order 3 or 6.*

(II)   *(a)* $n = 28$, $G = \mathrm{PSL}_2(8)$ *is of type* $(2,3,7)$, $(2,3,9)$, *or* $(2,2,2,3)$,
and $A = \mathrm{P\Gamma L}_2(8)$.

   (b) $n = 45$, $G = \mathrm{PSL}_2(9)$ *is of type* $(2,4,5)$, *and either* $A = \mathrm{M}_{10}$, *or*
$A = \mathrm{P\Gamma L}_2(9)$.

If we take a group G and a branch cycle description from this theorem, then Riemann's existence theorem implies the existence of a rational function over some number field K with G as geometric monodromy group, and branching given by the branch cycle description. However, two difficult arithmetic problems are left. First, it is not clear how small we may take K, in particular whether we may take $K = \mathbb{Q}$. This is the descent problem encountered in the inverse Galois problem. Secondly, it is difficult to get a hold on the arithmetic monodromy group A once we have fixed K and $f(X) \in K(X)$. So after having done all the group–theoretic work yielding the above theorem, the question remains whether there are indeed arithmetically exceptional functions with the data in the theorem.

The cases in (II) have been dealt with in [GMS97, Section 6] using variants of the rational rigidity theorem, and they are shown to appear over $\mathbb{Q}$. The cases (I)(a)(i) and (I)(a)(ii) are very easy to deal with, and basically lead to cyclic polynomials X^p and the Rédei functions (see [GMS97, Section 7]) in the first case, and the Dickson polynomials in the second case. Case (I)(a)(iii) appears also over the rationals, with the added feature that if $A = \mathrm{Alt}_4$, then $\hat{\mathbb{Q}}$ can be any cyclic cubic extension of $\mathbb{Q}$.

The four infinite series in (I)(b) are intimately connected with isogenies of elliptic curves. A careful analysis is contained in [GMS97]. We want to remark that, as an alternative to the treatment in [GMS97], one can also use the branch cycle argument Theorem 3.2 to show that the cases (I)(b)(ii), (iii), and (iv) do not occur over the rationals. Section 5 contains information concerning the first of these series, as we will need this setup to decide a question of Thompson.

As to (I)(c): Only (vi) has been shown to not occur at all over any number field K. Case (iv), which is the hardest, will be dealt with in Section 6.

In the following we decide the existence over the rationals in all other cases.

Cases (I)(c)(i) and (I)(c)(iii). In the first case the element of order 8 is not rational in A, and in the second case the element of order 4 is not rational in A, so these cases do not occur by the branch cycle argument Theorem 3.2.

Case (I)(c)(ii). The given tuple $(\sigma_1, \sigma_2, \sigma_3)$ is rigid in G. The elements σ_1 and σ_2 are rational in G, and σ_3 is rational in A. The values of the irreducible characters of G at σ_3 generate the cyclotomic field $K = \mathbb{Q}(\zeta_5)$, where ζ_5 is a primitive 5th root of unity. The rational rigidity criterion thus gives a regular Galois extension $L/K(t)$ with group G and branching data given by the σ_i. The four conjugacy classes in G of elements of order 10 are permuted transitively by $\mathrm{Aut}(G)$, because A already permutes them transitively. Thus L is Galois over $\mathbb{Q}(t)$, see [Völ96, Section 3.1.2]. One obtains that $A = \mathrm{Gal}(L/\mathbb{Q}(t))$. Let U be a subgroup of index 25 in A, and E the fixed field of U. Then $A = GU$, so $E/\mathbb{Q}(t)$ is regular. Further, the Riemann–Hurwitz formula (2) shows that E has genus 0, and is even rational because σ_2 has a unique fixed point, and so E has a rational place. Thus $E = \mathbb{Q}(x)$. Write $t = f(x)$, and f is the desired rational function.

Case (I)(c)(v). Here we have four different kinds of branching types. In the first one of type $(2, 4, 6)$, we obtain that the associated triple of the σ_i is rigid and consists of elements which are rational in G. So the usual rational rigidity criterion shows that we cannot have $A > G$.

Now suppose that we have branching type $(2, 2, 2, 4)$. Here indeed there are arithmetically exceptional functions with this data. It seems to be difficult to exactly write them all down. Instead we give just one example, and show in Appendix A how we got it (and how to possibly get others). Set

$$f(X) = \frac{X(X^4 - 8X^3 + 12X^2 - 48X - 28)^2}{(X^2 - 2)^4}$$

and let $F(X, Y) \in \mathbb{Q}[X, Y]$ be the numerator of $(f(X) - f(Y))/(X - Y)$ as a reduced fraction. One verifies easily that $f(X, 1)$ is irreducible over the rationals ($f(X, 1)$ is irreducible even modulo 3), so $F(X, Y)$ is irreducible over $\mathbb{Q}$, hence the arithmetic monodromy group A of f is doubly transitive. On the other hand, one easily checks that $F(X, Y)$ factors (into two factors of degree 4) over $K := \mathbb{Q}(\sqrt{2})$, so the geometric monodromy group G is not doubly transitive. There are only 3 doubly transitive groups of degree 9 with a subgroup of index 2 which is not doubly transitive, namely $\mathrm{AGL}_1(9)$, M_9, and $\mathrm{A\Gamma L}_1(9)$. From the branching over ∞ and 0 we see already that G contains elements of cycle types $1^1 4^2$ and $1^1 2^4$. Now let β be a root of $2Z^2 + 88Z + 343$. The numerator of $f(X) - \beta$ factors as $(25X^3 + 16X^2 + 3\beta X^2 + 100X + 56 - 2\beta)(25X^3 - 208X^2 - 14\beta X^2 + 702X + 16\beta X + 112 - 4\beta)^2$, so f has two more branch points, and the inertia generators have cycle type $1^3 2^3$. None of the index 2 subgroups of $\mathrm{AGL}_1(9)$ and M_9 has elements of this type, so we have $A = \mathrm{A\Gamma L}_1(9)$, and the only not doubly transitive subgroup of A containing elements of the previous types is $G = \mathrm{A\Sigma L}_1(9)$.

From this function f we immediately get also a function corresponding to the branching type $(2, 2, 2, 2, 2)$. Namely note that $f(X^2) = g(X)^2$ for $g \in \mathbb{Q}(X)$. It is easy to verify that f and g have the same pairs of arithmetic and geometric monodromy group, and that the branching of g is as claimed.

Next suppose that we have branching type $(2, 2, 2, 6)$. We are going to show that this does not occur. To start with we need

Lemma 4.2. *Let K be a field of characteristic 0, $\mathcal{E}$ be an elliptic curve and $\varphi : \mathcal{C} \to \mathcal{E}$ be a K-rational morphism of finite degree of an algebraic curve $\mathcal{C}$ defined over K of genus 1 to $\mathcal{E}$. Then $\mathcal{C}$ is an elliptic curve.*

Proof. Let J be the jacobian of $\mathcal{C}$, and $\Phi : \mathcal{C} \to J$ be the map such that $\Phi^\gamma \circ \Phi^{-1}$ is the translation map T_γ on J by a point P_γ depending on $\gamma \in \mathrm{Gal}(\bar{K}/K)$. Set $\Psi := \varphi \circ \Phi^{-1} : \mathcal{C} \to \mathcal{E}$. Without loss assume that Ψ maps a fixed K-rational point 0_J to a K-rational point $0_\mathcal{E}$, and that 0_J and $0_\mathcal{E}$ are the zero elements of the respective additions on the elliptic curves. We get

$$\Psi^\gamma \circ T_\gamma = \Psi.$$

So for Q a point on J, we get

$$\Psi(Q) = \Psi^\gamma(Q + P_\gamma) = \Psi^\gamma(Q) + \Psi^\gamma(P_\gamma).$$

But $Q = 0_J$ shows $\Psi^\gamma(P_\gamma) = 0_\mathcal{E}$, hence $\Psi = \Psi^\gamma$ for all $\gamma \in \mathrm{Gal}(\bar{K}/K)$, so Ψ and therefore also Φ is defined over K.   □

Now fix a branch cycle description $(\sigma_1, \sigma_2, \sigma_3, \sigma_4)$ of type $(2, 2, 2, 6)$, and let K be the field of constants. Without loss let ∞ correspond to the element σ_4 of order 6. Two of the involutions, say σ_1 and σ_2, are conjugate in G, whereas σ_3 is not conjugate to them. So the branch point corresponding to σ_3 first has to be K-rational, but it is even $\mathbb{Q}$-rational for otherwise (by the branch cycle argument) an element of $\mathrm{Gal}(\bar{\mathbb{Q}}/\mathbb{Q})$ would interchange both points corresponding to σ_1 and σ_2 with the one belonging to σ_3, because σ_1, σ_2, and σ_3 are conjugate in A.

The group A has a subgroup U of index 4 with $A = GU$, so the fixed field of U in L is a regular extension of $\mathbb{Q}(t)$. The action of A on A/U is the dihedral group action, so there is precisely one group W between U and A of index 2 in A. Using the Riemann–Hurwitz genus formula (2) we compute that the fixed fields L_U and L_W of U and W both have genus 1. As ∞ is ramified in $L_W/\mathbb{Q}(t)$, we get that $L_W = \mathbb{Q}(t, x)$, where $x^2 = q(t)$ for a cubic polynomial $q \in \mathbb{Q}[T]$. The zeros of q are the finite branch points of f, so in particular q has a rational root, so the elliptic curve $X^2 = q(T)$ has a rational point of order 2. The inclusion $L_W \subset L_U$ induces a rational morphism of a genus 1 curve with function field L_U to $X^2 = q(T)$ of degree 2. By the preceding lemma, L_U is thus the function field of an elliptic curve. Let (z, v) be a generic point on this curve with equation $z^2 = q'(v)$ for some cubic polynomial q'. Then $L_U = \mathbb{Q}(z, v)$. By the immediate part of [GMS97, Lemma 5.3], the isogeny of degree 2 gives $t = R(v)$, where $R \in \mathbb{Q}(V)$ has degree 2. But then we get the quadratic extension $\mathbb{Q}(v)$ of $\mathbb{Q}(t)$ inside L_U (and different from L_W, for instance because the genus is different), contrary to the fact that W is the unique group properly between U and A.

5   Rational functions with branching type $(2, 2, 2, 2)$

Throughout this section let K be a field of characteristic 0. We call a rational function a $(2,2,2,2)$–function if it has exactly 4 branch points, and branching type $(2,2,2,2)$. The following proposition gives a characterization of indecomposable $(2,2,2,2)$–functions of odd degree. The easy extension to the decomposable case (which we don't need here) is left to the reader.

Proposition 5.1. *Let $f \in K(X)$ be a $(2,2,2,2)$–function of odd degree n which is indecomposable over K. For a transcendental t let $A := \mathrm{Gal}(f(X) - t/K(t))$ and $G := \mathrm{Gal}(f(X) - t/\bar{K}(t))$ be the arithmetic and geometric monodromy group, respectively. Then f has degree p^m with a prime p and $m \in \{1, 2\}$, $A = \mathbb{F}_p^m \rtimes H$ with $H \le \mathrm{GL}_m(p)$, and $G = \mathbb{F}_p^m \rtimes <-1>$.*

Proof. We have $G = <\sigma_1, \sigma_2, \sigma_3, \sigma_4>$ with inertia generators σ_i of order 2, and $\sigma_1 \sigma_2 \sigma_3 \sigma_4 = 1$. From $\sigma_1 \sigma_2 = \sigma_4^{-1} \sigma_3^{-1}$ we see that the σ_i act by inversion on $\sigma_1 \sigma_2$. For $i \ne j$ set $T_{i,j} = <\sigma_i \sigma_j>$. We see as before that the σ_k act by inversion on $T_{i,j}$. Note that $T_{1,2} = T_{3,4}$, and so on. Also, $\sigma_1 \sigma_2$ commutes with $\sigma_1 \sigma_3$. From that we see that the normal subgroup N of G which is generated by the $T_{i,j}$ is abelian and actually generated by $\sigma_1 \sigma_2$ and $\sigma_1 \sigma_3$. Also, N has index 2 in G, and therefore is transitive (there are at most 2 orbits, they have the same lengths, but the degree is odd.) Note that f indecomposable implies that A is primitive, so every non–trivial normal subgroup of A is transitive. Hence N is even normal in A, for otherwise N would embed into the direct product of the groups G/N^a for a running through A, so N were a 2–group, contrary to odd degree. So N is a minimal and hence elementary abelian p–subgroup of A, which is generated by two elements. Thus A embeds naturally into the affine general linear group $\mathrm{AGL}_m(p)$ for $m = 1$ or 2, and the action of the σ_i on $\mathbb{F}_p^m$ is of the form $x \mapsto -x + t_i$ for $t_i \in \mathbb{F}_p^m$. (Note that by the Riemann–Hurwitz formula (1), each σ_i has exactly one fixed point.) In particular, $G = \mathbb{F}_p^m \rtimes <-1>$. $\qquad\square$

Remark 5.2. (A, G) is arithmetically exceptional if and only if H contains an element which has neither 1 nor -1 as eigenvalue.

We now relate the arithmetic monodromy group A to the arithmetic of elliptic curves. For the applications in this paper we need only the case where one of the branch points is K–rational. A linear fractional change of f over K can move this point to infinity, but does not affect A. Thus suppose that ∞ is one of the branch points of f. Also, we may and do assume that the unique simple point in the fiber $f^{-1}(\infty)$ is also ∞. Accordingly, write

$$f(X) = \frac{R(X)}{S(X)^2}$$

with $R, S \in K(X)$, $\deg R = n$, $\deg S = (n-1)/2$. We may assume that R and S are monic. Let λ_i be the 3 finite branch points of f, and μ_i be the simple point in the fiber of λ_i. Set $q_\lambda(X) = (X-\lambda_1)(X-\lambda_2)(X-\lambda_3) \in K[X]$ and $q_\mu(X) = (X - \mu_1)(X - \mu_2)(X - \mu_3) \in K[X]$. We have

$$f(X) - \lambda_i = (X - \mu_i)\frac{Q_i(X)^2}{S(X)^2}$$

for $Q_i(X) \in \bar{K}[X]$. The roots of the monic polynomial $Q_1(X)Q_2(X)Q_3(X)$ are the roots of the monic numerator of the derivative of f, thus

$$Q_1(X)Q_2(X)Q_3(X) = f'(X)S(X)^3,$$

hence

$$q_\lambda(f(X)) = q_\mu(X)f'(X).$$

From that we see that the morphism

$$(x, y) \mapsto (f(x), yf'(x))$$

induces a K–rational isogeny of degree n of the elliptic curve $E_\mu : Y^2 = q_\mu(X)$ to $E_\lambda : Y^2 = q_\lambda(X)$.

The interesting question in this context is the structure of A, or more precisely, the field of constants of f. In [GMS97, Proposition 5.4] we prove that the field of constants is generated over K by the X–coordinates of the finite points in the kernel of ϕ, where $\phi : E_\mu \to E_\lambda$ is the associated isogeny. But these X–coordinates are just the roots of the polynomial S. Hence we get

Proposition 5.3. *Let f be a $(2, 2, 2, 2)$–function as above, and $\hat{K}$ be the algebraic closure of K in a normal closure of $K(x)/K(f(x))$. Then $\hat{K}$ is the splitting field of $S(X)$ over K.*

Note that A_1/G_1 can be naturally identified with $\mathrm{Gal}(\hat{K}/K)$. On the other hand $\mathrm{Gal}(\hat{K}/K)$ acts on the elements of the kernel of ϕ, which is an $\mathbb{F}_p$–space of dimension m, thus $\mathrm{Gal}(\hat{K}/K)$ maps into $\mathrm{GL}_m(p)$. The induced action on the X–coordinates of these kernel elements is just the action of A_1 on $\mathbb{F}_p^m/<-1>$ (this follows from the proof of [GMS97, Proposition 5.4]), so we get the following

Corollary 5.4. *Let f be a $(2, 2, 2, 2)$–function as above. Then f is arithmetically exceptional if and only if $\mathrm{Gal}(S(X)/K)$ contains an element which does not fix a root of $S(X)$.*

Remark 5.5. If the degree of f is a prime p (so $m = 1$ in our notation), then the condition on S is easily seen to be equivalent to $S(X)$ having no root in K. For $m = 2$, a typical situation where the condition holds is when $S(X)$ is irreducible over K. Using Hilbert's irreducibility theorem and a theorem of Weber, one can easily construct for each odd p a function f of degree p^2 such that S is irreducible over K, see [GMS97, Proposition 5.6].

Let K be a number field. It is amusing to give a direct proof of the permutation property of the functions f from the Corollary without going the detour over the group–theoretic equivalence of arithmetic exceptionality. Note that if the Galois group of $S(X)$ contains an element which fixes no root, then $S(X)$ has no root modulo infinitely many primes by Chebotarëv's density theorem. Thus we get a direct proof of the Corollary from

Proposition 5.6. *Let K be a number field, and $\mathfrak{p}$ be a prime of K, such that f, S, and the associated elliptic curves can be reduced modulo $\mathfrak{p}$. (That of course is possible for all but finitely many primes.) Let $\bar{K}$ be the residue field of the prime $\mathfrak{p}$. If $S(x)$ modulo $\mathfrak{p}$ has no root in $\bar{K}$, then $f \bmod \mathfrak{p}$ is bijective on $\bar{K} \cup \{\infty\}$.*

Proof. We work over the field $\bar{K}$, and understand the coefficients of f, S and so on being reduced modulo $\mathfrak{p}$, so as being in $\bar{K}$. The hypothesis gives that $f(\infty) = \infty$, and $f(a) \neq \infty$ for $a \in \bar{K}$. So we only need to show injectivity on $\bar{K}$. By replacing $f(X)$ with $f(X + r)$ we may assume that $q_\mu(X) = X^3 + d_1 X + d_0$.

Suppose there is $a, b \in \bar{K}$, $a \neq b$, such that $f(a) = f(b)$. Then there are u, v in a quadratic extension of $\bar{K}$ such that $u^2 = q_\mu(a)$ and $v^2 = q_\mu(b)$. So by the previous development, the points $(f(a), uf'(a))$ and $(f(b), vf'(b))$ lie on the elliptic curve $Y^2 = q_\lambda(X)$. As $f(a) = f(b)$, we have $uf'(a) = \pm vf'(b)$, so by replacing possibly v by $-v$ we may assume that $uf'(a) = vf'(b)$. Also, $u^2, v^2 \in \bar{K}$, so $uv \in \bar{K}$. (Note that if $f'(a) = 0$, then $(f(a), 0)$ is a point of order 2 on $Y^2 = q_\lambda(X)$, so $u = 0$ as also (a, u) has order two by the odd degree of the isogeny induced by f.) Furthermore, the difference $(a, u) - (b, v)$ lies in the kernel of the isogeny ϕ. The X–coordinate of this difference hence is a root of S. The addition formula on elliptic curves gives:

$$(a, u) - (b, v) = (a, u) + (b, -v)$$

$$= \left(-a - b + \left(\frac{u + v}{a - b}\right)^2, \star\right),$$

but $(u + v)^2 = u^2 + v^2 + 2uv \in \bar{K}$, so we get a root of S in $\bar{K}$, contrary to the hypothesis.   □

5.1   Rational isogenies of degree 5

If $f \in \mathbb{Q}(X)$ is a function of branching type $(2, 2, 2, 2)$, and $n = \deg f$ is a prime, then we get a rational isogeny of an elliptic curve over $\mathbb{Q}$ of degree n. According to a result of Mazur [Maz78, Theorem 1], this can happen only for a few values of n. This is the reason for the short list of primes in Theorem 1.1(e). In Section 6 we have to look closer at the case $n = 5$. There are infinitely many elliptic curves admitting an isogeny of degree 5, however the possible j–invariants are restricted by the following: If $\phi : E \to E'$ is an isogeny of elliptic curves (everything defined over $\mathbb{Q}$), then there is an absolutely irreducible polynomial (modular equation) $F_n(J, J') \in \mathbb{Q}[J, J']$ of degree $n + 1$ in each variable, such that $F_n(j, j') = 0$, where j and j' is the j–invariant of E and E', respectively (see [Sil94, Chapter III,§6]). Now suppose that $n = 5$. Then F_5 admits a rational parametrization as follows, see [Fri22, Viertes Kapitel]:

$$J = \frac{(T^2 + 10T + 5)^3}{T}$$

$$J' = \frac{(T'^2 + 10T' + 5)^3}{T'} \quad \text{with } TT' = 125.$$

One can write $T = R(J, J')/S(J, J')$ with $R, S \in \mathbb{Q}[J, J']$. So, as we have to have rational values for j and j', the corresponding parameter is rational except for those pairs (j, j') for which R and S vanish. One computes that in these cases $j = j' \in \{1728, -32768, 287496, -884736\}$, and verifies directly (e.g. using the Maple package apecs for computations with elliptic curves [Con97]) that we cannot have rational isogenies of degree 5 in these cases. Conversely, if an elliptic curve has a j–invariant of the form $\frac{(\eta^2 + 10\eta + 5)^3}{\eta}$ for some non–zero rational η, then one can compute that the 5th division polynomial has a factor of degree 2. So the curve has a 5–division point P whose X–coordinate has degree at most 2 over $\mathbb{Q}$. One easily checks that the group generated by P is Galois invariant, so dividing by this group gives a rational isogeny of degree 5.

Summarizing, we get

Proposition 5.7. *Let j be the j–invariant of an elliptic curve over $\mathbb{Q}$. Then the curve admits a rational isogeny of degree 5 if and only if $j = \frac{(\eta^2 + 10\eta + 5)^3}{\eta}$ for some non–zero rational η.*

In order to determine the field of constants, we also need

Proposition 5.8. *Let $f(X) = R(X)/S(X)^2$ be a $(2, 2, 2, 2)$–function of degree 5 as above, $q_\lambda(X)$, $q_\mu(X)$ be the associated cubic polynomials. Let j be the j–invariant of $Y^2 = q_\lambda(X)$. Then $j = \frac{(\eta^2 + 10\eta + 5)^3}{\eta}$ for some non–zero rational η, and the field of constants of f is $\mathbb{Q}(\sqrt{5(\eta^2 + 22\eta + 125)})$.*

Proof. f induces an isogeny ϕ from $Y^2 = q_\mu(X)$ to $Y^2 = q_\lambda(X)$. Let $\hat{\phi}$ be the dual isogeny. Then $\Phi := \hat{\phi} \circ \phi$ is the multiplication by 5 map on $Y^2 = q_\mu(X)$. The kernel of ϕ of course is contained in the kernel of Φ, so $S(X)$ is a divisor of the 5–th division polynomial of $Y^2 = q_\mu(X)$. By the previous proposition, we may assume that $j = \frac{(\eta^2 + 10\eta + 5)^3}{\eta}$. Let j' be the j–invariant of $Y^2 = q_\mu(X)$. Then $j' = \frac{(\eta'^2 + 10\eta' + 5)^3}{\eta'}$ with $\eta' = 125/\eta$. Knowing j', we can compute the discriminant of $S(X)$. It is, up to a square factor, equal to $5(\eta^2 + 22\eta + 125)$. The result follows. $\qquad\square$

6   Application to a question of Thompson

As mentioned already in Section 4, it is usually a difficult problem to decide whether rational functions $f(Z) \in \mathbb{C}(Z)$ with specific geometric monodromy groups (and branching data) are defined over certain small fields. In [Tho90] Thompson proposes the project to determine the cases where f is already defined over the rationals, and adds " ... , an analysis which may require several years of hard work." Specifically, he investigates the case of a certain degree 25 function, but has to leave undecided the question about small fields of definition. The purpose of this section is to give a different approach and to settle this question. We give the method in reasonably complete detail, as it works also in many other instances. As this function also appears in Theorem 4.1(I)(c)(iv) as a possible candidate of an arithmetically exceptional function, we give more details in order to also show existence of examples with the correct pair of arithmetic and geometric monodromy group.

6.1   Thompson's question, group–theoretic preparation

Let $W = A\Gamma L(1, 25)$ be the affine semi–linear group acting on the elements of the finite field $\mathbb{F}_{25}$. Thus $W = T \rtimes (S \rtimes F)$, where T is the group of translations $x \mapsto x + t$, S is the group of scalar multiplications $x \mapsto ax$ for non–zero a, and the group F of order 2 is generated by the Frobenius map $x \mapsto x^5$.

For i a divisor of 24, denote by $S_{(i)}$ the subgroup of S of order i, and set $W_{(i)} := T \rtimes (S_{(i)} \rtimes F)$. So $W_{(24)} = W$. Set $G := W_{(6)}$, and note that W/G is cyclic of order 4.

Let β be an element of order 12 in $\mathbb{F}_{25}^*$, and define elements σ_1, σ_2, σ_3, $\sigma_4 \in G$ as follows:

$$\sigma_1 : x \mapsto \beta^{20} x$$
$$\sigma_2 : x \mapsto -x + \beta$$
$$\sigma_3 : x \mapsto -\beta^{20} x^5 + \beta$$
$$\sigma_4 : x \mapsto x^5$$

One immediately verifies that these σ_i generate G, and that $\sigma_1 \sigma_2 \sigma_3 \sigma_4 = 1$. (Here we write the action on $\mathbb{F}_{25}$ from the right.) The following table gives the cycle type and index of these elements:

	σ_1	σ_2	σ_3	σ_4
Cycle type	$1^1 3^8$	$1^1 2^{12}$	$1^5 2^{10}$	$1^5 2^{10}$
Ind	16	12	10	10

For natural action

Thus, by (2) we see that $(\sigma_1, \sigma_2, \sigma_3, \sigma_4)$ is a genus 0 system, so there is a rational function $f(Z) \in \mathbb{C}(Z)$ having G as geometric monodromy group and branch cycle description given by the σ_i. Thompson's question [Tho90] is whether we can have $f \in \mathbb{Q}(Z)$. We give an answer which also takes care of the different possibilities of the arithmetic monodromy group of f.

Definition 6.1. Let K be a field, and $f, \tilde{f} \in K(X)$ rational functions. We call f and $\tilde{f}$ *linearly equivalent over* K, if there are linear fractional functions $\ell_1, \ell_2 \in K(X)$ such that $f(X) = \ell_1(\tilde{f}(\ell_2(X)))$.

Theorem 6.2. *Let $f(Z) \in \mathbb{Q}(Z)$ be a rational function with geometric monodromy group G and branching data as above. Let $A := \mathrm{Gal}(f(Z) - t | \mathbb{Q}(t))$ be the arithmetic monodromy group. Then there are, up to linear equivalence, exactly one such functions with $A = G$, and exactly two with $A/G = C_2$. In the latter two cases, the field of constants is $\mathbb{Q}(\sqrt{5})$.*

The proof of this theorem is the subject of the following subsections.

Remark 6.3. The normalizer of G in the symmetric group S_{25} is W. We have $W/G = C_4$. Grouptheoretically, there is the third possibility that $A/G = C_4$. We have not been able to prove existence in this case, though we have very strong evidence for that.

Corollary 6.4. *Up to linear equivalence, there are exactly two arithmetically exceptional functions belonging to Theorem 4.1(I)(c)(iv).*

6.2   Passing to a different rational function

Let f be as in Theorem 6.2, and L be a splitting field of $f(Z) - t$ over $\mathbb{Q}(t)$. Denote by $\hat{\mathbb{Q}}$ the algebraic closure of $\mathbb{Q}$ in L. Then $G = \mathrm{Gal}(L|\hat{\mathbb{Q}}(t))$ and $A := \mathrm{Gal}(L|\mathbb{Q}(t))$, so $A/G = \mathrm{Gal}(\hat{\mathbb{Q}}|\mathbb{Q})$.

Thompson's idea [Tho90] was to look at the fixed field of T in $\mathbb{C}L$ over $\mathbb{C}(t)$. Then $\mathbb{C}L$ is an unramified Galois extension of this field (of genus 2) with Galois group $S_{(6)} \rtimes F$. Here we rather work with fixed fields which have genus 0, and involve rational functions of type $(2,2,2,2)$.

Recall that A is one of the groups $W_{(6)}$, $W_{(12)}$, $W_{(24)}$. If $A = W_{(i)}$, then A has, up to conjugation, the unique subgroup $A^{(3)} := W_{(i/3)}$ of index 3. Set

$G^{(3)} := A^{(3)} \cap G$. Let $L^{(3)}$ be the fixed field of $A^{(3)}$ in L. As $A = GA^{(3)}$, we get that $L^{(3)}$ is a regular extension of $\mathbb{Q}(t)$ of degree 3. The action of the σ_i on the coset space $G/G^{(3)}$ gives the following cycle types:

	σ_1	σ_2	σ_3	σ_4
Cycle type	3^1	1^3	$1^1 2^1$	$1^1 2^1$

For action on $G/G^{(3)}$

By the Riemann–Hurwitz genus formula (2), we see that $L^{(3)}$ has genus 0, furthermore $L^{(3)}$ is a rational field, because σ_3 has a unique fixed point on $G/G^{(3)}$.

Let H be an F–stable $\mathbb{F}_5$–subspace of $T = \mathbb{F}_{25}$. Then H is also stable under $S_{(2)}$ and $S_{(4)}$, so the semidirect product $H \rtimes (S_{(i/3)} \rtimes F)$ is a subgroup of index 15 of $W_{(i)}$ for $i = 6$ or $i = 12$.

Suppose $|A/G| \leq 2$, so $A = W_{(i)}$ for $i = 6$ or 12. Then let $A^{(15)}$ be the subgroup of index 15 in A constructed above, which is also a subgroup of index 5 of $A^{(3)}$. As $A = GA^{(15)}$, the fixed field $L^{(15)}$ of this subgroup is a regular extension of $\mathbb{Q}(t)$. The action of the σ_i on $G/G^{(15)}$ gives the following cycle types:

	σ_1	σ_2	σ_3	σ_4
Cycle type	3^5	$1^3 2^6$	$1^1 2^7$	$1^5 2^5$
Ind	10	6	7	5

For action on $G/G^{(15)}$

For the genus $g_{L^{(15)}}$ of $L^{(15)}$ we obtain

$$2(15 - 1 + g_{L^{(15)}}) = 10 + 6 + 7 + 5,$$

hence $g_{L^{(15)}} = 0$. As σ_3 has a unique fixed point on $G/G^{(15)}$, the field $L^{(15)}$ has a rational place, so this field is rational. As the cycle types of the σ_i in this action are all different, we get that the branch points of $L/\mathbb{Q}(t)$ are rational. Write $L^{(15)} = \mathbb{Q}(z)$, so $t = g(z)$ for a rational function $g(Z) \in \mathbb{Q}(Z)$. We get a decomposition $g(Z) = g_1(g_2(Z))$, with $g_i \in \mathbb{Q}(Z)$, $\deg g_1 = 3$, and $\deg g_2 = 5$, such that $L^{(3)} = \mathbb{Q}(g_2(z))$.

6.3   Rationality question for $|A/G| \leq 2$

We use the results from Section 5 to precisely pin down the function g_2 of degree 5. As the branch points of g are rational, we may make the following assumptions: ∞ corresponds to σ_1, and $0 = g_1^{-1}(\infty)$. Let 0 correspond to σ_3. The ramification information from above shows that the single point in the fiber $g_2^{-1}(0)$ is a branch point of g_2. Assume that this branch point is ∞. Without loss assume that $4/27$ is the branch point corresponding to σ_4, and that g_1 has monic numerator and denominator. This gives

$$g_1(X) = \frac{(X-1)^2}{X^3}.$$

Finally, let $1/\mu$ be the branch point corresponding to σ_2. The ramification information from above shows that $g_1^{-1}(1/\mu)$ consists of 3 different points, which are branch points of g_2. What we get is that g_2 is a $(2,2,2,2)$ function with branch points ∞ and the three roots of $X^3 - \mu(X-1)^2$. The elliptic curve associated to the branching data of g_2 thus is (see Section 5)

$$Y^2 = X^3 - \mu(X-1)^2. \tag{3}$$

The j-invariant of $\mathcal{E}$ is

$$j = 256\frac{\mu(\mu-6)^3}{4\mu-27}.$$

On the other hand, as $\mathcal{E}$ has a rational isogeny of degree 5, its j-invariant is of the form

$$j = \frac{(\eta^2 + 10\eta + 5)^3}{\eta}$$

for some non-zero rational η, see Proposition 5.7. This gives the algebraic curve relation

$$\mathcal{C}: \; 256\mu(\mu-6)^3\eta = (4\mu-27)(\eta^2+10\eta+5)^3.$$

The curve $\mathcal{C}$ is birationally equivalent to the elliptic curve

$$\mathcal{E}: \; V^2 = U^3 - 7U^2 - 144U = U(U+9)(U-16).$$

Using the MAPLE package [Hoe95] and some adhoc tricks, we get the following birational correspondence, where $\zeta := (\eta^2 + 10\eta + 5)/(\mu - 6)$:

$$\mu = \frac{27(864V + U^4 - 36U^3 + 4320U)}{4(U-36)U^3}$$

$$\eta = \frac{36V + VU - 13U^2 + 108U}{2U^2}$$

$$U = \frac{9\zeta^2 - 36(\eta+1)\zeta + 144(\eta-1)}{4(\eta^2+4\eta-1)}$$

$$V = \frac{9(2\eta+7)\zeta^2 - 36(\eta+9)\zeta - 144(3\eta+5)}{4(\eta^2+4\eta-1)}$$

Lemma 6.5. *The field of constants of f is the same one as the field of constants of g_2.*

Proof. Of course, the field of constants of g_2 is contained in the field of constants of f. One verifies that the index of the core of $A^{(15)}$ in $A^{(3)}$ is 10 if $A = W_{(6)}$, and 20 if $A = W_{(12)}$. So the degrees of the two fields of constants are the same. □

We are now going to study the rational points on $\mathcal{C}$ via the rational points on $\mathcal{E}$.

Lemma 6.6. *The finite rational points (u_0, v_0) on $\mathcal{E}$ are $(0,0)$, $(16,0)$, $(-9,0)$, $(-4,\pm20)$, and $(36,\pm180)$.*

Proof. As 7 does not divide the discriminant of $\mathcal{E}$, the rational torsion points of $\mathcal{E}$ map injectively to the $\mathbb{F}_7$–rational points of the reduction $\mathcal{E}$ modulo 7, see [Sil86, VII.3.1]. We compute that there are exactly 8 points modulo 7 (including the one at infinity), so we are done once we know that the Mordell–Weil rank of $\mathcal{E}$ is 0. This however is well–known. Namely the linear transformation $X = U/4 - 1$, $Y = (V - U)/8$ maps $\mathcal{E}$ to the curve

$$Y^2 + XY + Y = X^3 + X^2 - 10X - 10,$$

which is C15 (one of the 8 curves with conductor 15) in the notation of [Cre97], and shown to have rank 0 there, confer [Cre97, page 110]. □

Lemma 6.7. *The rational points (μ_0, η_0) on $\mathcal{C}$ are $(135/128, -25/8)$, $(-5/4, -25/2)$, $(-675/8, -40)$, and $(27/4, 0)$.*

Proof. As $\eta^2 + 4\eta - 1$ has no root in $\mathbb{Q}$, the above transformation equations show that the only possible rational points (μ_0, η_0) which are not mapped to finite points on $\mathcal{E}$ have $\mu_0 = 6$. But that leads to $\eta^2 + 10\eta + 5 = 0$, which has no rational solution. The finite points (u_0, v_0) on $\mathcal{E}$ which give points on $\mathcal{C}$ are those with $u_0 \neq 0, 36$. (Those with $u_0 = 0$ or 36 give points on the projective completion of $\mathcal{C}$.) □

We summarize:

Proposition 6.8. *Let C be the set of linear equivalence classes of rational functions $g \in \mathbb{Q}(X)$ such that $g(X) = g_1(g_2(X))$ with $g_1, g_2 \in \mathbb{Q}(X)$ and the following holds:*

(a) $\deg g_1 = 3$, *and g_1 has three rational branch points of branching type* 3^1, 1^12^1, *and* 1^12^1.

(b) $\deg g_2 = 5$. *Furthermore, g has, besides the three branch points of g_1, also the different rational branch point $1/\mu$, g_2 is a $(2,2,2,2)$–function with branch points $g_1^{-1}(1/\mu)$ and a simple point $g_1^{-1}(b)$, where b is a branch point of g_1 of type 1^12^1.*

Let $\hat{Q}$ be the field of contants of g_2. Then C has size 3, with g_1 and g_2 as above, where $\mu_0 = -675/8$ gives $\hat{Q} = \mathbb{Q}$, and $\mu_0 = 135/128$ or $-5/4$ gives $\hat{Q} = \mathbb{Q}(\sqrt{5})$.

Proof. The case $(\mu_0, \eta_0) = (27/4, 0)$ is nonsense, whereas the other rational points on C give examples as stated. The claim about $\hat{Q}$ follows from Proposition 5.8   □

6.4   Existence of f for $|A/G| \leq 2$

We now use the functions g_i from Proposition 6.8 in order to show that we get back the desired functions f whose existence we hypothetically assumed.

Let $\hat{Q}$ be the field of constants of g_2. It is clear that the geometric monodromy group of g is a transitive subgroup of the wreath product $D_5 \wr D_3$, where D_5 denotes the dihedral group of degree 5, and that in the case $\hat{Q} = \mathbb{Q}$ the arithmetic monodromy group is a subgroup of the same wreath product, whereas in the case $\hat{Q} = \mathbb{Q}(\sqrt{5})$ it is a subgroup of $(C_5 \rtimes C_4) \wr D_3$. Also, we know the cycle types of the branch cycle description of g. Using the computer algebra system GAP [S$^+$95], one verifies that such a 4–tuple generates $W_{(6)}$ in its action on 15 points, and that this group G is selfnormalizing in $D_5 \wr D_3$, and that the normalizer of G in $(C_5 \rtimes C_4) \wr D_3$ is $W_{(12)}$. So the normalizers in these wreath products are just the expected arithmetic monodromy groups A. Now let L be a splitting field of $g(Z) - t$ over $\mathbb{Q}(t)$, and E be a fixed field of a subgroup U of A of index 25. One verifies that the σ_i induce the expected action on $G/(G \cap U)$, also $A = GU$, so E has genus 0 and is regular over $\mathbb{Q}(t)$. Also, $E = \mathbb{Q}(y)$ is rational, because σ_1 has a unique fixed point in this degree 25 action. Write $t = f(y)$ for $f(Y) \in \mathbb{Q}(Y)$, and f is the desired function.

A   Computation of the $(2, 2, 2, 4)$–example

Let K be the proposed field of constants. In $G = \mathrm{Gal}(L/K(t))$ there is a subgroup U of index 6, such that the cycle types of σ_1, σ_2, σ_3, and σ_4 are $1^2 2^2$, $1^4 2^1$, 2^3, and $2^1 4^1$. Thus there is a rational function $r(X)$ of degree 6 over K such that the fixed field of U is $K(x)$ with $r(x) = t$. Let b_i be the branch point corresponding to σ_i. From the degree 9 action we see that b_1 and b_4 are rational. Without loss assume that $b_4 = \infty$, the 4–fold point over b_4 is ∞, and the other one is 0. A consideration similar to the one in the $(2, 2, 6)$–case, utilizing the regular degree 4–extension over $\mathbb{Q}(t)$, shows that b_2 and b_3 are algebraically conjugate and generate K over $\mathbb{Q}$. Without loss assume $b_2 = -\lambda$, $b_3 = \lambda$, with $\lambda^2 \in \mathbb{Q}$. If we make a further choice, namely assume that the double point above b_2 is 1, then r is given by

$$r(X) = -2\frac{\lambda(8X^3 + 8\beta X^2 + 12X + 8\beta X + \beta^2 X + 8\beta + 16)^2}{(36 + 24\beta + \beta^2)^2 X^2} + \lambda,$$

where $\beta \in K$. Furthermore, we compute

$$b_1 = -\frac{\lambda(\beta^4 - 80\beta^3 - 504\beta^2 - 1728\beta - 2160)}{\beta^4 + 48\beta^3 + 648\beta^2 + 1728\beta + 1296}.$$

If we write $\beta = u + v\lambda$ with $u, v \in \mathbb{Q}$, and use that b_1 has to be rational, we get a polynomial condition in u and v. This polynomial is the product of two genus 0 factors over $\mathbb{Q}$, and it is easy to find rational points on them. One of them gives $\lambda = \sqrt{2}$ and $b_1 = 44/25$. Of course, we have used only necessary conditions so far. Yet, nothing guarantees that the Galois closure of $K(x)/K(t)$ is Galois over $\mathbb{Q}(t)$. However, in the specific case one can use an "almost–argument" to verify that. Namely express the coefficients of r in terms of $\lambda = \sqrt{2}$, and denote by $\bar{r}$ the function where we replace λ by $-\lambda$. So the numerator of $(r(X) - t)(\bar{r}(X) - t)$ is in $\mathbb{Q}(t)[X]$ of degree 12, and the Galois group should have size $|A| = 144$. One now checks that using the computer algebra system KASH [DFK$^+$96] for various specializations of t. So we have a good candidate for the location of the branch points in order to compute the function $f \in \mathbb{Q}(X)$ of degree 9. The branching data gives polynomial equations for the coefficients of f. The resulting system is too big to be handled and solved by the usual Groebner basis packages. Instead, we use a MAPLE package by Raphael Nauheim [Nau95], which computes the solutions modulo a fixed prime, and lift them to p–adic numbers for sufficiently many digits in order to see periodicities and then guess the rational numbers. Once one has such a function, it is routine to verify that it has the desired properties, as we did in Section 4.

References

[Con97]   I. Connell, *Apecs (arithmetic of plane elliptic curves), a program written in maple*, available via anonymous ftp from math.mcgill.ca in /pub/apecs (1997).

[Cre97]   J. E. Cremona, *Algorithms for Modular Elliptic Curves*, Cambridge University Press (1997).

[DFK$^+$96] M. Daberkow, C. Fieker, J. Klüners, M. Pohst, K. Roegner, S. M., K. Wildanger, *KANT V4*, J. Symb. Comput. (1996), **24**(3), 267–283, KASH software available from ftp.math.tu-berlin.de in /pub/algebra/Kant/Kash/Binaries.

[FGS93]   M. Fried, R. Guralnick, J. Saxl, *Schur covers and Carlitz's conjecture*, Israel J. Math. (1993), **82**, 157–225.

[Fri22]   R. Fricke, *Die elliptischen Funktionen und ihre Anwendungen, Zweiter Teil*, B. G. Teubner, Leipzig, Berlin (1922).

[Fri70]   M. Fried, *On a conjecture of Schur*, Michigan Math. J. (1970), **17**, 41–55.

[Fri78]   M. Fried, *Galois groups and complex multiplication*, Trans. Amer. Math. Soc. (1978), **235**, 141–163.

[GMS97]   R. Guralnick, P. Müller, J. Saxl, *The rational function analogue of a question of Schur and exceptionality of permutation representations*, preprint.

[Hoe95]   M. v. Hoeij, *An algorithm for computing the Weierstraß normal form*, ISSAC '95 Proceedings (1995), Implementation to be found at http://klein.math.fsu.edu/~hoeij.

[Maz78]   B. Mazur, *Rational isogenies of prime degree*, Invent. Math. (1978), **44**, 129–162.

[MM]   G. Malle, B. H. Matzat, *Inverse Galois Theory*, Book manuscript.

[Mül97]   P. Müller, *A Weil–bound free proof of Schur's conjecture*, Finite Fields Appl. (1997), **3**, 25–32.

[Nau95]   R. Nauheim, *Algebraische Gleichungssysteme bei schlechter Reduktion*, Ph.D. thesis, Universität Heidelberg (1995), Software available from ftp.iwr.uni-heidelberg.de in /pub/nauheim.

[S⁺95]   M. Schönert, et al., *GAP – Groups, Algorithms, and Programming*, Lehrstuhl D für Mathematik, RWTH Aachen, Germany (1995).

[Sch23]   I. Schur, *Über den Zusammenhang zwischen einem Problem der Zahlentheorie und einem Satz über algebraische Funktionen*, S.-B. Preuss. Akad. Wiss., Phys.–Math. Klasse (1923), pp. 123–134.

[Sil86]   J. H. Silverman, *The Arithmetic of Elliptic Curves*, Springer–Verlag, New York (1986).

[Sil94]   J. H. Silverman, *Advanced Topics in the Arithmetic of Elliptic Curves*, Springer–Verlag, New York (1994).

[Tho90]   J. G. Thompson, *Groups of genus zero and certain rational functions*, in *Groups, Sel. Pap. Aust. Natl. Univ. Group Theory Program, 3rd Int. Conf. Theory Groups Rel. Top., Canberra/Aust. 1989*, vol. 1456 of *Lect. Notes Math.* (1990), 1990 pp. 185–190, Zbl. 749.12005.

[Tur95]   G. Turnwald, *On Schur's conjecture*, J. Austral. Math. Soc. Ser. A (1995), **58**, 312–357.

[Völ96]   H. Völklein, *Groups as Galois Groups – an Introduction*, Cambridge University Press, New York (1996).

IWR, UNIVERSITÄT HEIDELBERG, IM NEUENHEIMER FELD 368,
D-69120 HEIDELBERG, GERMANY
E-mail: Peter.Mueller@iwr.uni-Heidelberg.de

Tangential base points and Eisenstein power series

HIROAKI NAKAMURA

In this note, we discuss a Galois theoretic topic where the two subjects of the title intersect. Three co-related sections will be arranged as follows. In Part I, we review basic notion of tangential base points for etale fundamental groups of schemes of characteristic zero. Then, in Part II, we introduce 'Eisenstein power series' as a main factor of the Galois representation "of Gassner-Magnus type" arising from an affine elliptic curve with 'Weierstrass tangential base point'. Part III is devoted to examining the Eisenstein power series in the case of the Tate elliptic curve over the formal power series ring $\mathbb{Q}[[q]]$ (introduced in Roquette [R], Deligne-Rapoport [DR]). We deduce then a certain explicit relation (Th.3.5) between such Eisenstein power series and Ihara's Jacobi-sum power series [I1].

I

In [GR], A.Grothendieck invented Galois theory for general connected schemes. It is based on axiomatic characterization of a "Galois category" which models on the category $\mathrm{Rev}(X)$ of all finite etale covers of a scheme X. In this theory, the role of a base point of π_1 is played by a certain "Galois functor" $\mathrm{Rev}(X) \to \{\text{finite sets}\}$ which axiomatizes the functor of taking fibre sets over a "base point" for all covers in $\mathrm{Rev}(X)$. Then, a chain between two "base points" is by definition an invertible natural transformation between such Galois functors. In particular, the fundamental group based at a Galois functor Φ is the functorial automorphism group $\mathrm{Aut}(\Phi)$, or equivalently, the automorphism group of the coherent sequence of finite sets $\{\Phi(Y)\}_{Y \in \mathrm{Rev}(X)}$ (with maps induced from those in $\mathrm{Rev}(X)$) topologized naturally as a profinite group.

Recently, the notion of "base point at infinity" seems to be calling certain attentions of Galois-theorists, according as fascinating problems of Grothendieck [G] are known (cf. V.G.Drinfeld [Dr], L.Schneps&P.Lochak [SL].) This notion was founded rigorously by P.Deligne [De] as "tangential base point" for more general π_1-theory of motives (including Betti, de Rham, etale realizations etc.) Still in the original Galois context, G.Anderson and Y.Ihara [AI] initiated effective use of Puiseux power series to represent such a base point, which has led to a number of practical applications in Galois-Teichmüller theory ([IM], [IN], [Ma], [N2,3],...) Inspired from these works, in this paper, we shall employ the following simple definition of a tangential base point.

(1.1) Definition. Let X be a connected scheme, and $k((t))$ be the field of Laurent power series in t over a field k of characteristic zero. A k-rational tangential base point on X is, by definition, a morphism $\vec{v} : \operatorname{Spec} k((t)) \to X$. This amounts to giving a scheme-theoretic point x of X together with an embedding of the residue field of x into the field $k((t))$. (The point x may or maynot be a k-rational point of X; see examples below.)

A basic motivation to introduce the above definition is that it has been often the case that a certain role of a "base point at infinity" can be played by a generic point of a 1-dimensional subscheme with specified 1-parameter "t". Let us explain how such a tangential base point $\vec{v}$ could work in the study of Galois representations in fundamental groups. Following Anderson-Ihara [AI], we fix an algebraically closed overfield $\Omega = \bar{k}\{\{t\}\}$, the field of Puiseux power series in the symbols "$t^{1/n}$" with $(t^{1/mn})^m = t^{1/n}$ $(m, n \in \mathbb{N})$, which is the union of the Laurent power series fields $\bar{k}((t^{1/n}))$ for $n \in \mathbb{N}$. Given such a $\vec{v}$ (and $\Omega_{\vec{v}}$), for each cover $Y \in \operatorname{Rev}(X)$, we may associate the set of its $\Omega_{\vec{v}}$-valued points $Y(\Omega_{\vec{v}})$. This is a finite set as the fibre of the finite etale morphism $Y \to X$ over the geometric point $\vec{v}$ on X. Noticing also that every morphism $Y' \to Y$ in $\operatorname{Rev}(X)$ induces a natural map $Y'(\Omega_{\vec{v}}) \to Y(\Omega_{\vec{v}})$, we get a coherent sequence of finite sets $\{Y(\Omega_{\vec{v}})\}$ indexed by the objects $Y \in \operatorname{Rev}(X)$, or equivalently, a fibre functor $\Phi_{\vec{v}} : \operatorname{Rev}(X) \to \{\text{finite sets}\}$ $(Y \mapsto \Phi_{\vec{v}}(Y) = Y(\Omega_{\vec{v}}))$.

Now, the absolute Galois group $G_k = \operatorname{Gal}(\bar{k}/k)$ acts on $\Omega_{\vec{v}}$ by the co-efficientwise transformation of power series $\sum_{\alpha \in \mathbb{Q}} a_\alpha t^\alpha \mapsto \sum_{\alpha \in \mathbb{Q}} \sigma(a_\alpha) t^\alpha$, hence induces an automorphism of the sequence $\{Y(\Omega_{\vec{v}})\}_{Y \in \operatorname{Rev}(X)}$ coherently. Thus, we obtain a natural homomorphism

$$s_{\vec{v}} : G_k \to \pi_1(X, \vec{v}) := \operatorname{Aut}(\Phi_{\vec{v}}).$$

When X is defined to be geometrically connected over k, then $s_{\vec{v}}$ gives a splitting of the canonical exact sequence

$$(1.2) \qquad 1 \longrightarrow \pi_1(X_{\bar{k}}, \vec{v}) \longrightarrow \pi_1(X, \vec{v}) \overset{p_{X/k}}{\longrightarrow} G_k \longrightarrow 1.$$

By conjugation, $s_{\vec{v}}$ defines a Galois representation

$$\varphi_{\vec{v}} : G_k \to \operatorname{Aut}(\pi_1(X_{\bar{k}}, \vec{v}))$$

which lifts the exterior Galois representation

$$\varphi : G_k \to \operatorname{Out}(\pi_1(X_{\bar{k}}, \vec{v}))$$

induced from the exact sequence (1.2). Generally speaking, the group-theoretic character of φ is independent of the choice of base points, while that of $\varphi_{\vec{v}}$ is dependent on $\vec{v}$.

Example 0. Any k-rational point $x \in X(k)$ gives automatically a k-rational tangential base point via $k \hookrightarrow k((t))$.

Example 1. Let $X = \mathbf{P}^1_{\mathbb{Q}} - \{0, 1, \infty\}$ be the projective t-line over $\mathbb{Q}$ minus the three points $t = 0, 1, \infty$. Then, the residue field of the generic point x of X can be identified with the rational function field $\mathbb{Q}(t)$. The obvious embedding $\mathbb{Q}(t) \hookrightarrow \mathbb{Q}((t))$ determines a morphism

$$\operatorname{Spec} \mathbb{Q}((t)) \to X = \mathbf{P}^1_t - \{0, 1, \infty\},$$

whose target lies on the generic point x of X. We call the tangential base point given by this morphism the standard tangential base point on X, and denote it by $\overrightarrow{01}$. Note that the notion of $\overrightarrow{01}$ depends on the choice of the normalized coordinate t of $\mathbf{P}^1$ setting the 3 punctures to be $t = 0, 1, \infty$.

Now, we have a natural compactification $\mathbf{P}^1$ of X, with respect to which the above $\overrightarrow{01}$ can be extended to the morphism $\operatorname{Spec} \mathbb{Q}[[t]] \to \mathbf{P}^1_t$. This means that the "point" determined by $\overrightarrow{01}$ is not only the generic point x of X but also a uniformizer of the (completed) local ring at $t = 0$ on $\mathbf{P}^1_t$, i.e., a 1-dimensional tangent vector (consisting of 'direction' + 'speed') starting from $t = 0$. (If we change normalization (e.g., scale) of $t \in \mathbb{Q}((t))$ relative to the standard coordinate t of $\mathbf{P}^1$, the represented vector will differ from $\overrightarrow{01}$. In a few contexts where X is regarded as the elliptic modular curve of level 2, another tangential base point "$\frac{1}{16}\overrightarrow{01}$" plays a crucial role, as pointed out in [N2-3].) According to this realization, we usually picture $\overrightarrow{01}$ as a unit tangent vector rooted at 0 towards 1. Fix an embedding $\overline{\mathbb{Q}} \hookrightarrow \mathbb{C}$ so that the geometric fundamental group $\pi_1(X_{\overline{\mathbb{Q}}}, \overrightarrow{01})$ may be identified with the profinite completion of its natural Betti correspondent. Then, $\pi_1(X_{\overline{\mathbb{Q}}}, \overrightarrow{01})$ is a free profinite group $\hat{F}_2$ freely generated by the standard loops x, y running around the punctures $0, 1$ respectively.

It is known that the Galois representation $\varphi_{\overrightarrow{01}}$ embeds $G_{\mathbb{Q}}$ into $\operatorname{Aut}\hat{F}_2$ in such a way that

$$(1.3) \qquad \sigma(x) = x^{\chi(\sigma)}, \quad \sigma(y) = f_\sigma(x, y)^{-1} y^{\chi(\sigma)} f_\sigma(x, y) \quad (\sigma \in G_{\mathbb{Q}})$$

with $f_\sigma(x, y) \in [\hat{F}_2, \hat{F}_2]$ (G.V.Belyi), where $\chi : G_{\mathbb{Q}} \to \hat{\mathbb{Z}}^\times$ denotes the cyclotomic character. The pro-word f_σ is uniquely determined by the above formula, and plays a central role in the Grothendieck-Teichmüller theory.

Example 3. (Ihara-Matsumoto [IM]) Let X_n be the affine n-space $\mathbf{A}^n_{\mathbb{Q}}$ minus the discriminant locus D_n whose geometric fundamental group is isomorphic to the profinite Artin braid group $\hat{B}_n$ generated by $\tau_1, \ldots, \tau_{n-1}$ with relations $\tau_i \tau_j = \tau_j \tau_i$ ($|i - j| > 1$), $\tau_i \tau_{i+1} \tau_i = \tau_{i+1} \tau_i \tau_{i+1}$ ($1 \leq i < n - 1$). The covering space Y_n corresponding to the pure braid group $\hat{P}_n \subset \hat{B}_n$ can be naturally regarded as $\{(t_1, \ldots, t_n) \in \mathbf{A}_{\mathbb{Q}}^n \mid t_i \neq t_j \, (i \neq j)\}$. Define a tangential base point $\vec{v}' : \operatorname{Spec} \mathbb{Q}((t)) \rightarrow Y_n$ via $t \mapsto (t, t^2, \ldots, t^n)$, and let $\vec{v}$ be the projection image of $\vec{v}'$ on X_n. Then, it turns out that $\varphi_{\vec{v}} : G_{\mathbb{Q}} \rightarrow \operatorname{Aut} \hat{B}_n$ provides the Galois representation of the form:

$$(1.4) \qquad \begin{cases} \sigma(\tau_1) & = \tau_1^{\chi(\sigma)}, \\ \sigma(\tau_i) & = f_\sigma(y_i, \tau_i^2)^{-1} \tau_i^{\chi(\sigma)} f_\sigma(y_i, \tau_i^2) \qquad (1 \leq i \leq n - 1). \end{cases}$$

where $y_i = \tau_{i-1} \cdots \tau_1 \cdot \tau_1 \cdots \tau_{i-1}$. This Galois action is compatible with Drinfeld's formula discovered in the context of quasi-triangular, quasi-Hopf algebras ([Dr]).

Example 4. If a space X is given as a modular variety parametrizing certain types of objects, then to construct a tangential base point $\vec{v} : \operatorname{Spec} \mathbb{Q}((t)) \rightarrow X$ is equivalent to constructing such an object defined over $\mathbb{Q}((t))$. To do this, sometimes, formal patching method turns out to be useful in smoothing a specially degenerate object $Y_0/\mathbb{Q}$ over $\mathbb{Q}[[t]]$ whose generic fibre $Y_\eta/\mathbb{Q}((t))$ defines $\vec{v}$ with desired properties. We refer to [IN], [N2-3] for some of such examples of tangential base points constructed in the moduli spaces $M_{g,n}$ of the marked smooth curves. The method will also produce a "coalescing tangential base point" $\vec{v}(g)$ on the Hurwitz moduli space $\mathcal{H}(G; C_1, \ldots, C_r)$ associated to a Nielsen class $g \in Ni(G, C)$. Indeed, for any given transitive permutation group $G \subset S_n$ and for any generator system $g = (g_1, \ldots, g_r)$ with $g_1 \cdots g_r = 1$ (g_i lying in a conjugacy class $C_i \subset G$), define $r - 2$ triples $\{(x_i, y_i, z_i) \mid i = 1, \ldots, r - 2\}$ by setting $x_i = g_1 \cdots g_i$, $y_i = g_{i+1}$, $z_i = g_{i+2} \cdots g_r$. Then, since $x_i y_i z_i = 1$, each (x_i, y_i, z_i) (regarded as a branch cycle datum) defines a (not necessarily connected) branched cover $Y_i \rightarrow \mathbf{P}^1$ ramified only over $\{0, 1, \infty\}$. One obtains then an admissible cover $Y_s = \cup_i Y_i / \sim$ over a linear chain P of $\mathbf{P}_{01\infty}^1$ such that $\sim$ identifies the branch points on Y_i and Y_{i+1} according as the cycle orbits by $\langle z_i = x_{i+1}^{-1} \rangle$ in $\{1, \ldots, n\}$. One expects that suitable techniques for smoothing $Y \rightarrow P$ (cf. [HS], [W]) should yield a good $\vec{v}(g)$ on the Hurwitz moduli space, generalizing a prototype example given in [N2]. We hope to investigate some aspects of this construction in a circle of ideas of inverse Galois problems ([Fr]).

II

In [I1], Y.Ihara discovered deep aspects of arithmetic fundamental groups by interpolating complex multiplications of Fermat jacobians encoded in $\pi_1(\mathbf{P}^1 - \{0, 1, \infty\})$. Among other things, he introduced new l-adic power series $\mathcal{F}_\sigma \in \mathbb{Z}_l[[T_1, T_2]]$ ($\sigma \in G_{\mathbb{Q}}$) whose special values at roots of unity recover Jacobi sum grössencharacters. He also conjectured the explicit forms of Galois characters appearing in the coefficients of $\mathcal{F}_\sigma$ in terms of Soulé's l-adic cyclotomic elements, which was settled by Anderson [A], Coleman [C], Ihara-Kaneko-Yukinari [IKY] (see 3.6 below). Ihara [I2] also developed an l-adic theory of Fox's free differential calculus to control this power series $\mathcal{F}_\sigma$ in the framework of combinatorial group theory. This enables one to relate $\mathcal{F}_\sigma$ and f_σ of the previous section in a very simple way. We shall employ this treatment also here in a slightly more general setting applicable to higher genus curves.

Let X be a smooth projective curve of genus g over a field k of characteristic 0, S a non-empty closed subset of X with geometric cardinality n, and let $C = X - S$ be the affine complement curve. Practically, we shall be concerned with the "pure affine hyperbolic cases" of $(g, n) = (g, 1)$ or $(0, n)$ $(g \geq 1, n \geq 3)$, where the geometric fundamental group is a nonabelian free profinite group with its 1-st homology group being pure of weight -1 or -2 respectively.

Fix a rational prime p, and pick a k-rational tangential base point $\vec{v}$ on C. The Galois group G_k acts on $\pi_1(C_{\bar{k}}, \vec{v})$ and hence on its maximal pro-p quotient π. We shall write this action as $\varphi_{\vec{v}}^{(p)} : G_k \to \operatorname{Aut}(\pi)$. Note that π is a free pro-p group of rank $r := 2g + n - 1$ and that its abelianization H is canonically identified with the p-adic etale homology group $H_1(C_{\bar{k}}, \mathbb{Z}_p)$ which is a free $\mathbb{Z}_p$-module of rank r. Our interest will be concentrated on the kernel part of the composition map:

$$(2.1) \qquad \rho^{(p)} : G_k \xrightarrow{\varphi_{\vec{v}}^{(p)}} \operatorname{Aut}(\pi) \longrightarrow \operatorname{GL}(H).$$

The fixed field of the kernel of $\rho^{(p)}$ (resp. the kernel subgroup $\ker(\operatorname{Aut}(\pi) \to \operatorname{GL}(H))$ of $\operatorname{Aut}(\pi)$) will be denoted by $k(1)$ (resp. $\operatorname{Aut}_1 \pi$).

In order to analyze the restriction $\varphi_{\vec{v}}^{(p)}|_{G_{k(1)}}$ closely, we construct a certain combinatorial (anti-)representation

$$(2.2) \qquad \tilde{\mathfrak{A}} : \operatorname{Aut}_1 \pi \to \operatorname{GL}_r(\mathbb{Z}_p[[H]])$$

in analogy with the Gassner-Magnus representation in combinatorial group theory (cf. Ihara [I2], see also [Bi], [Mo] for topological aspects). Here, $\mathbb{Z}_p[[H]]$ is the abelianization of the complete group algebra $\mathbb{Z}_p[[\pi]]$ which is by definition the projective limit of the finite group rings $(\mathbb{Z}/p^n\mathbb{Z})[\pi/N]$ over the open normal subgroups $N \subset \pi$ and $n \in \mathbb{N}$. The Gassner-Magnus representation is a basic device to look at operations on the maximal meta-abelian quotient of π (cf. 2.7 below). To give its precise definition, we shall first introduce some terminology of free differential calculus.

If $\{x_1, \ldots, x_r\}$ is a free generator system of π, then, as shown by Lazard [La], $\mathbb{Z}_p[[\pi]]$ can be regarded as the ring of formal power series in non-commutative variables $t_i := x_i - 1$ $(i = 1, \ldots, r)$ over $\mathbb{Z}_p$. Each element

$\lambda \in \mathbb{Z}_p[[\pi]]$ is then written in the form

$$\lambda = \sum_w a_w \cdot w \qquad (a_w \in \mathbb{Z}_p),$$

where w runs over all finite words in $\{t_1, \ldots, t_r\}$ with non-negative exponents including the unity 1. We call a_1 the constant term of λ and denote it by $\varepsilon(\lambda)$. Classifying the other terms $a_w \cdot w$ ($w \neq 1$) of λ according to the right most letters, we may write uniquely $\lambda = \varepsilon(\lambda) + \sum_{i=1}^{r} \lambda_i t_i$ ($\lambda_i \in \mathbb{Z}_p[[\pi]]$). This λ_i is by definition the i-th free differential of λ, and will be denoted by $\partial\lambda/\partial x_i$:

$$(2.3) \qquad \lambda = \varepsilon(\lambda) + \sum_{i=1}^{r} \frac{\partial\lambda}{\partial x_i}(x_i - 1).$$

In the following, we use capital letters X_i, T_i to designate the images of $x_i, t_i \in \mathbb{Z}_p[[\pi]]$ in the abelianization $\mathbb{Z}_p[[H]]$ ($i = 1, 2$). Obviously, it follows that

$$\mathbb{Z}_p[[H]] = \mathbb{Z}_p[[T_1, \ldots, T_r]] \qquad (T_i = X_i - 1).$$

For a general element $\lambda \in \mathbb{Z}_p[[\pi]]$, we write λ^{ab} for its image in $\mathbb{Z}_p[[H]]$.

(2.4) Definition. For $\alpha \in \mathrm{Aut}_1\pi$, define its Gassner-Magnus matrix by

$$\bar{\mathfrak{A}}_\alpha := \left(\left(\frac{\partial\alpha(x_i)}{\partial x_j} \right)^{ab} \right)_{1 \leq i,j \leq r}.$$

(2.5) Proposition. *The mapping* $\bar{\mathfrak{A}} : \mathrm{Aut}_1\pi \to \mathrm{GL}_r(\mathbb{Z}_p[[H]])$ *($\alpha \mapsto \bar{\mathfrak{A}}_\alpha$) is an anti-representation, i.e.,* $\bar{\mathfrak{A}}_{\alpha\alpha'} = \bar{\mathfrak{A}}_{\alpha'}\bar{\mathfrak{A}}_\alpha$ *($\alpha, \alpha' \in \mathrm{Aut}_1\pi$).*

Proof. From direct computation, we have

$$\alpha\alpha'(x_i) = 1 + \sum_k \alpha\left(\frac{\partial\alpha'(x_i)}{\partial x_k} \right) \left(\sum_j \frac{\partial\alpha(x_k)}{\partial x_j}(x_j - 1) \right).$$

The result follows at once from the fact that α acts trivially on $\mathbb{Z}_p[[H]]$. $\square$

(2.6) We note that the above construction of $\bar{\mathfrak{A}}$ depends on the choice of the free basis $(x_1, \ldots, x_r)$ of π. In order to see its dependency on the choice, it would be convenient to introduce, more primitively, the Magnus matrices $\mathfrak{A}_\alpha$ for $\alpha \in \mathrm{Aut}(\pi)$ by

$$\mathfrak{A}_\alpha := \left(\left(\frac{\partial\alpha(x_i)}{\partial x_j} \right) \right)_{1 \leq i,j \leq r}.$$

with entries in the non-commutative algebra $\mathbb{Z}_p[[\pi]]$ satisfying the anti-1-cocycle property: $\mathfrak{A}_{\alpha\beta} = \alpha(\mathfrak{A}_\beta) \cdot \mathfrak{A}_\alpha$ $(\alpha, \beta \in \mathrm{Aut}(\pi))$. Then, one can show that if the free basis $(x_1, \ldots, x_r)$ is replaced by another basis $(x'_1, \ldots, x'_r)$, then the respective Magnus matrices $\mathfrak{A}_\alpha$, $\mathfrak{A}'_\alpha$ are related by "Jacobian matrices" as follows:

$$\mathfrak{A}_\alpha \cdot \frac{\partial(x_1, \ldots, x_r)}{\partial(x'_1, \ldots, x'_r)} = \alpha\left(\frac{\partial(x_1, \ldots, x_r)}{\partial(x'_1, \ldots, x'_r)}\right) \cdot \mathfrak{A}'_\alpha \qquad (\alpha \in \mathrm{Aut}(\pi)).$$

Since $\alpha \in \mathrm{Aut}_1\pi$ acts trivially on $\mathbb{Z}_p[[H]]$, the above implies that the Gassner-Magnus matrix $\bar{\mathfrak{A}}'_\alpha$ w.r.t. $(x'_1, \ldots, x'_r)$ is just the conjugation of $\bar{\mathfrak{A}}_\alpha$ by the Jacobian matrix $\left(\frac{\partial(x'_1, \ldots, x'_r)}{\partial(x_1, \ldots, x_r)}\right)^{ab}$.

The usefulness of the anti-1-cocycle representation of $\mathrm{Aut}(\pi)$ through Magnus matrices was shown by Anderson-Ihara [AI] Part 2 in their close study of Galois representations in $\pi_1(\mathbf{P}^1 - \{n \text{ points}\})$ (where a more 'geometric' variant was employed). Independently, Morita [Mo] presented effective applications of Magnus matrices in his theory of "traces" of topological surface mapping classes.

(2.7) We shall now briefly explain how the Gassner-Magnus representation looks at the meta-abelian quotient of π. Let π'' be the double commutator subgroup of π, i.e., $\pi'' = [\pi', \pi']$ where $\pi' = [\pi, \pi]$. Then, the pro-p version of Blanchfield-Lyndon theorem (cf. Brumer [Br] (5.2.2), Ihara [I2] Th.2.2) tells us an exact sequence of $\mathbb{Z}_p[[H]]$-modules:

$$(BL_p) \qquad 0 \to \pi'/\pi'' \xrightarrow{\partial} \mathbb{Z}_p[[H]] \otimes_{\mathbb{Z}_p[[\pi]]} I(\pi) \xrightarrow{\delta} I(H) \to 0,$$

where $I(*)$ denotes the augmentation ideal of $\mathbb{Z}_p[[*]]$, and the maps ∂, δ are defined by $\partial(a \bmod \pi'') = 1 \otimes (a-1)$, $\delta(b \otimes c) = b \cdot c^{ab}$. Since $I(\pi)$ is known to be the free $\mathbb{Z}_p[[\pi]]$-module of rank r with basis $x_i - 1$ $(i = 1, \ldots, r)$, one can identify the middle module with $\bigoplus_{i=1}^r \mathbb{Z}_p[[H]] \otimes (x_i - 1) \cong \mathbb{Z}_p[[H]]^{\oplus r}$ so that

$$\partial(a) = \left((\frac{\partial a}{\partial x_1})^{ab}, \ldots, (\frac{\partial a}{\partial x_r})^{ab}\right) \in \mathbb{Z}_p[[H]]^{\oplus r}.$$

Each automorphism $\alpha \in \mathrm{Aut}(\pi)$ acts on the modules of (BL_p) compatibly, especially on the middle one by $\alpha(b \otimes c) = \alpha(b) \otimes \alpha(c)$. In particular, $\alpha \in \mathrm{Aut}_1(\pi)$ acts on it $\mathbb{Z}_p[[H]]$-linearly with matrix representation given by the (transpose of) Gassner-Magnus matrix $\bar{\mathfrak{A}}_\alpha$. Noticing that π'/π'' is embedded there by ∂, one sees at least that the representation of $\mathrm{Aut}_1(\pi)$ in π'/π'' should be analyzed well by the Gassner-Magnus matrices.

Returning to the situation of Galois representation $\varphi_{\vec{v}}^{(p)} : G_k \to \mathrm{Aut}(\pi)$, our main concern thus turns to look at the composition with the Gassner-Magnus representation:

$$\bar{\mathfrak{A}}_{\vec{v}} = \bar{\mathfrak{A}} \circ \varphi_{\vec{v}}^{(p)}|_{G_{k(1)}} : G_{k(1)} \to \mathrm{GL}_r(\mathbb{Z}_p[[H]]).$$

In the remainder of this section, we review known results on the most basic two cases of $(g, n) = (0, 3), (1, 1)$. The former is Ihara's original case $C = \mathbf{P}^1 - \{0, 1, \infty\}$ ([I1,I2]), and the latter case is for $C =$ an elliptic curve minus one point, which was introduced/studied by Bloch [Bl], Tsunogai [T] and the author [N1].

Case 1: $C = \mathbf{P}^1 - \{0, 1, \infty\}$, $\vec{v} = \vec{01}$, $x_1 = x$, $x_2 = y$.

In this case, it follows from computations that

$$\bar{\mathfrak{A}}_{\vec{01}}(\sigma) = \begin{pmatrix} 1 & 0 \\ \left(\frac{\partial f_\sigma}{\partial x_1}(x_2 - 1)\right)^{ab} & \left(1 + \frac{\partial f_\sigma}{\partial x_2}(x_2 - 1)\right)^{ab} \end{pmatrix}$$

for $\sigma \in G_{\mathbb{Q}(\mu_{p^\infty})}$. (Note that $\mathbb{Q}(1) = \mathbb{Q}(\mu_{p^\infty})$ now). The power series

$$\mathcal{F}_\sigma(T_1, T_2) := \det \bar{\mathfrak{A}}_{\vec{01}}(\sigma) = \left(1 + \frac{\partial f_\sigma}{\partial x_2}(x_2 - 1)\right)^{ab}$$

is called the universal power series for Jacobi sums, or Ihara's power series ([I1,2], [Ic], [Mi]). In fact, Ihara showed that the mappings $\sigma \mapsto \mathcal{F}_\sigma(\zeta_{p^n}^a - 1, \zeta_{p^n}^b - 1)$ $(1 \le a, b < p^n)$ represent Jacobi sum grössencharacters over $\mathbb{Q}(\mu_{p^n})$, and also investigated the p-adic local behaviors of the coefficient characters. As in [I1], $\mathcal{F}_\sigma$ can be defined for all $\sigma \in G_\mathbb{Q}$, but in the present paper, we content ourselves with treating it only over $\mathbb{Q}(\mu_{p^\infty})$. By the above definition and (2.6), we see that $\mathcal{F}_\sigma$ in this range is determined only by the abelianization of the free basis (x_1, x_2) of π. In view of the above $(BL)_p$ specialized to this case, the module π'/π'' turns out to be a free $\mathbb{Z}_p[[H]]$-module of rank 1 generated by the class of $[x_1, x_2] = x_1 x_2 x_1^{-1} x_2^{-1} \bmod \pi''$. The image of ∂ is generated by $\partial([x_1, x_2]) = (-T_2, T_1) \in \mathbb{Z}_p[[H]]^{\oplus 2}$, and from this follows that $\bar{\mathfrak{A}}_{\vec{01}}(\sigma)$ (hence $\varphi_{\vec{01}}(\sigma)$) acts on $\mathrm{Im}(\partial) \cong \pi'/\pi''$ by multiplication by $\mathcal{F}_\sigma(T_1, T_2)$. This was in fact Ihara's original definition of $\mathcal{F}_\sigma$ in [I1].

Case 2: $C : y^2 = 4x^3 - g_2 x - g_3$ $(g_2, g_3 \in k)$, $\Delta = g_2^3 - 27g_3^2 \ne 0$.

In this case, we will take suitable generators x_1, x_2, z of π with $[x_1, x_2]z = 1$ so that z generates an inertia subgroup over the missed infinity point $O \in X$. For a tangential base point, we take $\vec{w} : \mathrm{Spec}\, k((t)) \to C$ defined by $t := -2x/y$ and call it the Weierstrass base point. The fixed field $k(1)$ of the kernel of G_k-action on $H = \pi/[\pi, \pi]$ is the field generated by all coordinates of the p-power division points of E over k. As shown in [N1] §6, there exists a unique power series $\mathcal{E}_\sigma(T_1, T_2) \in \mathbb{Z}_p[[T_1, T_2]]$ such that

$$\bar{\mathfrak{A}}_{\vec{w}}(\sigma) = \mathbf{1}_2 + \mathcal{E}_\sigma \cdot \begin{pmatrix} T_1 T_2 & -T_1^2 \\ T_2^2 & -T_1 T_2 \end{pmatrix}$$

for all $\sigma \in G_{k(1)}$. It is easy to see that $\mathcal{E}_\sigma$ depends only on the abelianization image $(\bar{x}_1, \bar{x}_2)$ of the free basis (x_1, x_2) of π. We shall call $\mathcal{E}_\sigma$ the Eisenstein power series associated to the Weierstrass equation $y^2 = 4x^3 - g_2 x - g_3$ and the basis $(\bar{x}_1, \bar{x}_2)$ of H. As in Case 1, again, the image of π'/π'' in $\mathbb{Z}_p[[H]]^{\oplus 2}$ via ∂ of (BL_p) is the free $\mathbb{Z}_p[[H]]$-module of rank 1 generated by $\partial(z) = (T_2, -T_1)$, but this time the action of $\tilde{\mathfrak{A}}_\alpha$ ($\alpha \in \text{Aut}_1 \pi$) on this image is trivial. This means that $\mathcal{E}_\sigma$ ($\sigma \in G_{k(1)}$) should be understood as an invariant of the 'unipotent' action of $\varphi_{\vec{w}}(\sigma)$ on the extension of (BL_p). In fact, using Bloch's construction described in [T],[N1], one can show more explicitly that

$$\varphi_{\vec{w}}(\sigma)(1 \otimes (x_i - 1)) = 1 \otimes (x_i - 1) + \mathcal{E}_\sigma(T_1, T_2) T_i \cdot \partial(z) \qquad (i = 1, 2)$$

holds for $\sigma \in G_{k(1)}$ in $\mathbb{Z}_p[[H]] \otimes_{\mathbb{Z}_p[[\pi]]} I(\pi)$.

(2.8) Remark. In [N1], we employed a special section $s : G_{k(1)} \to \pi_1(C)$ characterized by a certain group theoretical property instead of that induced from the above $\vec{w}$. The power series α_σ given in loc.cit. is the same as $\mathcal{E}_\sigma$ except that it misses constant term. If $p \geq 5$, the constant term of $\mathcal{E}_\sigma$ is $\frac{1}{12}\rho_\Delta(\sigma)$ where $\rho_\Delta : G_{k(1)} \to \mathbb{Z}_p$ is the Kummer character defined by the p-power roots of Δ.

In the next part III, we will show that $\mathcal{E}_\sigma$ for the Tate elliptic curve over $\mathbb{Q}((q))$ degenerates to a "logarithmic partial derivative" of Ihara's power series $\mathcal{F}_\sigma$.

III

In this section, we shall examine the Eisenstein power series arising from the Tate curve $T = \text{``}\mathbf{G}_m/q^{\mathbb{Z}}\text{''}$ over the rational power series ring $\mathbb{Q}[[q]]$ in one variable q. The affine equation defining T (minus the origin O) is

$$(3.1) \qquad\qquad y^2 + xy = x^3 + a_4(q)x + a_6(q),$$

where $a_4(q), a_6(q) \in \mathbb{Q}[[q]]$ are given by

$$a_4(q) = -5s_3(q), \quad a_6(q) = -\frac{1}{12}(5s_3(q) + 7s_5(q)),$$

$$s_k(q) := \sum_{n \geq 1} \sigma_k(n)q^n = \sum_{n \geq 1} \frac{n^k q^n}{1 - q^n} \qquad (k \geq 1).$$

The equation modulo q is $y^2 + xy = x^3$, hence T has a split multiplicative reduction. Indeed, the $\mathbb{Q}((q))$-rational points of $T - \{O\}$ are uniformized by $u \in \overline{\mathbb{Q}((q))}^\times \setminus q^{\mathbb{Z}}$ through the formulae:

$$x(u, q) = \sum_{n \in \mathbb{Z}} \frac{q^n u}{(1 - q^n u)^2} - 2s_1(q), \quad y(u, q) = \sum_{n \in \mathbb{Z}} \frac{(q^n u)^2}{(1 - q^n u)^3} + s_1(q),$$

and the special fibre T_s at $q = 0$ may be regarded as the nodal projective u-line with two points $u = 0, \infty$ identified (cf. [Si] V §3.) The completed local neighborhood at $u = 1$ in T can be identified as $\operatorname{Spec} \mathbb{Q}[[q, u - 1]]$ whose generic fibre is the spectrum of $\mathbb{Q}[[q, u - 1]] \otimes_{\mathbb{Q}[[q]]} \mathbb{Q}((q))$, the ring of formal power series in $u-1$ with bounded coefficients from $\mathbb{Q}((q))$. In the latter ring, we may arrange the mapping $q \mapsto t'$, $u - 1 \mapsto t'$ to define a tangential base point valued in $\mathbb{Q}((t'))$ on the generic fibre $T_\eta/\mathbb{Q}((q))$ of T minus the origin O_η. We write this tangential base point as $\vec{t} : \operatorname{Spec} \mathbb{Q}((t')) \to T_\eta - \{O_\eta\}$ and call it the Tate base point. In the following, we shall look at the Galois representation $\varphi_{\vec{t}} : G_{\mathbb{Q}} \to \operatorname{Aut} \pi_1(T_{\bar{\eta}} \setminus O)$, where $T_{\bar{\eta}} \setminus O$ denotes the generic geometric fibre of $T - \{O\}$.

First, let us connect the above $\mathbb{Q}$-rational base point $\vec{t}$ with the $\mathbb{Q}((q))$-rational Weiserstrass base point $\vec{w}$ on the generic elliptic curve T_η (introduced in the previous section). Indeed, we see that these two base points give essentially the same Galois action on $\pi_1(T_{\bar{\eta}} \setminus O)$ as follows. First, let us apply the change of variables "$X = x + \frac{1}{12}$, $Y = x + 2y$" to (3.1) to get the equation of Weierstrass form

$$(3.2) \qquad Y^2 = 4X^3 - g_2(q)X - g_3(q),$$

where

$$g_2(q) = 20(-\frac{B_4}{8} + \sum_{n \geq 1} \sigma_3(n)q^n),$$

$$g_3(q) = \frac{7}{3}(\frac{B_6}{12} - \sum_{n \geq 1} \sigma_5(n)q^n).$$

($B_4 = -1/30$, $B_6 = 1/42$ are the Bernoulli numbers.) Then, as explained in the previous section, the Weierstrass base point $\vec{w}$ on $T_\eta - O_\eta$ is defined as a tangential basepoint valued in $\mathbb{Q}((q))((t))$ by putting $t = -2X/Y$. Our claim here is that the Galois representation $\varphi_{\vec{t}} : G_{\mathbb{Q}} \to \operatorname{Aut} \pi_1(T_{\bar{\eta}} \setminus O)$ is essentially the same as the composite of $\varphi_{\vec{w}} : G_{\mathbb{Q}((q))} \to \operatorname{Aut} \pi_1(T_{\bar{\eta}} \setminus O)$ with the map $G_{\mathbb{Q}} \to G_{\mathbb{Q}((q))}$, where the last map is the one obtained from the coefficientwise $G_{\mathbb{Q}}$-action on the Puiseux power series in $\overline{\mathbb{Q}((q))} \hookrightarrow \overline{\mathbb{Q}}\{\{q\}\}$. Indeed, since x and y can be written respectively in the forms $(u-1)^{-2}(1 + \sum_m \alpha_m(u-1)^m)$, $-(u-1)^{-3}(1 + \sum_m \beta_m(u-1)^m)$ with $\alpha_m, \beta_m \in \mathbb{Q}[[q]]$ (cf. [Si] V §4), the coefficientwise $G_{\mathbb{Q}}$-actions on the two rings $\overline{\mathbb{Q}}[[q^{1/N}, t^{1/N}]]$, $\overline{\mathbb{Q}}[[q^{1/N}, (u-1)^{1/N}]]$ are compatible with their natural identification via $t = -2X/Y \equiv u - 1 \mod^{\times} (1 + (u-1)\mathbb{Q}[[q, u-1]])$. From this, the above relation of $\varphi_{\vec{w}}$ with $\varphi_{\vec{t}}$ follows.

Now, fix a rational prime p, and consider the p-adic Tate module $H = \varprojlim_m T_{\bar{\eta}}[p^m]$. As is well-known, in the Tate curve case, H is an extension of $\mathbb{Z}_p$ by $\mathbb{Z}_p(1)$. But in our case, we may split the extension in a natural way as follows. In fact, by the Tate uniformization by "$\mathbf{G}_m$" of $T_{\bar{\eta}}$, the p^m-division points $T_{\bar{\eta}}[p^m]$ may be identified with the subset $\{\zeta_{p^m}^a q^{b/p^m} \mid 0 \leq a, b < p^m\}$ of $\overline{\mathbb{Q}}\{\{q\}\}^\times$ with natural $G_{\mathbb{Q}((q))}$-action. (Here ζ_{p^m} is a primitive p^m-th roots of unity; we select those so that $\zeta_{p^n}^{p^m} = \zeta_{p^{n-m}}$ $(0 < m < n)$ once and for all.) Thus, we can take generators $\bar{x}_1, \bar{x}_2$ of H as projective sequences $\{q^{1/p^n}\}$, $\{\zeta_{p^m}\}$ respectively to obtain a splitting $H = \mathbb{Z}_p \bar{x}_1 \oplus \mathbb{Z}_p(1)\bar{x}_2$.

Let us then consider the maximal pro-p quotient π of $\pi_1(T_{\bar{\eta}} \setminus O, \vec{t})$ and the associated Galois representation $\varphi_{\vec{t}}^{(p)} : G_{\mathbb{Q}} \to \mathrm{Aut}(\pi)$. Then, with respect to the above basis $(\bar{x}_1, \bar{x}_2)$ of H, we have the Gassner-Magnus representation $\tilde{\mathfrak{A}}_{\vec{t}} : G_{\mathbb{Q}} \to \mathrm{GL}_2(\mathbb{Z}_p[[H]])$, which yields the (Tate-)Eisenstein power series $\mathcal{E}_\sigma^{\vec{t}}(T_1, T_2) \in \mathbb{Z}_p[[H]]$ $(\sigma \in G_{\mathbb{Q}(\mu_{p^\infty})})$ defined by

$$\bar{\mathfrak{A}}_{\vec{t}}(\sigma) = 1_2 + \mathcal{E}_\sigma^{\vec{t}} \cdot \begin{pmatrix} T_1 T_2 & -T_1^2 \\ T_2^2 & -T_1 T_2 \end{pmatrix}.$$

(3.3) Theorem. *Let* $U_i = \log(1 + T_i)$ $(i = 1, 2)$. *Then, in* $\mathbb{Q}_p[[U_1, U_2]]$, *we have*

$$\mathcal{E}_\sigma^{\vec{t}}(T_1, T_2) = \sum_{\substack{m \geq 2 \\ \text{even}}} \frac{\chi_{m+1}(\sigma)}{1 - l^m} \frac{U_2^m}{m!} \qquad (\sigma \in G_{\mathbb{Q}(\mu_{p^\infty})}).$$

Here $\chi_m : G_{\mathbb{Q}(\mu_{p^\infty})} \to \mathbb{Z}_p(m)$ *is the* m-*th Soulé character defined by the properties:*

$$\left(\prod_{\substack{1 \leq a < p^n \\ p \nmid a}} (1 - \zeta_{p^n}^a)^{a^{m-1}} \right)^{\frac{1}{p^n}(\sigma - 1)} = \zeta_{p^n}^{\chi_m(\sigma)} \qquad (\forall n \geq 1).$$

Proof. The statement follows from a more general formula given in [N1] which states that the coefficient $\kappa_{ij}(\sigma)$ of $U_1^i U_2^j / (1 - l^{i+j}) i! j!$ $((i, j) \neq (0, 0))$ is determined by the following Kummer properties:

$$\left(\prod_{\substack{0 \leq a, b < p^n \\ p \nmid (a, b)}} (\theta_{ab}^{(p^n)})^{a^i b^j} \right)^{\frac{1}{p^n}(\sigma - 1)} = \zeta_{p^n}^{12 \kappa_{ij}(\sigma)} \qquad (\forall n \geq 1),$$

where, for $0 \leq a, b < N = p^n$,

$$\theta_{ab}^{(N)} = q^{6 B_2(\frac{a}{N})} \zeta_N^{6b(\frac{a}{N} - 1)} \left[(1 - q^{\frac{a}{N}} \zeta_N^b) \prod_{n \geq 1} (1 - q^{n + \frac{a}{N}} \zeta_N^b)(1 - q^{n - \frac{a}{N}} \zeta_N^{-b}) \right]^{12}.$$

Here $B_2(T) = T^2 - T + \frac{1}{6}$ is the second Bernoulli polynomial. Observing that the coefficientwise $G_{\mathbb{Q}(\mu_{p^\infty})}$-action on the p-power roots of $\theta_{ab}^{(p^n)}$ is nontrivial only when $a = 0$, and noticing that $0^i = 1$ only when $i = 0$, we see that κ_{ij} occurs nontrivially only when $i = 0$, in which case it is equal to χ_{j+1}. The constant term turns out to vanish according to Remark(2.8) applied to $\Delta(q) = q \prod_{n \geq 1} (1 - q^n)^{12}$. $\square$

The above result (3.3) may also be deduced by an alternative method relating $\mathcal{E}_\sigma^{\vec{t}}$ explicitly with Ihara's power series $\mathcal{F}_\sigma$. We begin with

(3.4) Theorem. *One can take suitable generators x_1, x_2, z of $\pi_1(T_{\bar{\eta}} \setminus O, \vec{t})$ with $[x_1, x_2]z = 1$ such that (x_1, x_2) lifts $(\bar{x}_1, \bar{x}_2)$ above and that the Galois representation $\varphi_{\vec{t}} : G_{\mathbb{Q}} \to \operatorname{Aut} \pi_1(T_{\bar{\eta}} \setminus O, \vec{t})$ is expressed by the following formulae in terms of $(\chi(\sigma), f_\sigma)$ of §1 Example 1:*

$$
\begin{cases}
x_1 & \mapsto z^{\frac{1-\chi(\sigma)}{2}} f_\sigma(x_1 x_2 x_1^{-1}, z) x_1 f_\sigma(x_2^{-1}, z)^{-1}, \\
x_2 & \mapsto f_\sigma(x_2^{-1}, z) x_2^{\chi(\sigma)} f_\sigma(x_2^{-1}, z)^{-1} \\
z & \mapsto z^{\chi(\sigma)}
\end{cases}
$$

Proof. This assertion is essentially [N3] Cor.(4.5), except that the choice of generators differs from loc.cit. We first consider '$\mathbf{G}_m/q^{n\mathbb{Z}}$' $(n \geq 2)$ over $\mathbb{Q}[[q]]$ and realize the fundamental group of generic geometric fibre minus sections (one for each component) as a Van-Kampen composite of copies $\pi(i)$ $(i \in \mathbb{Z}/n\mathbb{Z})$ of $\pi_1(\mathbf{P}^1 - \{0, 1, \infty\})$. Identifying $\pi(i) = \langle 0_i, 1_i, \infty_i \mid 0_i 1_i \infty_i = 1 \rangle$, we compute the composite as the amalgamated product of the $\pi(i)$'s and $\langle e \rangle$ over the relations $\infty_i^{-1} = 0_{i+1}$ $(0 \leq i < n - 1)$, $\infty_{n-1} = e 0_0 e^{-1}$. Setting then standard generators $x_1 = e$, $x_2 = 0_0^{-1}$, $z_i = 1_{i-1}$ $(1 \leq i \leq n)$, we get the relation $[x_1, x_2] z_1 \cdots z_n = 1$. Then, [N3] Th.(3.15) computes the limit Galois representation $\varphi_{\vec{t}}$ on these generators in terms of the parameters $\chi(\sigma), f_\sigma$. The desired Galois representation follows from this computation after reducing $z_1 = z$, $z_2 = \cdots = z_n = 1$ (and checking its subtle independence of n). $\square$

Using the above, we shall compute $\mathcal{E}_\sigma^{\vec{t}}$ directly from the definition. Note that it suffices to look at $\frac{\partial \varphi^{(p)}(\sigma)(x_2)}{\partial x_2} - 1$ divided by $-T_1 T_2$ for $\sigma \in G_{\mathbb{Q}(\mu_{p^\infty})}$.

First, we compute

$$\frac{\partial \varphi_{\vec{t}}^{(p)}(\sigma)(x_2)}{\partial x_2} - 1 = \frac{\partial f_\sigma x_2 f_\sigma^{-1}}{\partial x_2} - 1 = (1 - f_\sigma x_2 f_\sigma^{-1})\frac{\partial f_\sigma}{\partial x_2} + f_\sigma - 1,$$

for $\sigma \in G_{\mathbb{Q}(\mu_{p^\infty})}$, which maps to $-T_2(\frac{\partial f_\sigma}{\partial x_2})^{\mathfrak{ab}} + 0$ in $\mathbb{Z}_p[[H]]$. But recalling $f_\sigma = f_\sigma(x_2^{-1}, z)$ here, we have

$$\frac{\partial f_\sigma}{\partial x_2} = \frac{\partial f_\sigma(x_2^{-1}, z)}{\partial x_2^{-1}}\frac{\partial x_2^{-1}}{\partial x_2} + \frac{\partial f_\sigma(x_2^{-1}, z)}{\partial z}\frac{\partial z}{\partial x_2},$$

where its first term must vanish in $\mathbb{Z}_p[[H]]$ because $\frac{\partial f_\sigma(x_2^{-1}, z)}{\partial x_2^{-1}}(x_2^{-1} - 1)$ is equal to $f_\sigma - 1 - \frac{\partial f_\sigma(x_2^{-1}, z)}{\partial z}(z - 1)$ which vanishes in $\mathbb{Z}_p[[H]]$. Then since $(\frac{\partial z}{\partial x_2})^{\mathfrak{ab}} = -T_1$, it follows that

$$\mathcal{E}_\sigma^{\vec{t}}(T_1, T_2) = -\Big(\lim_{\substack{w_1 \to x_2^{-1} \\ w_2 \to z}} \frac{\partial f_\sigma(w_1, w_2)}{\partial w_2}\Big)^{\mathfrak{ab}} = -\lim_{\substack{W_1 \to X_2^{-1} \\ W_2 \to 1}} \frac{\mathcal{F}_\sigma(W_1 - 1, W_2 - 1) - 1}{W_2 - 1}.$$

In terms of the variables $U_i = \log X_i$ ($i = 1, 2$), we conclude (after de l'Hospital's limit rule) the following relation between the Tate-Eisenstein power series and Jacobi sum power series.

(3.5) Theorem.

$$\mathcal{E}_\sigma^{\vec{t}}(T_1, T_2) = -\frac{\partial}{\partial T}\log \mathcal{F}_\sigma(S, T)\Big|_{\substack{S = \exp(-U_2) - 1, \\ T = 0}}$$

for $\sigma \in G_{\mathbb{Q}(\mu_{p^\infty})}$. □

This sort of relation between genus 1 and 0 was first expected by Takayuki Oda in his comments on a seminar talk by the author at RIMS, Kyoto University in 1993. The above formula was then obtained in the course of studies along [IN, N3] with Y.Ihara. Theorem (3.3) can then be deduced also by combining Theorem (3.5) with the following formula:

(3.6) Theorem. (Anderson [A], Coleman [C], Ihara-Kaneko-Yukinari [IKY])

$$\mathcal{F}_\sigma(T_1, T_2) = \exp\left(\sum_{\substack{m \geq 3 \\ odd}} \frac{\chi_m(\sigma)}{l^{m-1} - 1}\sum_{\substack{i+j=m \\ i,j \geq 1}} \frac{U_1^i U_2^j}{i!j!}\right) \qquad (\sigma \in G_{\mathbb{Q}(\mu_{p^\infty})}). \quad \square$$

The explicit formula (3.6) was proved by Anderson [A], Coleman [C] and Ihara-Kaneko-Yukinari [IKY] independently around 1985. Later Ichimura

[Ic], Miki [Mi] gave simplifications of the proof. All these proofs so far depended on the interpolation properties of the values $\mathcal{F}_\sigma(\zeta_{p^n}^a - 1, \zeta_{p^n}^b - 1)$ by Jacobi sums.

*   *   *

Recently, (in a more general profinite context) Ihara [I3], using the 5-cyclic relation of the Grothendieck-Teichmüller group, gave a purely algebraic proof of the factorization

$$\mathcal{F}_\sigma(T_1, T_2) = \Gamma_\sigma(T_1)\Gamma_\sigma(T_2)/\Gamma_\sigma((1 + T_1)(1 + T_2) - 1),$$

where $\Gamma_\sigma(T) \in \mathbb{W}_p[[T]]$ is Anderson's Gamma series [A] ($\mathbb{W}_p$: the ring of Witt vectors of $\bar{\mathbb{F}}_p$). In particular, $\mathcal{F}_\sigma$ has to be of the form

$$\mathcal{F}_\sigma = \exp\left(\sum_{\substack{m \geq 3 \\ odd}} c_m \sum_{i+j=m} \frac{U_1^i U_2^j}{i!j!}\right)$$

with some constants c_m. Then, he derived $c_m = \chi_m(\sigma)/(l^{m-1} - 1)$ by a direct method observing meta-cyclic covers of $\mathbf{P}^1 - \{0, 1, \infty\}$ (cf. also [De] §16 for the last technique). Thus, we now have a purely geometric proof of Theorem (3.6) without use of Jacobi sums.

Returning to our elliptic context, we see that combination of Theorems (3.3), (3.5) may also reconfirm the same values of c_m's independently, leading us to an elliptic interpretation of the logarithmic derivative of Anderson's Gamma series (with constant term dropped):

$$D \log \Gamma_\sigma(T_2) - D \log \Gamma_\sigma(0) = \mathcal{E}_\sigma^{\vec{t}}(T_1, T_2).$$

If p-adic Tate curves "$\mathbb{Q}_p^\times / q^{p^n \mathbb{Z}}$" ($n \in \mathbb{N}$) are employed instead of a single Tate curve over $\mathbb{Q}[[q]]$, then it can be shown that those Eisenstein power series $\{\mathcal{E}_\sigma^{(p^n)}(0, T_2)\}_{n \in \mathbb{N}}$ produce a $\mathbb{Q}_p$-valued distribution whose 'asymptotic expansion' gives a power series

$$2 \sum_{\substack{m \geq 2 \\ even}} E_{-m}^{(p)}(q)\varphi_{m+1}(\sigma)\frac{U_2^m}{m!}$$

for σ in the ramification subgroup $\mathcal{R}$ of $G_{\mathbb{Q}_p(\mu_{p^\infty})}$, where $\varphi_m : \mathcal{R} \to \mathbb{Z}_p$ represents the m-th Coates-Wiles homomorphism, and $E_s^{(p)}(q)$ is the p-adic Eisenstein series $\frac{1}{2}\zeta_p(1 - s) + \sum_n \sigma_{s-1}^{(p)}(n)q^n$ of weight s introduced by J.P.Serre [Se]. For this and other arithmetic aspects, we will have more discussions in subsequent works.

REFERENCES

[A]   G.Anderson, *The hyperadelic gamma function*, Invent. Math. **95** (1989), 63–131.

[AI]   G.Anderson, Y.Ihara, *Pro-l branched coverings of* $\mathbf{P}^1$ *and higher circular l-units, Part 1*, Ann. of Math. **128** (1988), 271–293; *Part 2*, Intern. J. Math. **1** (1990), 119–148.

[Bi]   J.S.Birman, *Braids, links, and mapping class groups*, Annals of Math. Studies, vol. 82, Princeton University Press, 1975.

[Bl]   S.Bloch, *letter to P.Deligne, May 1, 1984.*

[Br]   A.Brumer, *Pseudocompact algebras, profinite groups and class formations*, J. Algebra **4** (1966), 442–470.

[C]   R.Coleman, *Anderson-Ihara theory: Gauss sums and circular units*, Adv. Stud. in Pure Math. **17** (1989), 55–72.

[De]   P.Deligne, *Le groupe fondamental de la droite projective moins trois points*, The Galois Group over Q, ed. by Y.Ihara, K.Ribet, J.-P.Serre, Springer, 1989, pp. 79–297.

[Dr]   V.G.Drinfeld, *On quasitriangular quasi-Hopf algebras and a group closely connected with* $Gal(\bar{Q}/Q)$, Leningrad Math. J. **2(4)** (1991), 829–860.

[DR]   P.Deligne, M.Rapoport,, *Les schémas de modules de courbes elliptiques*, in "Modular functions of one variable II", Lecture Notes in Math., vol. 349, Springer, Berlin Heidelberg New York, 1973.

[F]   M.Fried, *Introduction to modular towers: generalizing dihedral group – modular curve connections*, Recent developments in the inverse Galois problem (M.Fried et al. eds), vol. 186, Contemp. Math. (AMS), 1995, pp. 111–171.

[GR]   A.Grothendieck, M.Raynaud, *Revêtement Etales et Groupe Fondamental (SGA1)*, Lecture Note in Math. **224** (1971), Springer, Berlin Heidelberg New York.

[G]   A.Grothendieck, *Esquisse d'un Programme, 1984*, in [SL-I], 7–48.

[HS]   D.Harbater, K.Stevenson, *Patching and Thickening Problems*, Preprint 1998.

[Ic]   H.Ichimura, *On the coefficients of the universal power series for Jacobi sums*, J. Fac. Sci. Univ. Tokyo IA **36** (1989), 1–7.

[I1]   Y.Ihara, *Profinite braid groups, Galois representations, and complex multiplications*, Ann. of Math. **123** (1986), 43–106.

[I2]   Y.Ihara, *On Galois representations arising from towers of coverings of* $\mathbf{P}^1 - \{0, 1, \infty\}$, Invent. Math. **86** (1986), 427–459.

[I3]   Y.Ihara, *On beta and gamma functions associated with the Grothendieck- Teichmüller modular group*, this volume.

[IKY]   Y.Ihara, M.Kaneko, A.Yukinari, *On some properties of the universal power seires for Jacobi sums*, Galois Representations and Arithmetic Algebraic Geometry, Advanced Studies in Pure Math., vol. 12, Kinokuniya Co. Ltd., North-Holland, 1987, pp. 65–86.

[IM]   Y.Ihara, M.Matsumoto, *On Galois actions on profinite completions of braid groups*, Recent developments in the inverse Galois problem (M.Fried et al. eds), vol. 186, Contemp. Math. (AMS), 1995, pp. 173–200.

[IN]   Y.Ihara, H.Nakamura, *On deformation of maximally degenerate stable marked curves and Oda's problem*, J. Reine Angew. Math. **487** (1997), 125–151.

[L]   M.Lazard, *Groupes analytiques p-adiques*, Publ. I.H.E.S **26** (1965), 389–603.

[Ma]   M.Matsumoto, *Galois group* $G_{\mathbb{Q}}$, *singularity* E_7, *and moduli* M_3, in [SL-II], 179–218.

[Mi]   H.Miki, *On Ihara's power series*, J. Number Theory **53** (1995), 23–38.

[Mo]   S.Morita, *Abelian quotients of subgroups of the mapping class groups of surfaces*, Duke Math. J. **70** (1993), 699–726.

[N1]   H.Nakamura, *On exterior Galois representations associated with open elliptic curves*, J. Math. Sci., Univ. Tokyo **2** (1995), 197–231.

[N2]   _____, *Galois representations in the profinite Teichmüller modular groups*, in [SL-I], 159–173.

[N3]   _____, *Limits of Galois representations in fundamental groups along maximal degeneration of marked curves I*, Amer. J. Math. (to appear).

[R]   P.Roquette, *Analytic theory of elliptic functions over local fields*, Hmburger Mathematische Einzelschriften, Vandenhoeck & Ruprecht in Göttingen, 1970.

[SL]   L.Schneps, P.Lochak (eds.), *Geometric Galois Actions I, II*, London Math. Soc. Lect. Note Ser., vol. 242, 243, Cambridge University Press, 1997.

[Se]   J.P.Serre, *Formes modulaires et fonctions zêta p-adiques*, Lect. Notes in Math. **350**, Springer-Verlag, 191–268.

[Si]   J.H.Silverman, *Advanced Topics in the Arithmetic of Elliptic Curves*, Graduate Texts in Math., vol. 151, Springer, 1994.

[T]   H.Tsunogai, *On the automorphism group of a free pro-l meta-abelian group and an application to Galois representations*, Math. Nachr **171** (1995), 315–324.

[W]   S.Wewers, *Deformation of tame admissible covers of curves*, Preprint 1998.

Department of Mathematics, Tokyo Metropolitan University,
Hachioji-shi, Tokyo 192-0397, JAPAN.
E-mail address: h-naka@comp.metro-u.ac.jp

Braid-abelian tuples in $\mathrm{Sp}_n(K)$

John Thompson [1] and Helmut Völklein [2], University of Florida

March 1997

Abstract: We study a class of generating systems of $\mathrm{PSp}_n(q)$ of length $m + 2$, where $n = 2m$. These systems are not quite rigid, but very close: The pure braid group induces an abelian group of permutations on inner classes of these tuples (actually, an elementary abelian group of order 2^{m-1}). The corresponding Hurwitz spaces are well under control. We show they are unirational varieties over $\mathbb{Q}$ in many cases. This yields Galois realizations over the rationals for $\mathrm{PSp}_n(q)$ under various conditions on n and q.

0 Introduction

Let $g_1, ..., g_r$ be generators of a finite group G with $g_1 \cdots g_r = 1$. We say they form a **rigid** generating system if for any generators $g_1', ..., g_r'$ of G with $g_1' \cdots g_r' = 1$ such that g_i' is conjugate g_i for all i, there is unique $g \in G$ with $g_i' = g^{-1} g_i g$ for all i. We say the system is **rational** if for each integer m prime to $|G|$ there is a permutation π of r letters such that g_i^m is conjugate $g_{\pi(i)}$ for all i. One version of the rigidity criterion (e.g., [V1], Cor. 3.13) says: If G has a rigid and rational system of r generators then G occurs as the Galois group of a regular extension of $\mathbb{Q}(x)$ with r branch points.

The rigidity criterion with $r = 3$ has been used extensively to obtain Galois realizations for various classes of (almost) simple groups, by Belyi, Thompson, Matzat's Heidelberg school and others. A comprehensive list of results can be found in the forthcoming book of Malle and Matzat [MM]. In [V2] it was shown that the groups $\mathrm{PGL}_n(q)$ and $\mathrm{PU}_n(q)$ have rigid generating systems of length $n+1$ (consisting of perspectivities). These systems (called Thompson tuples) are the only known rigid generating systems of a Lie type group (or any almost simple group) of length > 3. Now the question arises whether there are others.

[1] Partially supported by NSF grant DMS-9401399
[2] Partially supported by NSF grant DMS-9623199

The Thompson tuples are closely related to the Burau resp., Gassner representation of the Artin braid group over a finite field (see [V4]). A related representation of the symplectic braid group was studied in [MSV], and this led to Galois realizations over $\mathbb{Q}$ of the symplectic groups $\mathrm{Sp}_n(2^s)$ for $2^{s+1} \geq n$. For $\mathrm{Sp}_n(q)$, q odd, this method did not immediately apply, but the analogy with the case of the Artin braid group suggested there might be generators of $\mathrm{Sp}_n(q)$ with similar properties as the Thompson tuples. The present paper is devoted to the study of these tuples.

Section 1 studies the case $n = 4$, which in some sense is the heart of the matter. The general structure theorem for our tuples is Theorem 2.2. It contains the connection with Thompson tuples: If $(\sigma_1, ..., \sigma_m, \tau_1, \tau_2)$ is one of our tuples, then (excluding a certain degenerate case) the elements $\sigma_1, ..., \sigma_m$ leave two complementary totally isotropic subspaces W, W' invariant, and together with $(\sigma_1 \cdots \sigma_m)^{-1}$ induce a Thompson tuple in W and W'. Once this is proved, the existence and uniqueness theorem for Thompson tuples (see [V2]) can be used to complete the classification of our tuples.

The corresponding Hurwitz spaces are described in Theorem 5.1. Over $\mathbb{C}$, their structure is quite clear: They are obtained by adjoining the square roots of certain cross-ratios of branch points. The $\mathbb{Q}$-structure on these spaces is more difficult to pin down. Under certain conditions we can show that they are $\mathbb{Q}$-unirational. In one case, this requires looking at the boundary, which contains $\mathbb{Q}$-rational points coming from the Thompson tuples obtained by coalescing the last two branch points.

From this we obtain regular Galois realizations over $\mathbb{Q}$ for the groups $\mathrm{PSp}_n(q)$ under various conditions on n and q. This extends the results of [ThV].

NOTATION. Let K be a field, and $n \geq 1$. From section 2 on we assume $\mathrm{char}(K) \neq 2$. We let $K^* = K \setminus \{0\}$, and V is the space of column vectors of length n over $\bar{K}$. Here $\bar{K}$ is an algebraic closure of K. For $u, v, ... \in V$ we let $[u, v, ...]$ denote the subspace of V spanned by these elements. We view $\mathrm{GL}_n(K)$ as matrix group acting on V by left multiplication. Let I denote the identity matrix in $\mathrm{GL}_n(K)$.

For $\sigma \in \mathrm{GL}_n(K)$ let χ_σ be its characteristic polynomial multiplied by $(-1)^n$; i.e., $\chi_\sigma(x) = \prod_i (x - e_i)$, where $e_1, ..., e_n$ are the eigenvalues of σ, counted with multiplicity. We say σ is a **perspectivity** (resp., **bi-perspectivity**) if it has an eigenspace of dimension $n - 1$ (resp., of dimension $\geq n - 2$). Then the product of two perspectivities is a bi-perspectivity. A tuple $(\sigma_1, ..., \sigma_r)$ of elements of $\mathrm{GL}_n(K)$ is called **irreducible** if the elements generate an irreducible subgroup of $\mathrm{GL}_n(K)$.

1   Triples of bi-perspectivities in $\mathrm{GL}_4(K)$

In this section we take $n = 4$. Let $u_1, ..., u_4$ be the standard basis vectors in V; i.e., $u_1 = (1, 0, 0, 0)^t$ etc..

Lemma 1.1 *Let* $a, c \in K^*$. *Let* α_1, α_2 *be matrices in* $\mathrm{GL}_4(K)$ *of the form*

$$\alpha_1 = \begin{pmatrix} a_1 & 0 & x_1 & x_2 \\ 0 & b_1 & x_3 & x_4 \\ 0 & 0 & 1 & 0 \\ 0 & 0 & 0 & 1 \end{pmatrix}, \quad \alpha_2 = \begin{pmatrix} 1 & 0 & 0 & 0 \\ 0 & 1 & 0 & 0 \\ y_1 & y_2 & a_2 & 0 \\ y_3 & y_4 & 0 & b_2 \end{pmatrix}$$

satisfying $\mathrm{rank}(\alpha_1\alpha_2 - cI) \leq 2$ *and* $(\alpha_1\alpha_2 - aI)(u) = 0$ *for some* $u \neq 0$ *in* V. *Assume* $a_i \neq b_i$ *for* $i = 1, 2$. *Let* $H = <\alpha_1, \alpha_2>$.

(i) If $u = u_1 + ... + u_4$ *and* $a \neq c$ *then the* x_j, y_k *are uniquely determined by* a, c, a_i, b_i. *If further* $c = -1$ *and* $a_i b_i = 1$ *for* $i = 1, 2$ *then* H *leaves a non-zero symmetric bilinear form invariant; this form is non-degenerate if* $a \neq 1$ *and* $\mathrm{char}(K) \neq 2$.

(ii) If $u = u_1 + u_2 + u_3$ *and* $a_2 \neq c \neq b_2$ *then* $a = c$ *and* $(\alpha_1\alpha_2 - cI)^2 \neq 0$.

(iii) We cannot have $u = u_1 + u_2$. *If* $u = u_2 + u_3$ *then* H *fixes the span of* u_2 *and* u_3.

(iv) If $a = 1$ *then* H *fixes the vector* u.

(v) If $a_i \neq c \neq b_i$ *then* H *fixes none of the 1-spaces* $[u_j]$.

Proof : Write $u = e_1 u_1 + ... + e_4 u_4$ with $e_j \in K$. Comparing the third and fourth coordinates on both sides of the equation $\alpha_1\alpha_2\, u = a\, u$ yields

$$(1) \qquad e_1 y_1 + e_2 y_2 \;=\; e_3(a - a_2), \quad e_1 y_3 + e_2 y_4 \;=\; e_4(a - b_2)$$

Using this and the condition on the first two coordinates yields

$$(2) \qquad e_3 x_1 + e_4 x_2 \;=\; e_1(a - a_1)/a, \quad e_3 x_3 + e_4 x_4 \;=\; e_2(a - b_1)/a$$

If $u = u_1 + u_2$ then (2) yields $a_1 = a = b_1$, contradicting the hypothesis. If $u = u_2 + u_3$ then (1) gives $y_4 = 0$ and (2) gives $x_1 = 0$. This proves (iii). If $a = 1$ then $\alpha_2\, u = u$ by (1) and the shape of the first two rows of α_2; also $\alpha_1\alpha_2\, u = a\, u = u$. This proves (iv).

Now we use the condition $\mathrm{rank}(\alpha_1\alpha_2 - cI) \leq 2$. In the matrix $\alpha_1\alpha_2 - cI$, subtract x_1 times the third row plus x_2 times the fourth row from the first row; and subtract x_3 times the third row plus x_4 times the fourth row from the second row. This yields the matrix

$$(3) \qquad \begin{pmatrix} a_1 - c & 0 & cx_1 & cx_2 \\ 0 & b_1 - c & cx_3 & cx_4 \\ y_1 & y_2 & a_2 - c & 0 \\ y_3 & y_4 & 0 & b_2 - c \end{pmatrix}$$

This matrix has rank ≤ 2, hence all its 3×3-minors are zero. If α_1 fixes the 1-space $[u_3]$ then $x_1 = x_3 = 0$; hence vanishing of the left upper 3×3-minor yields $(a_1 - c)(b_1 - c)(a_2 - c) = 0$. This, coupled with suitable re-labeling of $u_1, ..., u_4$, proves (v).

Proof of (ii): Assume $u = u_1 + u_2 + u_3$ and $a_2 \neq c \neq b_2$. Adding the first column to the second in the matrix (3) and using (1), (2) yields the matrix

$$\begin{pmatrix} a_1 - c & a_1 - c & (a - a_1)c/a & cx_2 \\ 0 & b_1 - c & (a - b_1)c/a & cx_4 \\ y_1 & a - a_2 & a_2 - c & 0 \\ y_3 & 0 & 0 & b_2 - c \end{pmatrix}$$

Vanishing of the right lower 3×3-minor yields

$$(b_2 - c)\,(a - c)\,(b_1 a_2 - ac) \;=\; 0$$

Vanishing of the minor obtained by deleting the first column and second row yields

$$(b_2 - c)\,(a - c)\,(a_1 a_2 - ac) \;=\; 0$$

The last two equations imply $a = c$ (because $a_1 \neq b_1$). The last two columns of the matrix $\alpha_1 \alpha_2 - cI$ are

$$v = \begin{pmatrix} a_2 x_1 \\ a_2 x_3 \\ a_2 - c \\ 0 \end{pmatrix} \quad \text{and} \quad w = \begin{pmatrix} b_2 x_2 \\ b_2 x_4 \\ 0 \\ b_2 - c \end{pmatrix}$$

They are linearly independent, hence span $\mathrm{Im}(\alpha_1 \alpha_2 - cI)$. If $(\alpha_1 \alpha_2 - cI)^2 = 0$ it follows that $\ker(\alpha_1 \alpha_2 - cI) = \mathrm{Im}(\alpha_1 \alpha_2 - cI) = [v, w]$. This space contains $u = (1, 1, 1, 0)^t$, hence u is a multiple of v. Hence $x_1 = x_3$. On the other hand, (2) gives $x_1 = (a - a_1)/a$ and $x_3 = (a - b_1)/a$. Contradiction, since $a_1 \neq b_1$.

Proof of (i): Assume $u = u_1 + ... + u_4$. In the matrix (3), add the first column to the second and the fourth to the third, and multiply the third column by a. This together with (1), (2) gives the matrix

$$\begin{pmatrix} a_1 - c & a_1 - c & (a - a_1)c & cx_2 \\ 0 & b_1 - c & (a - b_1)c & cx_4 \\ y_1 & a - a_2 & (a_2 - c)a & 0 \\ y_3 & a - b_2 & (b_2 - c)a & b_2 - c \end{pmatrix}$$

Again we use that its 3×3-minors are zero. The upper left 3×3-minor yields y_1. Other minors yield y_3, x_2 and x_4. The remaining x_j, y_k are then given by (1) and (2). The result is:

$$(4) \qquad x_1 = \frac{(a_2 - c)\,(ac - a_1 b_2)}{ac\,(a_2 - b_2)}, \qquad x_2 = \frac{(c - b_2)\,(ac - a_1 a_2)}{ac\,(a_2 - b_2)}$$

$$x_3 = \frac{(a_2 - c)\,(ac - b_1 b_2)}{ac\,(a_2 - b_2)}, \qquad x_4 = \frac{(c - b_2)\,(ac - b_1 a_2)}{ac\,(a_2 - b_2)}$$

$$y_1 = \frac{(a_1 - c)\,(ac - b_1 a_2)}{c\,(a_1 - b_1)}, \qquad y_2 = \frac{(c - b_1)\,(ac - a_1 a_2)}{c\,(a_1 - b_1)}$$

$$y_3 = \frac{(a_1 - c)\,(ac - b_1 b_2)}{c\,(a_1 - b_1)}, \qquad y_4 = \frac{(c - b_1)\,(ac - a_1 b_2)}{c\,(a_1 - b_1)}$$

This proves the first assertion in (i). Now assume $c = -1$, $a_i b_i = 1$. Using (4) we check that the symmetric matrix

$$S = \begin{pmatrix} 0 & (a_1 - 1)\,(a_2 - 1)\,a & -a_1 a_2 a - 1 & a_1 a + a_2 \\ (a_1 - 1)\,(a_2 - 1)\,a & 0 & a_1 + a_2 a & -a_1 a_2 - a \\ -a_1 a_2 a - 1 & a_1 + a_2 a & 0 & (a_1 - 1)\,(a_2 - 1) \\ a_1 a + a_2 & -a_1 a_2 - a & (a_1 - 1)\,(a_2 - 1) & 0 \end{pmatrix}$$

satisfies $\alpha_i^t \, S \, \alpha_i = S$ for $i = 1, 2$. Thus α_1, α_2 leave the symmetric bilinear form associated with S invariant. This form is non-zero since $a_1 \neq 1 \neq a_2$. Since

$$\det(S) = -4\,a_1 a_2 a\,(a_1 - 1)^2\,(a_2 - 1)^2\,(a - 1)^2$$

we see the form is non-degenerate iff $a \neq 1$ and $\mathrm{char}(K) \neq 2$.

Remark 1.2 *The proof actually shows that given* $a, c, a_i, b_i \in K^*$ *with* $a \neq c$, $a_i \neq b_i$, *there are unique matrices* α_1, α_2 *of the shape given in Lemma 1.1 with* $\mathrm{rank}(\alpha_1 \alpha_2 - cI) \leq 2$ *and* $(\alpha_1 \alpha_2 - aI)(u_1 + \ldots + u_4) = 0$. *It can be shown that they generate an irreducible group if and only if* $a \neq 1 \neq c$, $a_1 a_2 b_1 b_2 \neq ac^2$, $a_i \neq c \neq b_i$ *and* $ac \notin \{a_1 a_2, b_1 b_2, a_1 b_2, b_1 a_2\}$. *This is not needed here. It can be used to produce rigid triples of bi-perspectivities in certain groups* $PGL_4(K)$ *and* $PU_4(K)$.

2   The tuples in $\mathrm{Sp}_n(K)$

From now on we assume $\mathrm{char}(K) \neq 2$. Let $n = 2m \geq 4$ even and $G = \mathrm{Sp}_n(K)$, the subgroup of $\mathrm{GL}_n(K)$ fixing a non-degenerate symplectic form $(,)$ on V defined over K. For elements or subspaces U, W of V we write $U \perp W$ iff $(U, W) = 0$; and $U^\perp = \{v \in V : v \perp U\}$. Let Σ be the set of $\sigma \in G$ with $\mathrm{rank}(\sigma - I) = 2$ having two distinct eigenvalues $\neq 1$. Then these two eigenvalues are of the form a, a^{-1} with $a \neq \pm 1$, and $\chi_\sigma = (x^2 - tx + 1)(x - 1)^{n-2}$ with $t = a + a^{-1} \neq \pm 2$. We call the eigenspaces $C = \mathrm{ker}(\sigma - aI)$ and $C' = \mathrm{ker}(\sigma - a^{-1}I)$ the **centers** of σ. These are 1-spaces in V with $C \not\perp C'$. The map $\sigma \mapsto t$ induces a bijection between the set of conjugacy classes of G contained in Σ and the $t \in K$ with $t \neq \pm 2$.

Lemma 2.1 Let $\sigma_1, ..., \sigma_{m+1} \in G$ with $\sigma_i \in \Sigma$ for $i \leq m$, $\mathrm{rank}(\sigma_{m+1} - I) \leq 2$ and $\sigma_1 \cdots \sigma_{m+1} = -I$. If $n = 4$ assume additionally that σ_3 is not similar $\mathrm{diag}(1, 1, -1, -1)$. Let $1 \leq j < k \leq m$. Then we can label the centers of σ_j (resp., σ_k) as C_j, C_j' (resp., C_k, C_k') such that $C_j \perp C_k$, $C_j' \perp C_k'$, $C_j \not\perp C_k'$ and $C_j' \not\perp C_k$. Further, the group $< \sigma_j, \sigma_k >$ fixes $C_j + C_k$ and $C_j' + C_k'$, but none of C_j, C_j', C_k, C_k'.

Proof : By braiding the tuple $(\sigma_1, ..., \sigma_m)$ (see section 3), we can move σ_j, σ_k into the first two positions. Thus it suffices to do the case $j = 1$, $k = 2$.

Let $X_i = (\sigma_i - I)(V)$. Then $\dim X_i = 2$ for $i \leq m$, and $\dim X_{m+1} \leq 2$. We first prove that:

$$(5) \qquad V = X_1 \oplus ... \oplus X_{m-1} \oplus X_{m+1}$$

Let $E = X_1 + ... + X_{m-1} + X_{m+1}$. Then E is invariant under all σ_i, and σ_m acts in V/E as -1. Since $\sigma_m \in \Sigma$ this implies $E = V$, hence (5). It follows that the space $U = X_1 + X_2$ has dimension 4 if $n > 4$. Also, U is invariant under σ_1 and σ_2.

Let $\sigma_0 = \sigma_1 \sigma_2 = -(\sigma_3 \cdots \sigma_{m+1})^{-1}$. Then σ_0 acts trivially in V/U, hence $\ker(\sigma_0 + I) \subset U$. We have $\ker(\sigma_3 - I) \cap ... \cap \ker(\sigma_{m+1} - I) \subset \ker(\sigma_0 + I)$, hence

$$(6) \qquad \dim U \cap \ker(\sigma_0 + I) \geq 2 \quad \text{and} \quad (\sigma_0 - I)(V) \subset U$$

Recall that $\dim U = 4$ if $n > 4$. If $\cdot n = 4$ then by (6) and the extra hypothesis in this case we see that $\dim U \geq 3$.

Case 1: $U \cap U^\perp \neq 0$.

Let $R = U \cap U^\perp$. The elements σ_1, σ_2 act trivially in V/U, hence in $V/(U + U^\perp) = V/(R^\perp)$. It follows that σ_1, σ_2 act trivially in R. Thus $R \neq U$, and so $\dim R = \dim U - 2$.

Since σ_0 acts trivially in R, it follows from (6) that σ_0 acts as -1 in U/R. Thus σ_1 and $-\sigma_2^{-1}$ induce the same transformation in U/R. Hence the centers of σ_1 coincide with those of σ_2 modulo R. Thus we can label the centers of σ_1 (resp., σ_2) as C_1, C_1' (resp., C_2, C_2') such that $C_1 \equiv C_2$ and $C_1' \equiv C_2'$ mod R. Then they satisfy the orthogonality relations in the Lemma. If $C_1 = C_2$ then $\dim U < 4$, hence $n = 4$; then σ_1, σ_2 act trivially in the 2-space $C_1^\perp / C_1$, hence so does σ_0 and thus σ_0 is similar to $\mathrm{diag}(1, 1, -1, -1)$. This is excluded by hypothesis, hence $C_1 \neq C_2$. Analogously, $C_1' \neq C_2'$.

Thus $C_1 + C_2 = C_1 + Y = C_2 + Y$ for some 1-space $Y \leq R$. Hence σ_1 and σ_2 fix $C_1 + C_2$ (and analogously, $C_1' + C_2'$). Finally, σ_1 acts on $C_1 + Y$ as $\mathrm{diag}(s, 1)$ with $s \neq 1$, hence fixes no 1-space on $C_1 + C_2 = C_1 + Y$ except C_1 and Y. Thus $\sigma_1(C_2) \neq C_2$. This (and its analogues) proves the Lemma in Case 1. Thus we may assume from now on that U is non-degenerate. Then in particular, $\dim U = 4$.

Connection to section 1:

Set $\alpha_i = \sigma_i|_U$ for $i = 1, 2$. Let u be an eigenvector of $\alpha_1\alpha_2 = \sigma_0|_U$, for some eigenvalue a. Since $U = X_1 + X_2$, and X_i is spanned by the centers of σ_i (for $i = 1, 2$), there is a basis $u_1, ..., u_4$ of U such that the matrices of α_1, α_2 with respect to this basis are as in Lemma 1.1, with $c = -1$, $a_i b_i = 1$ and $a_i, b_i \neq \pm 1$ for $i = 1, 2$. In particular, the centers of σ_1 (resp., σ_2) are $[u_1]$ and $[u_2]$ (resp., $[u_3]$ and $[u_4]$). None of these 1-spaces is fixed by $< \sigma_1, \sigma_2 >$, by part (v) (of Lemma 1.1).

If $a = 1$ then by (iv) we get $\ker(\alpha_1 - I) \cap \ker(\alpha_2 - I) \neq 0$, hence $\mathrm{Im}(\alpha_1 - I) + \mathrm{Im}(\alpha_2 - I) \neq U$. Contradiction, since $X_1 + X_2 = U$. Thus $a \neq 1$.

Case 2: $a \neq \pm 1$.

Then the eigenvalues of $\alpha_1\alpha_2$ are $-1, -1, a, a^{-1}$ with $a \neq a^{-1}$. Thus the a-eigenspace of $\alpha_1\alpha_2$ is 1-dimensional, hence equals $< u >$.

Replacing each u_ν by a scalar multiple we may assume $u = e_1 u_1 + ... + e_4 u_4$, $e_\nu \in \{0, 1\}$. Assume first that all $e_\nu = 1$. Then we are in case (i) of Lemma 1.1. Thus there is a non-zero symmetric form $(,)'$ on U invariant under the α_i. We can write it as $(v, w)' = (\lambda(v), w)$, $v, w \in U$, where $\lambda \in \mathrm{End}(U)$ commutes with the α_i. Then λ fixes all eigenspaces of α_i and $\alpha_1\alpha_2$; in particular, it fixes all $< u_j >$ and $< u >$, where $u = u_1 + ... + u_4$. Hence λ is a scalar. Thus $(,)'$ is symmetric as well as symplectic, hence is zero — a contradiction.

Thus not all $e_\nu = 1$. By (ii),(iii),(v) and their analogues we see that $e_1 \neq e_2$ and $e_3 \neq e_4$. Thus we may assume $u = u_2 + u_3$ (by suitable re-labeling of indices). Hence by (iii) the α_i fix the 2-space $W = C_1 + C_2$, where $C_1 = < u_2 >$, $C_2 = < u_3 >$. Then α_2 acts in W with eigenvalues 1 and $a_2 \neq 1$, which forces W to be degenerate; i.e., $C_1 \perp C_2$.

Now let u' be an a^{-1}-eigenvector of $\alpha_1\alpha_2$. It follows analogously that $u' \in W' = C_1' + C_2'$, where C_i' is a 1-dimensional eigenspace of α_i, the space W' is invariant under the α_i and $C_1' \perp C_2'$. The space $[u, u'] = \mathrm{Im}(\alpha_1\alpha_2 + I)$ is non-degenerate, hence cannot equal W. Thus $W \neq W'$. Thus if $W \cap W'$ is non-zero then it equals either C_1 or C_2, hence C_1 or C_2 is fixed by both α_i. This contradicts (v), hence $W \cap W' = 0$.

If $C_i \perp C_j'$ for $i \neq j$ then $C_j' = C_i^\perp \cap W'$ is fixed by α_1 and α_2. Again a contradiction to (v). This proves the claim.

Case 3: $\alpha_1\alpha_2$ has -1 as only eigenvalue.

Analogously to (5) we get $U = \mathrm{Im}(\alpha_1 - I) \oplus \mathrm{Im}(\alpha_1\alpha_2 + I)$. Thus $\mathrm{Im}(\alpha_1\alpha_2 + I)$ is 2-dimensional. It cannot be non-degenerate, since $\alpha_1\alpha_2 + I$ is nilpotent in Case 3. Thus $\mathrm{Im}(\alpha_1\alpha_2 + I)$ equals its perpendicular space $\ker(\alpha_1\alpha_2 + I)$. In particular, $(\alpha_1\alpha_2 + I)^2 = 0$.

By the previous paragraph, we can write $u_3 = u'' + u'$ with $u'' \in \mathrm{Im}(\alpha_1 - I)$ and $u' \in \mathrm{Im}(\alpha_1\alpha_2 + I) = \ker(\alpha_1\alpha_2 + I)$. Since $\mathrm{Im}(\alpha_1 - I) = [u_1, u_2]$ we have $u' = e_1 u_1 + e_2 u_2 + u_3$ for certain $e_\nu \in K$. In particular, $u' \neq 0$. Thus we can take $u = u'$ in Lemma 1.1. Replacing u_1 and u_2 by suitable scalar multiples we may assume $e_\nu \in \{0, 1\}$.

Since $(\alpha_1\alpha_2 - cI)^2 = (\alpha_1\alpha_2 + I)^2 = 0$ we have $u \neq u_1 + u_2 + u_3$ by (ii). Thus $u = u_1 + u_3$ or $u = u_2 + u_3$. Hence the α_i fix $[u_1, u_3]$ or $[u_2, u_3]$ by (iii) (and its analogues). Replacing u_3 by u_4 yields that the α_i also fix $[u_1, u_4]$ or $[u_2, u_4]$. The rest is as in Case 2. ∎

Let T be the set of transvections in $G = \mathrm{Sp}_n(K)$ (elements τ with $\mathrm{rank}(\tau - I) = 1$). The sets T and Σ are invariant under the normalizer $\hat{G} = \mathrm{GSp}_n(K)$ of G in $\mathrm{GL}_n(K)$, and $\hat{G}$ acts transitively on T. If K is finite then T consists of two G-orbits.

Let $\mathcal{N}$ be the set of tuples $(\sigma_1, ..., \sigma_m, \tau_1, \tau_2)$ with $\sigma_i \in \Sigma$, $\tau_j \in T$ and $\sigma_1 \cdots \sigma_m \tau_1 \tau_2 = -I$. Let $\hat{G}$ act component-wise on $\mathcal{N}$, and let $\hat{\mathcal{N}}$ be the set of $\hat{G}$-orbits on $\mathcal{N}$.

Theorem 2.2 *Let* $\mathbf{n} = (\sigma_1, ..., \sigma_m, \tau_1, \tau_2) \in \mathcal{N}$.

(i) The centers of all the σ_i *are mutually distinct, hence form a set of cardinality* $n = 2m$. *Orthogonality* $\perp$ *is an equivalence relation on this set, and there are exactly two equivalence classes, of the form* $\{C_1, ..., C_m\}$ *and* $\{C_1', ..., C_m'\}$, *where* C_i, C_i' *are the centers of* σ_i. *Let* W *(resp.,* W'*) be the sum of the* C_i *(resp.,* C_i'*). Then* W *and* W' *are maximal totally isotropic subspaces of* V, *invariant under all* σ_i, *and* $\dim(W \cap W') \leq 1$. *The group* $< \sigma_1, ..., \sigma_m, \tau_1, \tau_2 >$ *acts irreducibly in* V.

(ii) Let a_i *be the eigenvalue of* σ_i *on* C_i. *If* $a_1 \cdots a_m \neq (-1)^{m-1}$ *then* $V = W \oplus W'$. *In this case, the restrictions of* $\sigma_1, ..., \sigma_m$ *and of* $(\sigma_1 \cdots \sigma_m)^{-1}$ *to* W *form a Thompson tuple (in the sense of [V2]); in particular, the group* $< \sigma_1, ..., \sigma_m >$ *acts irreducibly in* W *(and* W'*).*

(iii) Consider another tuple $\tilde{\mathbf{n}} = (\tilde{\sigma}_1, ..., \tilde{\sigma}_m, \tilde{\tau}_1, \tilde{\tau}_2) \in \mathcal{N}$, *and define the* $\tilde{a}_i$ *as in (ii). (They depend on the choice of one of the two equivalence classes associated with* $\tilde{\mathbf{n}}$*). Then* $\mathbf{n}$ *and* $\tilde{\mathbf{n}}$ *are conjugate under* $\hat{G}$ *if and only if either* $\tilde{a}_i = a_i$ *for all* i *or* $\tilde{a}_i = a_i^{-1}$ *for all* i.

(iv) If in the tuple $(\sigma_1, ..., \sigma_m, \tau_1, \tau_2)$ *we replace the* m-*th entry by* $\tilde{\sigma}_m = \sigma_m^{\tau_1}$ *and the* $(m+1)$-*th entry by* $\tilde{\tau}_1 = \tau_1^{\sigma_m \tau_1}$ *then we obtain another tuple* $\tilde{\mathbf{n}}$ *in* $\mathcal{N}$. *Associated* $\tilde{a}_i$ *are given by* $\tilde{a}_i = a_i$ *for* $i = 1, ..., m-1$ *and* $\tilde{a}_m = a_m^{-1}$.

(v) Let $L = K(a_1, ..., a_m)$. *Then* $[L : K] \leq 2$. *Let* $S = K^*$ *if* $L = K$, *otherwise* $S = \{s \in L : N_{L/K}(s) = 1\}$. *Then all* a_i *lie in* S. *Now suppose* K *is finite. Then* τ_1 *and* τ_2 *are* G-*conjugate if and only if the element* $(-1)^{m-1}a_1 \cdots a_m$ *is a square in* S.

Proof : Set $\sigma_{m+1} = \tau_1\tau_2$. Then σ_{m+1} is not similar $\mathrm{diag}(-1, -1, 1, ..., 1)$ (because the latter matrix cannot be the product of two transvections). Hence $\sigma_1, ..., \sigma_{m+1}$ are as in Lemma 2.1. This implies the first assertion in (i).

For the second assertion, let i, j, k be distinct indices from $\{1, ..., m\}$. Let C_i, C_j, C_k be centers of $\sigma_i, \sigma_j, \sigma_k$, respectively, satisfying $C_i \perp C_j$ and $C_j \perp$

C_k. Let C_i', C_j', C_k' be the other center of σ_i, σ_j, σ_k, respectively. Assume $C_i \not\perp C_k$. Then $C_i \perp C_k'$ (by the Lemma). We consider the case $i < j < k$; the other cases can be done by obvious modifications. By braiding action, the elements σ_j, $\sigma_i^{\sigma_j}$, and σ_k occur in a tuple with the same properties as $(\sigma_1, ..., \sigma_{m+1})$. The centers of $\sigma_i^{\sigma_j}$ are $C = \sigma_j^{-1}(C_i)$ and $C' = \sigma_j^{-1}(C_i')$. We know that $C + C_i = C_i + C_j$, and either $C \perp C_k$ or $C \perp C_k'$. If $C \perp C_k$ then $(C + C_j) \perp C_k$, hence $C_i \perp C_k$, contradicting the assumption. If $C \perp C_k'$ then $(C + C_i) \perp C_k'$, hence $C_j \perp C_k'$, also a contradiction (since $C_j \perp C_k$). This proves that orthogonality is an equivalence relation on the set of centers. By Lemma 2.1, there are exactly two equivalence classes, each of which contains one center of each σ_i, $i \leq m$. Hence we can label these classes as claimed in (i).

The spaces W, W' are clearly totally isotropic, and by Lemma 2.1 they are invariant under all σ_i. Let B_1 be the 1-space $(\tau_1 - I)(V)$. Then $W + W' + B_1$ is invariant under all σ_i and τ_1, and these elements act trivially in $V/(W + W' + B_1)$. Hence τ_2 acts as -1 in $V/(W + W' + B_1)$, which implies $W + W' + B_1 = V$. Hence $\dim (W + W') \geq n - 1$. Since W, W' are totally isotropic this implies $\dim W \geq m - 1$. If $\dim W = m - 1$ then $\sigma_1, ..., \sigma_m$ act trivially in the 2-space $W^\perp/W$, hence σ_{m+1} acts as -1. Thus σ_{m+1} is similar $\mathrm{diag}(-1, -1, 1, ..., 1)$, a contradiction. Hence $\dim W = \dim W' = m$. This proves all assertions in (i) except the last one, which will be proved after the proof of (iv).

(ii) Set $C_{m+1} = (\sigma_{m+1} - I)(W)$. We claim that $\dim C_{m+1} = 1$ ($= \dim C_i$, $i \leq m$), and σ_i acts trivially in W/C_i for $i = 1, ..., m + 1$. The space $W \cap W'$ is orthogonal to all C_i, C_i', hence it contains none of the C_i, C_i' for $i \leq m$. This implies that σ_i acts trivially in W/C_i for $i = 1, ..., m$. If $V = W \oplus W'$ the group $< \sigma_1, ..., \sigma_{m+1} >$ acts dually in W and W'; since $\mathrm{rank}(\sigma_{m+1} - I) = 2$ by (5) it follows that $\dim C_{m+1} = 1$ and σ_{m+1} acts trivially in W/C_{m+1}.

Now assume $\dim (W \cap W') = 1$. The σ_i, $i \leq m$, act trivially in $V/(W + W')$, hence in $R := W \cap W'$ (because of the dual pairing between $V/(W+W')$ and $W \cap W' = (W + W')^\perp$). Thus σ_{m+1} acts as -1 in $V/(W + W')$ and in R. Since $\mathrm{rank}(\sigma_{m+1} - I) \leq 2$, it follows that σ_{m+1} acts trivially in W/R; and $R = C_{m+1}$. This proves the claim in the previous paragraph.

We further claim that $C_1, ... C_{m+1}$ form a frame in W; i.e., any m of them are linearly independent. Since $W = C_1 \oplus ... \oplus C_m$ we only need to show that C_{m+1} does not lie in the span of $m - 1$ of the others. If, say, $C_{m+1} \subset U := C_1 + ... + C_{m-1}$ then U is invariant under all σ_i, and σ_m acts as -1 in W/U. Contradiction, since $\sigma_m \in \Sigma$.

In the terminology of [V2], it follows that the restrictions of $\sigma_1, ..., \sigma_m, -\sigma_{m+1}$ to W form a $(1, ..., 1, -1; a_1, ..., a_m, (-1)^{m-1} a_1^{-1} \cdots a_m^{-1})$-tuple in $\mathrm{GL}(W)$, adapted to the frame $C_1, ... C_{m+1}$. If $a_1 \cdots a_m \neq (-1)^{m-1}$ then this tuple is irreducible by [V2], Lemma 2, hence is a Thompson tuple; further, $W \cap W' = 0$ since this space is invariant under the σ_i. This proves (ii).

(iv) It is clear that $\tilde{\mathbf{n}} \in \mathcal{N}$. The centers of $\tilde{\sigma}_m = \sigma_m^{\tau_1}$ are $\tau_1^{-1}(C_m)$ and $\tau_1^{-1}(C_m')$. Since $\tilde{\mathbf{n}} \in \mathcal{N}$ we know by (i) that exactly one of $\tau_1^{-1}(C_m) \perp C_1 + \ldots + C_{m-1}$ or $\tau_1^{-1}(C_m') \perp C_1 + \ldots + C_{m-1}$ holds. We need to rule out the former case. First we prove:

$$(7) \qquad \sigma_{m+1}(C_i) \neq C_i \quad \text{for} \quad i = 1, \ldots, m$$

By braiding it suffices to prove this for $i = m$. Assume $\sigma_{m+1}(C_m) = C_m$. Then the element $\sigma_1 \cdots \sigma_{m-1} = (-\sigma_m \sigma_{m+1})^{-1}$ fixes both summands in the direct sum decomposition $W = C_m \oplus (C_1 + \ldots + C_{m-1})$. Hence this element acts trivially in C_m (since it does so in $W/(C_1 + \ldots + C_{m-1})$). Also σ_{m+1} acts trivially in C_m since $C_m \neq C_{m+1}$. Thus σ_m acts as -1 in C_m. Contradiction, since $\sigma_m \in \Sigma$. This proves (7).

If $(\tau_1 - I)(V) \subset W$ then $W \subset (\tau_1 - I)(V)^\perp = \ker(\tau_1 - I)$, hence $\tau_1|_W = \mathrm{id}$ and so $\sigma_{m+1}^{-1}|_W = \tau_2^{-1}|_W$; then $(\tau_2^{-1} - I)(V) \subset W$, hence also $\tau_2^{-1}|_W = \mathrm{id}$ and so $\sigma_{m+1}^{-1}|_W = \mathrm{id}$, contradicting (7). This and its analogue proves

$$(8) \qquad (\tau_j - I)(V) \not\subset W \quad \text{for} \quad j = 1, 2$$

If $\tau_1(C_i) = C_i$ for some $i \leq m$ then $\tau_2^{-1}(C_i) = \sigma_{m+1}^{-1}(C_i) \subset W$, hence $\tau_2^{-1}(C_i) = C_i$ by (8) and so $\sigma_{m+1}(C_i) = \tau_1 \tau_2(C_i) = C_i$, contradicting (7). This proves

$$(9) \qquad \tau_1(C_i) \neq C_i \quad \text{for} \quad i = 1, \ldots, m$$

Now let $v \neq 0$ in C_m. Then $u := (\tau_1^{-1} - I)(v) \neq 0$ by (9). If $\tau_1^{-1}(C_m) \perp C_1 + \ldots + C_{m-1}$ then $u \perp C_1 + \ldots + C_{m-1}$, hence $C_1 + \ldots + C_{m-1} \subset u^\perp = (\tau_1 - I)(V)^\perp = \ker(\tau_1 - I)$, contradicting (9). Thus $\tau_1^{-1}(C_m) \not\perp C_1 + \ldots + C_{m-1}$. This concludes the proof of (iv).

Now we prove the remaining part of (i): irreducibility of the group $< \sigma_1, \ldots, \sigma_m, \tau_1, \tau_2 >$. By (iv) we may assume $a_1 \cdots a_m \neq (-1)^{m-1}$. Then $< \sigma_1, \ldots, \sigma_m >$ acts irreducibly and inequivalently in W and W'. Hence it suffices to show that τ_1 fixes neither W nor W'. This follows from (8) and (9) (and analogues).

(iii) Set $b = (-1)^m a_1^{-1} \cdots a_m^{-1}$. By the proof of (ii), σ_{m+1} acts on W as $\mathrm{diag}(b, 1, \ldots, 1)$ if $b \neq 1$, otherwise as transvection. Further, C_{m+1} and $C_{m+1}' = (\sigma_{m+1} - I)(W')$ are 1-spaces. Thus $Z := C_{m+1} + C_{m+1}' = (\sigma_{m+1} - I)(V)$ is a 2-space. The space $Y = (\tau_1 - I)(V) + (\tau_2 - I)(V)$ has dimension ≤ 2 and the element $\sigma_{m+1} = \tau_1 \tau_2$ acts trivially in V/Y. Hence $Y = Z$.

Now consider some $\tilde{\mathbf{n}}$ as in (iii). The "only if"-part of the claim is clear. To prove the "if"-part, we may assume $K = \bar{K}$. Indeed, if $\mathbf{n}$ and $\tilde{\mathbf{n}}$ are conjugate under $\mathrm{GSp}_n(\bar{K})$ then they are also conjugate under $\mathrm{GSp}_n(K)$ by the argument in [V2] after Theorem 3.

By (iv) we may further assume $a_1 \cdots a_m \neq (-1)^{m-1}$ (i.e., $b \neq -1$). Then $V = W \oplus W'$ by (ii), and similarly for $\tilde{\mathbf{n}}$. By transitivity of G on pairs of complementary maximal isotropic subspaces of V we may assume W (resp., W') is spanned by the a_i-eigenspaces (resp., a_i^{-1}-eigenspaces) of the $\tilde{\sigma}_i$, $i = 1, ..., m$. The uniqueness theorem for Thompson tuples (see [V2]) implies now the existence of $g \in \mathrm{GL}(W)$ with $\tilde{\sigma}_i|_W = (\sigma_i|_W)^g$ for $i = 1, ..., m$. This g extends to a unique element of G fixing W and W'. Via conjugating with this element we may assume $\tilde{\sigma}_i = \sigma_i$ for all i.

Assume first $b \neq 1$. Then σ_{m+1} acts on Z as $\mathrm{diag}(b, b^{-1})$ and acts trivially in $Z^\perp$, hence Z is non-degenerate. Thus τ_1 and τ_2 act non-trivially in Z, hence induce transvections in Z; same for $\tilde{\tau}_1$ and $\tilde{\tau}_2$. A simple calculation in 2×2-matrices (see [ThV]) now shows that $\tau_1|_Z$ and $\tilde{\tau}_1|_Z$ are conjugate under an element of $\mathrm{GL}(Z)$ fixing C_{m+1} and C'_{m+1}. This element extends (uniquely) to some $h \in \hat{G}$ acting as a scalar in W and in W'. This h conjugates τ_1 to $\tilde{\tau}_1$ and centralizes the σ_i, hence conjugates $\mathbf{n}$ to $\tilde{\mathbf{n}}$.

Now assume $b = 1$. Then σ_{m+1} acts trivially in Z, hence Z is totally isotropic and thus also τ_1, τ_2 act trivially in Z (since $(\tau_j - I)(V) \subset Z$). Choose $e_3 \in W \setminus (C'_{m+1})^\perp$. Then $e_1 := \sigma_{m+1}(e_3) - e_3 \in C_{m+1} \setminus \{0\}$ (since $e_3 \notin Z^\perp = \ker(\sigma_{m+1} - I)$). Choose $e_4 \in W'$ with $(e_1, e_4) = 1$ and $(e_3, e_4) = 0$. Then $e_2 := \sigma_{m+1}(e_4) - e_4 \in C'_{m+1} \setminus \{0\}$. Further, $(e_1, e_2) = 0$ and $(e_2, e_3) = (\sigma_{m+1}(e_4), e_3) = (e_4, \sigma_{m+1}^{-1}(e_3)) = (e_4, e_3 - e_1) = -(e_4, e_1) = 1$. Thus the 4-space $X = [e_1, e_2, e_3, e_4]$ is non-degenerate, and σ_{m+1}, τ_1, τ_2 fix X and act trivially in $X^\perp$.

Any $\rho \in \mathrm{GL}(X)$ with $(\rho - I)(V) \subset Z \subset \ker(\rho - I)$ is given in the basis e_1, e_2, e_3, e_4 by a matrix of the form

$$\begin{pmatrix} I_2 & A \\ 0 & I_2 \end{pmatrix}$$

where I_2 is the 2×2 identity matrix and $A \in M_2(K)$. Such ρ lies in $\mathrm{Sp}(X)$ iff the diagonal entries of A are equal; and ρ is a transvection iff $\det(A) = 0$. Further, σ_{m+1} corresponds to $A = I_2$. Let τ_1 correspond to $\begin{pmatrix} a & c \\ d & a \end{pmatrix}$. Then $a^2 = cd$, and τ_2 corresponds to $I_2 - \begin{pmatrix} a & c \\ d & a \end{pmatrix}$. The condition that the latter matrix has zero determinant yields $a = 1/2$. Thus $d \neq 0$. Since $(\tau_1 - I)(V) = [(1/2)e_1 + de_2]$, it follows that we can conjugate τ_1 by an element $h \in \hat{G}$ as above to assume $(\tau_1 - I)(V) = (\tilde{\tau}_1 - I)(V)$. This forces the 2×2 matrices corresponding to τ_1 and $\tilde{\tau}_1$ to be equal. Hence $\tau_1 = \tilde{\tau}_1$. This concludes the proof of (iii).

(v) Again we may assume $a_1 \cdots a_m \neq (-1)^{m-1}$, hence $V = W \oplus W'$. Each a_i satisfies an equation $x^2 - t_i x + 1 = 0$ with $t_i \in K$, hence if $a_i \notin K$ then $[K(a_i) : K] = 2$ and $\mathrm{N}_{K(a_i)/K}(a_i) = 1$. Thus $L = K(a_1, ..., a_m)$ is a finite Galois extension of K, and $\mathrm{Gal}(L/K)$ permutes C_i and C'_i for each

$i = 1, ..., m$. Also $\mathrm{Gal}(L/K)$ permutes W and W', since these are the only non-zero, proper subspaces of V invariant under all σ_i (and W, W' are defined over L). The stabilizer of W in $\mathrm{Gal}(L/K)$ fixes each a_i, hence is trivial. Thus $[L : K] \leq 2$, and if $[L : K] = 2$ then the non-trivial element of $\mathrm{Gal}(L/K)$ maps each a_i to a_i^{-1}. Hence all $a_i \in S$.

Now assume K is finite. We use the notation from the proof of (iii). Note that Z is defined over K, i.e., spanned by $Z_K = Z \cap K^n$. Further, $b \neq -1$ by assumption. Assume first $b \neq 1$. Then $\tau_1|_Z$ and $\tau_2|_Z$ are non-commuting transvections, hence by [V1], Lemma 3.27 there is a basis of Z_K in which these transvections take matrix form

$$T_1 = \begin{pmatrix} 1 & 1 \\ 0 & 1 \end{pmatrix} \quad \text{and} \quad T_2 = \begin{pmatrix} 1 & 0 \\ u & 1 \end{pmatrix}$$

with $u \in K^*$. Then $u + 2 = \mathrm{trace}(T_1 T_2) = \mathrm{trace}(\sigma_{m+1}|_Z) = b + b^{-1}$, hence $u = b + b^{-1} - 2$. It is easy to see that τ_1 and τ_2 are G-conjugate if and only if T_1 and T_2 are conjugate in $\mathrm{SL}_2(K)$. The latter holds if and only if $-u$ is a square in K^*, see [V1], Lemma 3.27.

We have seen that τ_1 and τ_2 are G-conjugate if and only if $2 - b - b^{-1}$ is a square in K^*. It remains to show that $2 - b - b^{-1}$ is a square in K^* if and only if $-b$ is a square in S. If $b \in K$ (i.e., $S = K^*$) this follows from the identity $2 - b - b^{-1} = (-b)(1 - b^{-1})^2$. Now assume $b \notin K$, i.e., $[L : K] = 2$. If $2 - b - b^{-1} = k^2$ for some $k \in K^*$ then

$$\mathrm{N}_{L/K}\left(\frac{b-1}{k}\right) = \frac{2 - b - b^{-1}}{k^2} = 1$$

and

$$-b = \frac{(b-1)^2}{2 - b - b^{-1}} = \left(\frac{b-1}{k}\right)^2 \in S^2.$$

Conversely, if $-b = s^2$ for some $s \in S$ then

$$2 - b - b^{-1} = (s + s^{-1})^2 \in K^2.$$

This concludes the case $b \neq 1$. The case $b = 1$ can be reduced to the previous case using (iv) and braiding unless $a_i^2 = -1$ for all i. In that case the claim can be shown using the set-up of (iii). We omit the details.

3   Braiding action on the tuples

The Artin braid group $\mathcal{B}_r$ on r strings has a presentation with generators $Q_1, ..., Q_{r-1}$ and relations

$$Q_i Q_{i+1} Q_i = Q_{i+1} Q_i Q_{i+1} \text{ for } i = 1, ..., r-2, \quad Q_i Q_j = Q_j Q_i \text{ for } |i-j| > 1.$$

Mapping Q_i to the transposition $(i, i+1)$ yields a homomorphism $\kappa : \mathcal{B}^{(r)} \rightarrow S_r$. The kernel of κ is called the pure braid group, and denoted by $\bar{\mathcal{B}}^{(r)}$. It is generated by the

$$Q_{ij} \; = \; Q_i \cdots Q_{j-2}\, Q_{j-1}^2\, Q_{j-2}^{-1} \cdots Q_i^{-1}, \quad 1 \le i < j \le r.$$

The braid group $\mathcal{B}_r$ acts from the right on tuples $(g_1, ..., g_r)$ in G^r by the rule that Q_i maps $(g_1, ..., g_r)$ to

$$(g_1, \ldots, g_{i-1},\; g_{i+1},\; g_{i+1}^{-1} g_i g_{i+1}\,, \ldots, g_r).$$

Now let $r = m + 2$ in the set-up of the previous section. We get an induced action of $< \bar{\mathcal{B}}^{(r)}, Q_1, ..., Q_{m-1}, Q_{m+1} >$ on $\mathcal{N}$ and $\hat{\mathcal{N}}$. From Theorem 2.2 (ii), (iii) we see that Q_{m+1} and the Q_{ij} with $1 \le i < j \le m$ act trivially on $\hat{\mathcal{N}}$. Since $Q_{i,m+2} = Q_{i,m+1}^{Q_{m+1}}$ for $1 \le i \le m$ it follows that $Q_{i,m+1}$ and $Q_{i,m+2}$ induce the same permutation on $\hat{\mathcal{N}}$. Further, Q_i $(i < m)$ acts as transposition $(i, i+1)$ on the tuple $(a_1, ..., a_m)$ associated with an element of $\mathcal{N}$ (see Theorem 2.2 (ii)). Finally, $Q_{i,m+1}$ $(i \le m)$ maps $\mathbf{n} \in \mathcal{N}$ associated with $(a_1, ..., a_m)$ to $\tilde{\mathbf{n}} \in \mathcal{N}$ associated with

$$(10) \qquad\qquad (a_1, ..., a_{i-1}, a_i^{-1}, a_{i+1}, ..., a_m).$$

For $i = m$ this is (iv) of Theorem 2.2. If true for some $1 < i \le m$ it follows for $i - 1$ because $Q_{i-1,m+1} = Q_{i,m+1}^{Q_{i-1}}$. This proves (10). Combined with Theorem 2.2(iii) this implies:

Corollary 3.1

(i) *Two elements of* $\hat{\mathcal{N}}$, *represented by the tuples* $(\sigma_1, ..., \sigma_m, \tau_1, \tau_2)$ *and* $(\tilde{\sigma}_1, ..., \tilde{\sigma}_m, \tilde{\tau}_1, \tilde{\tau}_2)$, *lie in the same* $\bar{\mathcal{B}}^{(r)}$-*orbit if and only if* $\mathrm{trace}(\sigma_i) = \mathrm{trace}(\tilde{\sigma}_i)$ *for all* i.

(ii) *All elements of* $\hat{\mathcal{N}}$ *have the same stabilizer in* $\bar{\mathcal{B}}^{(r)}$. *This group* $\mathcal{K}$ *is generated by the squares in* $\bar{\mathcal{B}}^{(r)}$ *together with the elements*

$$Q_{ij}, \quad 1 \le i < j \le m$$

$$Q_{i,m+1}\, Q_{i,m+2}^{-1}, \quad 1 \le i \le m$$

$$Q_{1,m+1} \cdots Q_{m,m+1} \quad \text{and} \quad Q_{m+1,m+2}.$$

Thus $\bar{\mathcal{B}}^{(r)}/\mathcal{K}$ *is an elementary abelian group of order* 2^{m-1} *acting freely on* $\hat{\mathcal{N}}$. *In particular, all* $\bar{\mathcal{B}}^{(r)}$-*orbits on* $\hat{\mathcal{N}}$ *have length* 2^{m-1}.

4   Existence of the tuples, and generated subgroup

Let $t_1, ..., t_m \in K \setminus \{\pm 2\}$ and $\mathbf{t} = (t_1, ..., t_m)$. Let $\mathcal{N}(\mathbf{t})$ be the set of all $(\sigma_1, ..., \sigma_m, \tau_1, \tau_2) \in \mathcal{N}$ with $\mathrm{trace}(\sigma_i) = t_i + n - 2$. Then σ_i is similar to $\mathrm{diag}(a_i, a_i^{-1}, 1, ..., 1)$, where $a_i + a_i^{-1} = t_i$. Let $\hat{\mathcal{N}}(\mathbf{t})$ be the set of $\hat{G}$-orbits on $\mathcal{N}(\mathbf{t})$.

First we show that if $a_i \in K$ for all i then $\mathcal{N}(\mathbf{t}) \neq \emptyset$. We may assume $a_1 \cdots a_m \neq (-1)^{m-1}$. Let W_0 be a K-vector space of dimension m. By the existence theorem for Thompson tuples (see [V2]) there is a Thompson tuple $(\sigma_1^0, ..., \sigma_m^0, -\sigma_{m+1}^0)$ in $GL(W_0)$ such that σ_i^0 $(i \leq m)$ has $\mathrm{diag}(a_i, 1, ..., 1)$ as a representing matrix, and $\mathrm{rank}(\sigma_{m+1} - I) = 1$. Let W_0' be the dual of W_0, and $<,>: W_0 \times W_0' \to K$ the dual pairing. Extend σ_i^0 to an element σ_i of $GL(W_0 \oplus W_0')$ by its natural action on W_0'. Then the σ_i preserve the non-degenerate symplectic form $f((w, w'), (u, u')) = <w, u'> - <u, w'>$. As in the proof of Theorem 2.2 (iii) we see that σ_{m+1} can be written as $\sigma_{m+1} = \tau_1 \tau_2$ for transvections τ_1, τ_2 that also preserve f. Then $(\sigma_1, ..., \sigma_m, \tau_1, \tau_2) \in \mathcal{N}(\mathbf{t})$.

Lemma 4.1 *Suppose* $K = \mathbb{F}_q$ *is finite. Then* $\mathcal{N}(\mathbf{t}) \neq \emptyset$ *if and only if* $a_i \in K$ *for all* i *or* $a_i \notin K$ *for all* i.

Proof : The condition is necessary by Theorem 2.2 (v). The remarks preceding the Lemma show that there is a tuple $\mathbf{n} = (\sigma_1, ..., \sigma_m, \tau_1, \tau_2)$ in $Sp_n(\bar{K})$ with the desired properties. Let α be the automorphism of $Sp_n(\bar{K})$ raising each matrix entry to the power q. Then $\mathbf{n}^\alpha$ is a tuple with the same properties, hence by Theorem 2.2 (iii) there is $h \in Sp_n(\bar{K})$ with $\mathbf{n}^\alpha = \mathbf{n}^h$. By Lang's theorem there is $g \in Sp_n(\bar{K})$ with $h = g^{-1}g^\alpha$. Then $(\mathbf{n}^{g^{-1}})^\alpha = \mathbf{n}^{g^{-1}}$. Hence $\mathbf{n}^{g^{-1}} \in \mathcal{N}(\mathbf{t})$. ∎

Lemma 4.2 *Suppose* $n \geq 10$ *and* $K = \mathbb{F}_q$, *where* q *is a power of the prime* p. *Let* $(\sigma_1, ..., \sigma_m, \tau_1, \tau_2) \in \mathcal{N}$, *with* σ_i *similar* $\mathrm{diag}(a_i, a_i^{-1}, 1, ..., 1)$. *Let* $t_i = a_i + a_i^{-1}$.

(1) The group $G = Sp_n(q)$ *is generated by* $\sigma_1, ..., \sigma_m, \tau_1, \tau_2$ *if and only if* $\mathbb{F}_p(t_1, ..., t_m) = \mathbb{F}_q$.

(2) Assume $a_1 \cdots a_m \neq (-1)^{m-1}$, *and set* $H = <\sigma_1, ..., \sigma_m>$. *If* $\mathbb{F}_p(t_1, ..., t_m) = \mathbb{F}_q$ *and the* a_i *lie in* $\mathbb{F}_q$ *(resp., the* a_i *lie not in* $\mathbb{F}_q$*) then* H *is isomorphic to a group between* $SL_m(q)$ *and* $GL_m(q)$ *(resp., between* $SU_m(q^2)$ *and* $U_m(q^2)$ *). The images of* $\sigma_1, ..., \sigma_m$ *and* $(\sigma_1 \cdots \sigma_m)^{-1}$ *under this isomorphism form a Thompson tuple in* $GL_m(q)$, *resp.,* $GL_m(q^2)$ *whose* i-*th element* $(i \leq m)$ *is a matrix similar to* $\mathrm{diag}(a_i, 1, ..., 1)$, *and whose last element* σ *satisfies* $\mathrm{rank}(\sigma + I) = 1$.

Proof : We first prove (b). By Theorem 2.2 (ii) we know that restriction gives an injection $H \longrightarrow GL(W)$, and the images of $\sigma_1, ..., \sigma_m$ and

14

$(\sigma_1 \cdots \sigma_m)^{-1}$ form a Thompson tuple in $\mathrm{GL}(W)$ with $\sigma_i|_W$ $(i \leq m)$ and $(\sigma_1 \cdots \sigma_m)|_W^{-1}$ of the claimed form. Let $L = \mathbb{F}_p(a_1, ..., a_m)$ $(= \mathbb{F}_q$ or $\mathbb{F}_{q^2})$. Then W is defined over L, hence the $\sigma_i|_W$ lie actually in $\mathrm{GL}(W \cap L^n) \cong \mathrm{GL}_m(L)$, and their eigenvalues generate L. Now (b) follows from the classification theorem for Thompson tuples (see [V2]).

Part (a) follows from (b) as in [MSV], Theorem 1. (Actually, since we are now in the case $\mathrm{char}(K) \neq 2$ the proof is much easier than in [MSV]).

5   The resulting Galois realizations

From now on we let $K = \mathbb{F}_q$, and so $G = \mathrm{Sp}_n(q)$. Let $\bar{G} = G/\{\pm I\} = \mathrm{PSp}_n(q)$. Fix $t_1, ..., t_m \in K \setminus \{\pm 2\}$ and set $\mathbf{t} = (t_1, ..., t_m)$. Choose $a_1, ..., a_m \in \mathbb{F}_{q^2}$ with $a_i + a_i^{-1} = t_i$. Assume that either all $a_i \in \mathbb{F}_q$ or all $a_i \notin \mathbb{F}_q$. In the former case, set $S = \mathbb{F}_q^*$, and in the latter case let S be the group of elements of $\mathbb{F}_{q^2}$ of norm 1 over $\mathbb{F}_q$. Consider the class of G consisting of all matrices similar $\mathrm{diag}(a_i, a_i^{-1}, 1, ..., 1)$, and let C_i be the image of this class in $\bar{G}$ (for $i = 1, ..., m$). Fix one class of transvections in $\bar{G}$, and call it C_{r-1} (where again $r = m+2$). Here we call an element of $\bar{G}$ a transvection if it is the image of a transvection of G. Let $\mathbf{C}(\mathbf{t}) = (C_1, ..., C_r)$, where $C_r = C_{r-1}$ if $(-1)^{m-1}a_1 \cdots a_m$ is a square in S, and otherwise C_r is the other class of transvections in $\bar{G}$. We say this tuple is **rational** if $C_1^k, ..., C_r^k$ is a permutation of $C_1, ..., C_r$ for each integer k prime to the order of $\bar{G}$. Similarly, we call the tuple $(a_1, ..., a_m)$ rational if $a_1^k, ..., a_m^k$ is a permutation of $a_1, ..., a_m$ for each integer k prime to $q^2 - 1$. We order the tuple such that $C_1 = ... = C_{r_1}$, $C_{r_1+1} = ... = C_{r_1+r_2}$, etc., and there are no further equalities between the C_i.

We further assume there are generators $g_1, ..., g_r$ of $\bar{G}$ with $g_1 \cdots g_r = 1$ and $g_i \in C_i$ for $i = 1, ..., r$. Let $\bar{\mathcal{N}}(\mathbf{t})$ be the set of $\bar{G}$-orbits on the set of those tuples $(g_1, ..., g_r)$. Each g_i, $i \leq m$ lifts to a unique $\sigma_i \in \Sigma$; and g_{r-1}, g_r lift to unique transvections $\tau_1, \tau_2 \in G$. Then $\sigma_1 \cdots \sigma_m \tau_1 \tau_2 = -I$ because by Scott's formula [Sc] there is no such tuple generating G with product $= I$. Since the elements of $\hat{G} \setminus G$ switch the two classes of transvections, it follows by Theorem 2.2(v) that $\bar{\mathcal{N}}(\mathbf{t})$ can naturally be identified with the above set $\hat{\mathcal{N}}(\mathbf{t})$. This identification is compatible with the natural action of $\mathcal{B}^{(r)}$. Hence by Corollary 3.1 the action of $\mathcal{B}^{(r)}$ on $\bar{\mathcal{N}}(\mathbf{t})$ is transitive, and the kernel of this action is the group $\mathcal{K}$. Since $\mathcal{B}^{(r)}/\mathcal{K}$ is abelian, $\mathcal{K}$ also equals the stabilizer in $\mathcal{B}^{(r)}$ of each element of $\bar{\mathcal{N}}(\mathbf{t})$. Let $\mathcal{B}_r(\mathbf{t})$ be the stabilizer in $\mathcal{B}_r$ of $\bar{\mathcal{N}}(\mathbf{t})$, and let $\mathcal{K}_2$ be the stabilizer in $\mathcal{B}_r(\mathbf{t})$ of some element of $\bar{\mathcal{N}}(\mathbf{t})$. Then $\mathcal{K} = \mathcal{K}_2 \cap \mathcal{B}^{(r)}$ and $\mathcal{B}_r(\mathbf{t}) = \mathcal{K}_2 \mathcal{B}^{(r)}$ (since $\mathcal{B}^{(r)}$ acts transitively on $\bar{\mathcal{N}}(\mathbf{t})$). The map $\kappa : \mathcal{B}_r \to S_r$ induces an isomorphism $\mathcal{B}_r(\mathbf{t})/\mathcal{B}^{(r)} \to S_{r_1} \times S_{r_2} \times ....$

Let $x_1, ..., x_r \in \mathbb{C}$ be algebraically independent over $\mathbb{Q}$. Let S_r act on the field $\mathbb{Q}(x_1, ..., x_r)$ by permuting $x_1, ..., x_r$. The fixed field of S_r is $\mathbb{Q}(s_1, ..., s_r)$, where $s_1, ..., s_r$ are the elementary symmetric functions in

$x_1, ..., x_r$. Let $L(t)$ be the fixed field of the subgroup $S_{r_1} \times S_{r_2} \times ....$ Then $L(t) = \bar{\mathbb{Q}}(u_1, ..., u_r)$, where $u_1, ..., u_{r_1}$ are the elementary symmetric functions in $x_1, ..., x_{r_1}$ etc.

Let $\Omega_0/\mathbb{C}(x)$ be a Galois extension with branch points $x_1, ..., x_r$ and Galois group isomorphic to $\bar{G}$, such that C_i corresponds to the class of distinguished inertia group generators associated with x_i for $i = 1, ..., r$. Such Ω_0 exists by Riemann's existence theorem because $\bar{\mathcal{N}}(t) \neq \emptyset$ (see [V1] Thm. 2.13). By [V1], Prop. 7.12, this Ω_0 has a unique minimal field of definition $L_1 \subset \mathbb{C}$; i.e., there is $\Omega_1 \subset \Omega_0$, Galois over $L_1(x)$ and regular over L_1 such that restriction gives an isomorphism $G(\Omega_0/\mathbb{C}(x)) \to G(\Omega_1/L_1(x))$ (and L_1 is minimal with this property). This L_1 is finite over $\mathbb{Q}(s_1, ..., s_r)$. Further, L_1 is regular over $\mathbb{Q}$ if the tuple $\mathbf{C}(t)$ is rational (because $\mathcal{B}^{(r)}$ acts transitively on $\bar{\mathcal{N}}(t)$, see [FV], Thm. 1 or [V3], Thm. 3.9). Each $\alpha \in \mathrm{Aut}(\mathbb{C}/L_1)$ extends to an automorphism of Ω_0 fixing x and centralizing $G(\Omega_0/\mathbb{C}(x))$. By the branch cycle argument ([V1] Lemma 2.8), α permutes $x_1, ..., x_r$ such that if $\alpha(x_i) = x_j$ then $C_i = C_j^k$, where k is an integer such that α acts on the $|G|$-th roots of unity as $\zeta \mapsto \zeta^k$.

Let $L_2 = L_1\bar{\mathbb{Q}} = $ minimal field of definition of Ω_0 containing $\bar{\mathbb{Q}}$. The extensions $L_2/\bar{\mathbb{Q}}(s_1, ..., s_r)$ and $\bar{\mathbb{Q}}(x_1, ..., x_r)/\bar{\mathbb{Q}}(s_1, ..., s_r)$ are unramified over the complement of the discriminant locus in affine r-space with coordinates $s_1, ..., s_r$. Let Π be the maximal extension of $\bar{\mathbb{Q}}(s_1, ..., s_r)$ unramified over this complement. Then $G(\Pi/\bar{\mathbb{Q}}(s_1, ..., s_r))$ can be identified with the profinite completion of $\mathcal{B}_r$. Thus under the Galois correspondence, L_2 and $\bar{\mathbb{Q}}(x_1, ..., x_r)$ correspond to certain subgroups of finite index of $\mathcal{B}_r$; actually, $\bar{\mathbb{Q}}(x_1, ..., x_r)$ corresponds to $\mathcal{B}^{(r)}$ and L_2 corresponds to (a conjugate of) $\mathcal{K}_2$ by [V1], Cor. 10.21. Thus $L := L_2\bar{\mathbb{Q}}(x_1, ..., x_r)$ corresponds to $\mathcal{K}_2 \cap \mathcal{B}^{(r)} = \mathcal{K}$, and $L_2 \cap \bar{\mathbb{Q}}(x_1, ..., x_r)$ corresponds to $\mathcal{K}_2\mathcal{B}^{(r)} = \mathcal{B}_r(t)$. Thus $L_2 \cap \bar{\mathbb{Q}}(x_1, ..., x_r) = L(t) = \bar{\mathbb{Q}}(u_1, ..., u_r)$.

Let $\Gamma = G(L_2/L_1)$. Then restriction gives an isomorphism $\Gamma \to G(\bar{\mathbb{Q}}/\mathbb{Q})$ if the tuple $\mathbf{C}(t)$ is rational (because then L_1 is regular over $\mathbb{Q}$). Fix some $\gamma \in \Gamma$, and extend it some way to an automorphism α of $\mathbb{C}$. This α permutes $x_1, ..., x_r$ as described above. Thus γ permutes $u_1, ..., u_r$ accordingly, permuting the blocks $\{u_1, ..., u_{r_1}\}, \{u_{r_1+1}, ..., u_{r_1+r_2}\}$,etc. Hence we can extend $\gamma|_{L(t)}$ to $\bar{\mathbb{Q}}(x_1, ..., x_r)$ by permuting $x_1, ..., x_r$ in any way compatible with the action on these blocks. This further extends γ to $L = L_2\bar{\mathbb{Q}}(x_1, ..., x_r)$ because L_2 and $\bar{\mathbb{Q}}(x_1, ..., x_r)$ are linearly disjoint over $L(t)$.

Theorem 5.1 *Let $t_1, ..., t_m \in \mathbb{F}_q \setminus \{\pm 2\}$ and set $\mathbf{t} = (t_1, ..., t_m)$. Let $\mathbf{C}(t) = (C_1, ..., C_r)$ be the associated tuple of conjugacy classes of $\bar{G} = PSp_n(q)$ (where $r = m + 2$). Assume $\bar{\mathcal{N}}(t) \neq \emptyset$. Let $x_1, ..., x_r \in \mathbb{C}$ be algebraically independent over $\mathbb{Q}$. Let $\Omega_0/\mathbb{C}(x)$ be a Galois extension with branch points $x_1, ..., x_r$ and Galois group isomorphic to $\bar{G}$, such that C_i corresponds to the class of distinguished inertia group generators associated with x_i for $i = 1, ..., r$. Let L_1 be the minimal field of definition of Ω_0, let $L_2 = L_1\bar{\mathbb{Q}}$ and $L = L_1\bar{\mathbb{Q}}(x_1, ..., x_r)$. Then the following holds.*

(i) We have $L = \bar{\mathbb{Q}}(y_1, ..., y_{m-1}, x_m, x_{m+1}, x_{m+2})$, *where*

$$y_i^2 \;=\; \frac{x_{m+1} - x_i}{x_{m+1} - x_m} \;:\; \frac{x_{m+2} - x_i}{x_{m+2} - x_m}, \quad i = 1, ..., m-1$$

the cross ratio of $x_i, x_m, x_{m+1}, x_{m+2}$. *Let* $\tilde{L} = \bar{\mathbb{Q}}(z_1, ..., z_m, x_{m+1}, x_{m+2})$, *where*

$$z_i^2 \;=\; \frac{x_{m+1} - x_i}{x_{m+2} - x_i}, \quad i = 1, ..., m$$

Then $[\tilde{L} : L] = 2$ *and* $\tilde{L} = L(z_m)$.

(ii) Assume the tuple $(C_1, ..., C_r)$ *is rational and also the class* C_m *is rational in* $\bar{G}$. *If either* q *is a square or* m *is odd and* $C_{2\nu} = C_{2\nu-1}$ *for* $2\nu \leq m$ *then we can extend the action of* $\Gamma = G(L_2/L_1)$ *to an action on* L *such that* L^Γ *is a rational function field over* $\mathbb{Q}$.

(iii) Assume there are $a_1, ..., a_m \in \mathbb{F}_{q^2}^*$ *with* $a_i + a_i^{-1} = t_i$ *such that* $(a_1, ..., a_m)$ *is a rational tuple. We further assume* m *is even and* $a_{2\nu} = a_{2\nu-1}$ *for* $2\nu \leq m$, *as well as* $C_{r-1} \neq C_r$. *Then we can extend the action of* Γ *to* $\tilde{L}$ *such that* $\tilde{L}^\Gamma$ *is a rational function field over* $\mathbb{Q}$.

Proof : (i) The above identification of $G(\Pi/\bar{\mathbb{Q}}(s_1, ..., s_r))$ with the profinite completion of $\mathcal{B}_r$ induces an identification of $G(\Pi/\bar{\mathbb{Q}}(x_1, ..., x_r))$ with the profinite completion of $\mathcal{B}^{(r)}$. Under this identification, Q_{ij} becomes the generator of an inertia group over the place $x_i = x_j$ of $\bar{\mathbb{Q}}(x_1, ..., x_r)$, for $1 \leq i < j \leq r$ (see [Ma], 3.2 or [SV, Prop. 1 and section 3.2]). Adjoining $\sqrt{x_k - x_\ell}$, $1 \leq k < \ell \leq r$ yields an extension of $\bar{\mathbb{Q}}(x_1, ..., x_r)$ such that the place $x_i = x_j$ ramifies in this extension iff $\{i, j\} = \{k, \ell\}$. Since $\sqrt{x_k - x_\ell} \in \Pi$ we get

Claim 1: The element Q_{ij} maps $\sqrt{x_i - x_j}$ to $-\sqrt{x_i - x_j}$, and fixes all $\sqrt{x_k - x_\ell}$ with $\{k, \ell\} \neq \{i, j\}$.

Using Claim 1 we check that the generators of $\mathcal{K}$ given in Corollary 3.1 fix $y_1, ..., y_{m-1}$. Thus the fixed field L of $\mathcal{K}$ contains $y_1, ..., y_{m-1}$. Now (i) follows because $[\bar{\mathbb{Q}}(y_1, ..., y_{m-1}, x_m, x_{m+1}, x_{m+2}) : \bar{\mathbb{Q}}(x_1, ..., x_r)] = 2^{m-1} = [\mathcal{B}^{(r)} : \mathcal{K}] = [L : \bar{\mathbb{Q}}(x_1, ..., x_r)]$. (The assertion about $\tilde{L}$ is clear).

(ii) Assume first that q is a square. Then the classes C_{m+1} and C_{m+2} (of transvections) are rational in $\bar{G}$. Since also C_m is assumed to be rational, it follows from the remarks before the Theorem that we can extend the action of Γ to L fixing x_m, x_{m+1}, x_{m+2} (and permuting the other x_i). Then Γ permutes the $\pm y_i$, hence fixes the linear span of the transcendence basis of L given in (i). Thus L^Γ is rational over $\mathbb{Q}$ by [V1], Lemma 8.7.

Now assume q is not a square. Let p be the prime dividing q. Let C' be the field of p-th roots of unity in $\bar{\mathbb{Q}}$, and C'' the field of $(q^2 - 1)$-th roots of unity. Then for $C = C'C''$ we have $G(C/\mathbb{Q}) = G(C/C') \times G(C/C'')$.

Define a permutation action of $G(C/\mathbb{Q})$ on the x_i as follows: $G(C/C'')$ interchanges $x_{2\nu}$ and $x_{2\nu-1}$ for $2\nu \leq m$, as well as x_{m+1} and x_{m+2}. (Since

$G(C/C'')$ is cyclic, there is only one such action). Further, $G(C/C')$ fixes x_{m+1}, x_{m+2}, and $\beta \in G(C/C')$ maps x_i to x_j as follows: We have $C_j^k = C_i$ where k is an integer such that β acts on the (q^2-1)-th roots of 1 as $\zeta \mapsto \zeta^k$. To distinguish j among the $\tilde{j}$ with $C_{\tilde{j}} = C_j$, let $\tilde{i}$ (resp., $\tilde{j}$) be the smallest index with $C_{\tilde{i}} = C_i$ (resp., $C_{\tilde{j}} = C_j$). Then $j - \tilde{j} = i - \tilde{i}$. Since these actions of $G(C/C')$ and $G(C/C'')$ commute, they combine to give an action of $G(C/\mathbb{Q})$.

If $k \in \mathbb{Z}$ is a square (resp., not a square) mod p then the two classes C_{m+1} and C_{m+2} of transvections are fixed (resp., interchanged) by taking them to the k-th power. It follows by the remarks before the Theorem that the action of Γ extends to L such that Γ permutes $x_1, ..., x_r$ via its quotient $G(C/\mathbb{Q})$ as in the previous paragraph.

Now assume m is odd. Then $G(C/\mathbb{Q})$ fixes x_m, hence $G(C/C'')$ interchanges $\pm y_{2\nu}$ and $\pm y_{2\nu-1}^{-1}$. Since $G(C/C')$ fixes the sets $\{x_1, x_3, ..., x_{m-2}\}$ and $\{x_2, x_4, ..., x_{m-1}\}$, it follows that Γ fixes the linear span of the transcendence basis $x_m, x_{m+1}, x_{m+2}, y_1, y_2^{-1}, y_3, y_4^{-1}, ...$ of L. Thus the claim follows again by [V1], Lemma 8.7.

(iii) Assume the hypothesis of (iii). Then we can refine the action of $G(C/\mathbb{Q})$ on the x_i such that if $\beta \in G(C/C')$ maps x_i to x_j then $a_j^k = a_i$, with k as above; and further, if $\tilde{i}$ (resp., $\tilde{j}$) is the smallest index with $a_{\tilde{i}} = a_i$ (resp., $a_{\tilde{j}} = a_j$) then $j - \tilde{j} = i - \tilde{i}$.

This again extends the action of Γ to L. Below we extend it further from L to $\tilde{L}$. Assume this for the moment. Then we see as before that Γ fixes the linear span of $x_{m+1}, x_{m+2}, z_1, z_2^{-1}, z_3, z_4^{-1}, ...$ The rest is as for (ii).

Rational points of L over the discriminant locus:

Identify $x_1, ..., x_r$ with the coordinate functions on $\bar{\mathbb{Q}}^r$, which gives the usual identification of the points of $\bar{\mathbb{Q}}^r$ with the maximal ideals of $\bar{\mathbb{Q}}[x_1,...,x_r]$. This gives an action of Γ on $\bar{\mathbb{Q}}^r$ (different from the standard $G_{\mathbb{Q}}$-action). Define a (finite) point of L, $\tilde{L}$ etc. to be a maximal ideal of the integral closure of $\bar{\mathbb{Q}}[x_1, ..., x_r]$ in L, $\tilde{L}$ etc. We need

Claim 2: L has a Γ-fixed point **p** over each Γ-fixed point of $\bar{\mathbb{Q}}^r$ of the form $(p_1, ..., p_r)$ with $p_{r-1} = p_r$ and $p_i \neq p_j$ for $1 \le i < j \le r-1$.

Proof : Pick a base point p_0 on the Riemann sphere $\mathbb{P}^1$ different from the p_i, $i \ge 1$. Consider loops in $\mathbb{P}^1 \setminus \{p_1, ..., p_{r-1}\}$, based at p_0. Choose simply closed disjoint loops γ_i around p_i, $i = 1, ..., r-2$, and γ around $p_{r-1} = p_r$, satisfying the usual condition that $\gamma_1 \cdots \gamma_{r-2}\gamma$ is null-homotopic. Let $p_r' \neq p_r$ be close to p_r and inside γ (hence different from $p_1, ..., p_{r-1}$). Choose loops γ_{r-1} and γ_r inside γ around p_{r-1} and p_r', respectively, such that $\gamma_{r-1}\gamma_r$ is homotopic to γ.

The Galois covers φ' of $\mathbb{P}^1$ with group $\bar{G}$, branch points $p_1, ..., p_{r-1}, p_r'$ and associated classes $C_1, ..., C_r$ (see [V1], Ch. 4 and 5) correspond to the elements of $\bar{\mathcal{N}}(\mathbf{t})$, where a tuple $(g_1, ..., g_r)$ corresponding to φ' consists of the natural images of $\gamma_1, ..., \gamma_r$ in $\mathrm{Deck}(\varphi') \cong \bar{G}$. Fix such $g_1, ..., g_r$

having the property that their lifts $\sigma_1, ..., \sigma_m, \tau_1, \tau_2 \in G$ (see the beginning of section 5) correspond via Theorem 2.2 to the tuple $(a_1, ..., a_m)$ used to define the action of $G(C/\mathbb{Q})$ on L.

We consider the behavior of the covers φ' (corresponding to this fixed tuple $g_1, ..., g_r$) as p'_r moves into p_r: Then φ' collapses into a cover φ of $\mathbb{P}^1$ with group $\bar{H} =< g_1, ..., g_{r-2} >$ and branch points $p_1, ..., p_{r-1}$; the images in $\mathrm{Deck}(\varphi) \cong \bar{H}$ of $\gamma_1, ..., \gamma_{r-2}, \gamma$ are $g_1, ..., g_{r-2}, g$, where $g = g_{r-1}g_r$. Note that $\bar{H} = H/\{\pm 1\}$, where $H =< \sigma_1, ..., \sigma_m >$ is as in Lemma 4.2; in particular, the image of σ_i under the natural embedding $H \to \mathrm{GL}_m(q^2)$ is similar $\mathrm{diag}(a_i, 1, ..., 1)$.

The field extension $\Omega/\mathbb{C}(x)$ corresponding to φ is obtained by specializing the generic extension Ω_0. This specialization maps x_i to p_i, hence yields a specialization of L corresponding to a point $\mathbf{p}$ of L over $(p_1, ..., p_r)$.

If an element of Γ maps $\mathbf{p}$ to another point $\mathbf{p}'$ of L over $(p_1, ..., p_r)$, then it transforms Ω into the extension of $\mathbb{C}(x)$ associated with $\mathbf{p}'$. But we know how Γ acts on the ramification type of an extension of $\mathbb{C}(x)$: via the branch cycle argument [V1] Lemma 2.8. The class of $\bar{H}$ associated with the branch point p_i, $i \leq m$ of $\Omega/\mathbb{C}(x)$ is represented by $\mathrm{diag}(a_i, 1, ..., 1)$. By definition of the action of Γ on $x_1, ..., x_r$, hence on $p_1, ..., p_r$, and the branch cycle argument it follows that the extension of $\mathbb{C}(x)$ associated with $\mathbf{p}'$ corresponds to the same choice of $(a_1, ..., a_m)$ (since the tuple $(a_1, ..., a_m)$ is rational). But the 2^{m-1} points of L over $(p_1, ..., p_r)$ correspond bijectively to the 2^{m-1} distinct choices for $a_1, ..., a_m$ modulo inversion. It follows that Γ fixes $\mathbf{p}$. Hence Claim 2.

Extending the action of Γ from L to $\tilde{L}$:

Let $\tilde{\mathbf{p}}$ be a point of $\tilde{L}$ over $\mathbf{p}$, where $\mathbf{p}$ is as in Claim 2. Since $\tilde{\mathbf{p}}$ lies over the point $(p_1, ..., p_r)$ which is unramified in the extension $\tilde{L}/\bar{\mathbb{Q}}(x_1, ..., x_r)$, the point $\tilde{\mathbf{p}}$ gives rise to a splitting of the basic exact sequence

$$ 1 \quad \to \quad G(\tilde{L}/\bar{\mathbb{Q}}(x_1, ..., x_r)) \quad \to \quad G(\tilde{L}/\bar{\mathbb{Q}}(x_1, ..., x_r)^{\Gamma}) \quad \to \quad \Gamma \quad \to \quad 1 $$

in the usual way: The stabilizer $\tilde{\Gamma}$ of $\tilde{\mathbf{p}}$ in $G(\tilde{L}/\bar{\mathbb{Q}}(x_1, ..., x_r)^{\Gamma})$ effects the splitting. The induced isomorphism $\tilde{\Gamma} \to \Gamma$ extends the action of Γ from $\bar{\mathbb{Q}}(x_1, ..., x_r)$ to $\tilde{L}$. This action agrees with the action of Γ on L constructed above because the latter fixes $\mathbf{p}$.

Corollary 5.2 *The group* $PSp_n(q)$, *q odd, occurs regularly over* $\mathbb{Q}$ *if* $n \geq 4(q+1)$ *and one of the following holds:*

(1)   q is a square.

(2)   n/2 is divisible by 4 and $q \equiv 5$ *or* $7 \bmod 12$.

(3)   n/2 is odd and $q \equiv 3$ *or* $5 \bmod 8$.

Proof : Take $S = \mathbb{F}_q^*$ (a cyclic group of order $q - 1$) or $S = \{a \in \mathbb{F}_{q^2} : \mathrm{N}_{\mathbb{F}_{q^2}/\mathbb{F}_q}(a) = 1\}$ (a cyclic group of order $q + 1$) such that if q is a square then S has an element of order 4, and if q is not a square and $m = n/2$ is even (resp., odd) then S has an element of order 3, but none of order 4 (resp., S has an element of order 4, but none of order 8). Such S exists in each of the cases (1)-(3).

If $n \geq 4(q+1)$ then there are elements $a_1, ..., a_{2\ell} \in S \backslash \{\pm 1\}$, where $2\ell = m$ or $2\ell = m - 1$, that form a rational tuple, such that the $a_i + a_i^{-1}$ generate the field $\mathbb{F}_q$. Then $a_1 \cdots a_{2\ell} = 1$. If m is odd let a_m be of order 4. Then in any case, if q is not a square then $(-1)^{m-1}a_1 \cdots a_m$ is not a square in S. Let $t_i = a_i + a_i^{-1}$ and $\mathbf{t} = (t_1, ..., t_m)$. Then the associated tuple $\mathbf{C}(\mathbf{t}) = (C_1, ..., C_r)$ is rational. By Lemma 4.1 there is a tuple $(\sigma_1, ..., \sigma_m, \tau_1, \tau_2)$ in $\mathcal{N}(\mathbf{t})$. Let $g_1, ..., g_r$ be the images of these elements in $\bar{G} = \mathrm{PSp}_n(q)$. We may assume $g_{r-1} \in C_{r-1}$. Then $g_r \in C_r$ by Theorem 2.2(v). By Lemma 4.2, $\bar{G}$ is generated by $g_1, ..., g_r$. Hence $\mathcal{N}(\mathbf{t}) \neq \emptyset$, and so there is an extension $\Omega_0/\mathbb{C}(x)$ as in Theorem 5.1.

We can further modify $a_1, ..., a_m$ such that the hypothesis of Theorem 5.1 (ii) resp. (iii) holds. Then the field $M = L^\Gamma$ resp. $M = \tilde{L}^\Gamma$ is rational over $\mathbb{Q}$, say $M = \mathbb{Q}(t_1, ..., t_r)$. Since M contains the minimal field of definition L_1 of Ω_0, there is $\Omega_M \subset \Omega_0$ such that restriction gives an isomorphism $G(\Omega_0/\mathbb{C}(x)) \to G(\Omega_M/M(x)) = G(\Omega_M/\mathbb{Q}(t_1, ..., t_r, x))$. Since $\bar{G} \cong G(\Omega_0/\mathbb{C}(x))$ it follows that $\bar{G}$ occurs regularly over $\mathbb{Q}$.

References

[FV]   M. FRIED AND H. VÖLKLEIN, The inverse Galois problem and rational points on moduli spaces, Math. Annalen **290** (1991), 771-800.

[MSV] K. MAGAARD, K. STRAMBACH AND H. VÖLKLEIN, Finite quotients of the pure symplectic braid group, Israel J. of Math. 106 (1998), 13–28.

[Ma]   B. H. MATZAT, Zöpfe und Galoissche Gruppen, J. Reine und Angew. Math. **420** (1991), 99–159.

[Sc]   L. SCOTT, Matrices and cohomology, *Ann. Math.* 105 (1977), 473-492.

[SV]   K. STRAMBACH AND H. VÖLKLEIN, Generalized braid groups and rigidity, J. Algebra **175** (1995), 604–615.

[ThV] J.G. THOMPSON AND H. VÖLKLEIN, Symplectic groups as Galois groups, *J. Group Theory* 1 (1998), 1–58.

[V1]   H. VÖLKLEIN, Groups as Galois Groups – an Introduction, Cambr. Studies in Adv. Math. 53, Cambridge Univ. Press 1996.

[V2]   H. VÖLKLEIN, Rigid generators of classical groups, Math. Annalen 311 (1998), 421–438.

[V3]   H. VÖLKLEIN, Moduli spaces for covers of the Riemann sphere, Israel J. Math. **85** (1994), 407–430.

[V4]   H. VÖLKLEIN, Braid group action, embedding problems and the groups $PGL(n,q)$, $PU(n,q^2)$, Forum Math. **6** (1994), 513–535.

Author's adresses:
   John Thompson (e-mail: thompson@math.ufl.edu) and Helmut Völklein (e-mail: helmut@math.ufl.edu), Dept. of Mathematics, University of Florida, Gainesville, FL 32611, USA

Deformation of tame admissible covers of curves

Stefan Wewers

wewers@exp-math.uni-essen.de

Abstract

Let X be a semistable curve over a complete local ring and let $\bar{\rho} : \bar{Y} \to \bar{X}$ be a tame admissible cover of the special fiber. To lift $\bar{\rho}$ to a tame admissible cover $\rho : Y \to X$, it suffices to lift $\bar{\rho}$ locally in small neighborhoods of the singular points. The present paper gives a proof of this result using formal patching. As an application in the case of smooth curves, a proof of Grothendieck's Theorem on the tame fundamental group of smooth projective curves in positive characteristic is included.

Introduction

Let X be a semistable curve over a complete local ring and let $\bar{\rho} : \bar{Y} \to \bar{X}$ be a tame admissible cover of the special fiber. Choose a horizontal divisor $D \subset X$ lifting the branch locus $\bar{D} \subset \bar{X}$ of $\bar{\rho}$. To deform $\bar{\rho}$ to a tame admissible cover $\rho : Y \to X$, ramified along D, it suffices to lift $\bar{\rho}$ locally in small neighborhoods of the singular points. To do so, one has to choose roots of local parameters of the singular points of X. Hence, deformations of $\bar{\rho}$ exist in general only after a tamely ramified extension of the base ring and are in general not unique.

In the case of smooth curves, however, deformation of tame admissible covers is always possible and is unique. This result was first proved by Grothendieck and used in his theory of specialization of fundamental groups, [8]. In [9] deformation of mock covers is studied and applied to tame fundamental groups. These results can be reformulated in terms of deformation of tame admissible covers which are unramified over the singular points. In [23], deformation of admissible covers over complete discrete valuation rings are described. This is used to construct a specialization morphism of fundamental groups of curves with semistable reduction. The fundamental group of the special fiber, classifying admissible covers, is described by a graph of groups. There are also many results on deformation of covers of curves which are not admissible. They are all proved using some version of either formal or rigid patching. Let us only mention the result of Harbater [10] that every finite

1

group is a Galois group over $\mathbb{Q}_p(x)$ and the proof of Abhyankar's conjecture in [22] and [11].

The deformation theorem proved in this paper is used in [13], [19] and [25] to compactify Hurwitz spaces. For this application it is important to have a precise uniqueness statement and to work over quite general complete local rings. Even though rigid patching has often shown to be the more flexible approach, in particular for covers with wild ramification, in the present situation formal patching seems to be more appropriate. The potential and the mechanisms of formal patching are certainly well known to algebraic geometers. But there seems to be no reference for this particular result which is reasonably self contained and accessible for a wider audience.

Therefore, the present paper has two goals. First, to give a rigorous proof of the general deformation theorem of admissible covers. Second, to make formal patching and its application to fundamental groups more accessible to non-specialists.

The main result

Let R be a complete noetherian local ring with separably closed residue field k and let X/R be a projective *nodal curve*. By this we mean that the special fiber $\bar{X} := X \times_R k$ has at worst ordinary double points $x_1, \ldots, x_n \in \bar{X}$ as singularities and their complete local rings on X are of the form

$$\mathcal{O}_{X,x_i} \cong R[[u_i, v_i \mid u_i v_i = t_i]],$$

where the isomorphism is induced by elements $u_i, v_i \in \mathcal{O}_{X,x_i}$ with $t_i := u_i v_i \in R$. The elements u_i, v_i can even be chosen from the henselian local ring $\mathcal{O}_{X,x_i} \subset \mathcal{O}_{X,x_i}$. In other words, X is locally around x_i (in the étale topology) isomorphic to the standard nodal curve $\operatorname{Spec} R[u_i, v_i|u_i v_i = t_i]$.

Let $D \subset X$ be a *mark* on X, i.e. a horizontal divisor which is étale over $\operatorname{Spec} R$ and does not meet the singular points. A *tame admissible cover* $\rho : Y \to (X, D)$ is a finite morphism between nodal curves, which is tamely ramified along D, étale over $X^{\mathrm{sm}} - D$ and verifies the following condition over the singular points. Let $y_j \in Y$ be a point lying over one of the singular points x_i. Then y_j is a singular point of Y/R, i.e. $\mathcal{O}_{Y,y_j} = R[[s_j, s_j|r_j s_j = \tau_j]]$ with $\tau_j := r_j s_j \in R$. Moreover, we can choose $r_j, s_j \in \mathcal{O}_{Y,y_j}$ such that $r_j^{n_j} = u_j$ and $s_j^{n_j} = v_j$ for an integer n_j prime to the characteristic of k and $u_i, v_i \in \mathcal{O}_{X,x_i}$ as above.

Suppose we are given a tame admissible cover $\bar{\rho} : \bar{Y} \to (\bar{X}, \bar{D})$ of the special fiber $\bar{X} := X \times_R k$ of X. For every point $y_j \in \bar{Y}$ lying over a singular point $x_i \in X$ we can choose $\bar{r}_j, \bar{s}_j \in \mathcal{O}_{\bar{Y},y_j}$ with $\bar{r}_j \bar{s}_j = 0$, $\bar{r}_j^{n_j} = \bar{u}_i$ and $\bar{s}_j^{n_j} = \bar{v}_j$, where u_i, v_i are as before and $\bar{u}_i, \bar{v}_i$ denote their restrictions to the special fiber. A *deformation* of $\bar{\rho}$ to R is a tame admissible cover $\rho : Y \to X$ with $\rho \times_R k = \bar{\rho}$. For every deformation ρ of $\bar{\rho}$ there are unique lifts $r_j, s_j \in \mathcal{O}_{Y,y_j}$ of $\bar{r}_j, \bar{s}_j$ with $r_j^{n_j} = u_i$, $s_j^{n_j} = v_j$ and $\tau_j := r_j s_j \in R$. Thus the deformation ρ determines a tuple $(\tau_j)_j$ of elements of R with $\tau_j^{n_j} = t_i$. Let us call the tuple (τ_j) the *deformation datum* for $\bar{\rho}$ corresponding to the deformation ρ. Our main result can be stated as follows.

Theorem: *The assignment*

$$\rho \longmapsto (\tau_j := r_j s_j)_j$$

induces a bijection between isomorphism classes of deformations of $\bar{\rho}$ to R and the set of tuples $(\tau_j)_j$ of elements of R with $\tau_j^{n_j} = t_i$. Moreover, the isomorphism between two deformations with the same deformation datum $(\tau_j)_j$ is unique.

This theorem appears as a claim in [13] (page 61 f) for simple covers and in [19], §3.23 in the same generality as above. In the case that R is a complete discrete valuation ring and $D = \emptyset$ it is proved in [23]. The surjectivity of the map in the theorem is proved in [12] for the case $R = k[[\tau_1, \ldots, \tau_n]]$. Other special cases of the theorem appear in various papers dealing with Galois action on fundamental groups, e.g. in [15].

Outline

Section 1 serves an an introduction to formal patching. We explain the general idea of the proof and present the necessary tools, namely étale localization, descent and Grothendieck's Existence Theorem. In Section 2 we show that every tame admissible cover is étale locally isomorphic to a cover of a certain standard shape. To show this, we introduce local coordinate systems of a nodal curve at an ordinary double point. Section 3 contains the proof of the main result. In Section 4 we give a proof of Grothendieck's theorem on the tame fundamental group of curves in positive characteristic. The appendix contains some results about étale ring extensions and henselization which are used in this paper.

Throughout, we assume that the reader is familiar with the definition of a scheme. However, we have tried to keep the references as accessible as possible. Most proofs make only references to Hartshorne's book or a standard textbook on commutative algebra.

I thank David Harbater and Helmut Völklein for helpful discussions about the content of this paper.

1 Formal patching

This section is an introduction to formal patching, as it is understood in this paper. The reader familiar with this circle of ideas can skip it without problems. The general reference for this section is [8].

We start in 1.1 with an outline of the proof of our main result from a more general point of view. More precisely, we explain those ideas of the proof which are independent of the special case of admissible covers. This outline motivates the following subsections, where the tools we need to do formal patching are presented. These are étale localization, étale descent and Grothendieck's Existence Theorem.

1.1   Outline of the proof

1.1.1   Let R be a complete local ring with residue field k. Let X be a scheme over R. We write $\bar{X} := X \times_R k$ for the special fiber. Let P be a property of morphisms of schemes which is local, in an appropriate sense. To fix ideas, we assume that a morphism with property P is finite.

Problem 1.1.1 Suppose we are given a morphism $\bar{\rho} : \bar{Y} \to \bar{X}$ with property P. Does there exist a morphism $\rho : Y \to X$ with property P such that $\bar{Y} = Y \times_R k$?

The present paper deals with this problem in the case that a morphism with property P is an admissible cover of curves. In the rest of this section we show that, under certain general conditions on the scheme X and on the property P, Problem 1.1.1 can be solved using standard techniques of algebraic geometry.

1.1.2   The main idea is to solve Problem 1.1.1 first locally and then to glue the local solutions together to a global solution. To do this, we choose an open covering $(U_i)_{i \in I}$ of X. Then $(\bar{U}_i := U_i \cap \bar{X})_i$ is an open covering of $\bar{X}$ and $(\bar{V}_i := \bar{\rho}^{-1}(\bar{U}_i))_i$ is an open covering of $\bar{Y}$. Moreover, the maps $\bar{\rho}_i : \bar{V}_i \to \bar{U}_i$ induced by $\bar{\rho}$ have property P (since P is a local property). For $i, j \in I$ we let $U_{i,j} := U_i \cap U_j$, $\bar{U}_{i,j} := \bar{U}_i \cap \bar{U}_j$ and $\bar{V}_{i,j} := \bar{V}_i \cap \bar{V}_j$.

Condition 1.1.2 If the covering $(U_i)_i$ is chosen sufficiently fine, then the following holds.

(i) The morphisms $\bar{\rho}_i : \bar{V}_i \to \bar{U}_i$ can be lifted to morphisms $\rho_i : V_i \to U_i$ with property P.

(ii) For $i \neq j$, let U be an open subset of $U_{i,j}$, and let $\bar{U} := U \cap \bar{X}$ and $\bar{V} := \bar{\rho}^{-1}(\bar{U})$. Given two morphism $\rho_1 : V_1 \to U$ and $\rho_2 : V_2 \to U$ lifting $\bar{\rho}|_{\bar{V}} : \bar{V} \to \bar{U}$ and having property P, there exists a unique isomorphism $\alpha : V_1 \xrightarrow{\sim} V_2$ with $\rho_2 \circ \alpha = \rho_1$ and $\alpha|_{\bar{V}} = \mathrm{Id}_{\bar{V}}$.

Provided we use the right notion of an open covering, Condition 1.1.2 is sufficient to solve Problem 1.1.1. Given the local lifts $\rho_i : V_i \to U_i$ of (i), Condition (ii) makes sure there are unique isomorphisms

$$\alpha_{i,j} : \rho_j^{-1}(U_{i,j}) \xrightarrow{\sim} \rho_i^{-1}(U_{i,j}) \tag{1}$$

with $\rho_i \circ \alpha_{i,j} = \rho_j$ and $\alpha_{i,i} = \text{Id}$. Their uniqueness forces the $\alpha_{i,j}$ to verify the cocycle relation

$$\alpha_{i,j}\big|_{\rho_j^{-1}(U_{i,j,k})} \circ \alpha_{j,k}\big|_{\rho_k^{-1}(U_{i,j,k})} = \alpha_{i,k}\big|_{\rho_k^{-1}(U_{i,j,k})}, \tag{2}$$

where $U_{i,j,k} := U_i \cap U_j \cap U_k$ for $i, j, k \in I$. In this situation we can glue the schemes V_i along the isomorphisms $\alpha_{i,j}$ and obtain a scheme Y together with a morphism $\rho : Y \to X$ such that $\rho_i = \rho|_{V_i}$. It follows that ρ is a lift of $\bar{\rho}$ with property P, solving Problem 1.1.1.

In the case that the U_i are Zariski open subsets of X, this gluing process is given as an exercise in [14] II, Exercise 2.12. But it is in general very difficult to choose a sufficiently fine Zariski covering such that Condition 1.1.2 holds. An elegant solution for this problem is to replace the Zariski topology by the finer *étale topology*. This means that we replace open subsets $U_i \subset X$ by étale morphisms $U_i \to X$ and the intersections $U_i \cap U_j$ by the fiber products $U_i \times_X U_j$ (for the moment we will keep the old notation). In the context of the étale topology, the gluing process described above can be accomplished with the theory of *descent*. This theory is a generalization of both Zariski gluing and Galois descent. We will see in Section 1.2 that the descent theorem we need here reduces immediately to an algebraic lemma dealing with faithfully flat descent of modules.

1.1.3   Whether we can find a covering $(U_i)_i$ of X verifying Condition 1.1.2 depends of course on the property P. In Section 2.3 we will show that if P means being an admissible cover of curves, then we can find an étale covering of X of a certain standard shape. For such a covering Condition 1.1.2 (i) is then easily verified. This is the only part of our formal patching process dealing with the special situation of admissible covers. In our version of formal patching, Condition 1.1.2 (ii) depends on the following étaleness condition.

Condition 1.1.3 There is a dense open subset $U_0 \subset X$ such that $\bar{\rho} : \bar{Y} \to \bar{X}$ is étale over $U_0 \cap \bar{X}$ and all lifts $\rho : Y \to X$ of $\bar{\rho}$ with property P are étale over U_0. Moreover, for a sufficiently fine open covering $(U_i)_i$ of X and for $i \neq j$ we may assume that $U_{i,j} \subset U_0$.

In the case of admissible covers of curves, one can take for U_0 the complement of the branch locus inside the smooth locus of the curve. Then Condition 1.1.3 holds if k is algebraically closed.

Assume that Condition 1.1.3 holds. Then the morphisms $\bar{\rho}_{i,j} : \bar{V}_{i,j} \to \bar{U}_{i,j}$ are étale and any lift of $\bar{\rho}_{i,j}$ to $\rho_{i,j} : V_{i,j} \to U_{i,j}$ has to be étale. The following lemma shows that if R is artinian, then Condition 1.1.3 implies Condition 1.1.2 (ii).

Lemma 1.1.4 *Let A be a ring with a nilpotent ideal I; let $\bar{A} := A/I$. Then every finite étale $\bar{A}$-algebra $\bar{B}$ lifts uniquely to a finite étale A-algebra B.*

If R is artinian, then $\mathfrak{m}^n = 0$ for $n \gg 0$. We may assume that the open subset $U \subset U_{i,j}$ of Condition 1.1.2 (ii) is affine, $U = \operatorname{Spec} A$. Then $I := \mathfrak{m}A$ is a nilpotent ideal of A and $\bar{V} = \bar{\rho}^{-1}(\bar{U}) = \operatorname{Spec} \bar{B}$ for a finite $\bar{A}$-algebra. Assume moreover that Condition 1.1.3 holds. Then we may assume that $\bar{B}$ is étale over $\bar{A}$, and Condition 1.1.2 (ii) follows from Lemma 1.1.4.

1.1.4   We have seen that we can solve Problem 1.1.1 if R is artinian and the Conditions 1.1.2 (i) and 1.1.3 hold. We would like to extend this result to the case that R is a complete noetherian local ring. The problem is that we can not apply Lemma 1.1.4 in the same way as before. Note that even if A is a finitely generated R-algebra, A is in general not complete with respect to the ideal $I := \mathfrak{m}A$. It is easy to see that the analogous version of Lemma 1.1.4 is actually false in this situation. To get around this difficulty we need a further condition.

Condition 1.1.5 The ring R is noetherian and complete and X is a projective scheme over R.

For $n \geq 0$ let $R_n := R/\mathfrak{m}^{n+1}$ and $X_n := X \times_R R_n$. Assume that the Conditions 1.1.2 (i) and 1.1.3 hold if we replace the R-scheme X by the R_n-schemes X_n. Then we can use Lemma 1.1.4 and étale descent to construct a sequence of finite morphisms $\rho_n : Y_n \to X_n$ with property P such that $Y_n = Y_{n+1} \times_{R_{n+1}} R_n$ and $\bar{Y} = Y_0$. In this situation and under Condition 1.1.5 we can apply Grothendieck's Existence Theorem (see Section 1.4). This theorem shows that there is a finite morphism $\rho : Y \to X$ such that $Y_n = Y \times_R R_n$ for all $n \geq 0$. It remains to show that ρ has property P. Since this depends strongly on P, we formulate it as the last condition.

Condition 1.1.6 If the morphisms $\rho_n : Y_n \to X_n$ all have property P, then $\rho : Y \to X$ has property P.

1.2   Etale localization

We are going to introduce some terminology related to the étale topology of a scheme. All the definitions are restricted to *affine* étale morphisms. This is all we will need and it reduces the technical background. Throughout, X denotes a separated scheme.

1.2.1   A morphism of schemes $\varphi : U \to X$ is called **affine étale**, if $U = \operatorname{Spec} A$ is affine, its image $\varphi(U)$ is contained in some affine open subset $\operatorname{Spec} B \subset X$ and the ring extension $B \to A$ induced by φ is étale (see the Appendix for a definition of 'étale').

Lemma 1.2.1 *Let $\varphi : U = \operatorname{Spec} A \to X$ be an affine étale map. Then for any affine open subset $\operatorname{Spec} B \subset X$ containing $\varphi(U)$, $B \to A$ is étale.*

Moreover, if $\varphi' : U' = \operatorname{Spec} A' \to X$ *is another affine étale map, the same is true for the projection of the fibered product* $U \times_X U'$ *to* X *and for any morphism* $U' \to U$ *of* X-*schemes.*

Proof: The first claim follows from the fact that 'étale' is an open condition (see the Appendix). For the second claim, take $V := \operatorname{Spec} B \cap \operatorname{Spec} B'$, where $\operatorname{Spec} B \subset X$ (resp. $\operatorname{Spec} B' \subset X$) contains $\varphi(U)$ (resp. $\varphi(U')$). By [14] II, Ex. 4.3, $V = \operatorname{Spec} B''$ is again affine (here we use that X is separated). By the construction of fibred products in [14] II, Thm. 3.3, we get $U \times_X U' = (U \cap \phi^{-1}V) \times_V (U' \cap (\phi')^{-1}V) = \operatorname{Spec} A''$ with $A'' = (A \otimes_B B'') \otimes_{B''} (B'' \otimes_{B'} A)$. The B''-algebra A'' is étale by Lemma 5.1.2 (i), proving the second claim.

If there is any X-morphism $f : U' \to U$, then we can take one $\operatorname{Spec} B \subset X$ containing both $\varphi(U)$ and $\varphi'(U')$, and then f is given by a B-algebra morphism $A' \to A$, which must be étale by Lemma 5.1.2 (ii). This proves the last claim. ∎

An **(affine étale) covering**[1] $\mathcal{U}$ of X is a family $(\varphi_i : U_i \to X)_{i \in I}$ of affine étale morphisms whose images cover X.

Given a covering $\mathcal{U} = (U_i \to X)_{i \in I}$, we will frequently use the following notation.

$$U_{i,j} := U_i \times_X U_j, \qquad U_{i,j,k} := U_i \times_X U_j \times_X U_k, \qquad i, j, k \in I.$$

Note that, if the U_i are Zariski open subsets of X, then we actually have $U_{i,j} = U_i \cap U_j$, $U_{i,j,k} = U_i \cap U_j \cap U_k$. We have lots of natural maps:

$$U \xleftarrow{} \underset{i}{U_i} \overset{}{\underset{i,j}{\rightleftarrows}} \underset{i,j}{U_{i,j}} \overset{}{\underset{i,j,k}{\Lleftarrow}} U_{i,j,k}. \tag{3}$$

The two arrows in the middle we call $p^{(1)}_{i,j}$ (projection to the first factor, U_i) and $p^{(2)}_{i,j}$ (projection to U_j). On the right hand side, we have three maps $q^{(l)}_{i,j,k}$, $l = 1, 2, 3$, for the projection leaving out the l-th factor.

By Lemma 1.2.1, $U_{i,j} = \operatorname{Spec} A_{i,j}$, $U_{i,j,k} = \operatorname{Spec} A_{i,j,k}$. Therefore (3) corresponds to a complex of ring morphisms:

$$A' := \bigoplus_i A_i \rightrightarrows A'' := \bigoplus_{i,j} A_{i,j} \rightrightarrows A''' := \bigoplus_{i,j,k} A_{i,j,k}. \tag{4}$$

Lemma 1.2.2 *Assume* $X = \operatorname{Spec} A$ *affine. Then:*

(i) *Canonically,* $A'' = A' \otimes_A A'$ *and* $A'' = A' \otimes_A A' \otimes_A A'$

(ii) *The natural morphism* $A \to A'$ *is faithfully flat*

Proof: By the construction of fibered products in the affine case, $A_{i,j} = A_i \otimes_A A_j$ and $A_{i,j,k} = A_i \otimes_A A_j \otimes_A A_k$. This proves (i). $A \to A'$ is étale, therefore flat. It is faithfully flat because $\operatorname{Spec} A' = \coprod_i U_i \to X = \operatorname{Spec} A$ is surjective ([17] 4.C (iii)). ∎

---

[1]note that we give 'covering' (recouvrement, Überdeckung) a different meaning than 'cover' (revêtement, Überlagerung)

1.2.2   Let k be a field. A **geometric point** of X is a scheme morphism $x : \operatorname{Spec} k \to X$ such that k is an algebraically closed field. Let $x : \operatorname{Spec} k \to X$ be a geometric point. An **(affine étale) neighborhood** of x is a pair (U, x'), where $U \to X$ is an affine étale morphism $U \to X$ and $x' : \operatorname{Spec} k \to U$ a lift of x to U. Frequently, we will write U instead of (U, x') and call it simply a neighborhood of x. Let $\mathsf{Et}(X, x)$ denote the category of neighborhoods of x. A morphism $(U_1, x_1) \to (U_2, x_2)$ is a commutative diagram

$$
\begin{array}{ccccc}
\operatorname{Spec} k & = & \operatorname{Spec} k & = & \operatorname{Spec} k \\
\downarrow {\scriptstyle x_1} & & \downarrow {\scriptstyle x_2} & & \downarrow {\scriptstyle x} \\
U_1 & \longrightarrow & U_2 & \longrightarrow & X
\end{array}
$$

If a morphism from U_1 to U_2 exists, we will say that U_1 is **smaller** than U_2. Let $\mathsf{Et}'(X, x)$ be the full subcategory of all connected neighborhoods.

It follows from Lemma 5.2.1 that $\mathsf{Et}'(X, x)$ is a filtered inverse system. Therefore we can define the **(strict) henselian local ring** of X at x as

$$
\mathcal{O}_{X,x} := \varinjlim A, \qquad U = \operatorname{Spec} A \in \mathsf{Et}'(X, x) \tag{5}
$$

Of course, for every neighborhood $U = \operatorname{Spec} A$ of x, $\mathcal{O}_{X,x}$ is the henselization of $A \to k$, as defined in the Appendix. We define the **(strict) complete local ring** $\widehat{\mathcal{O}}_{X,x}$ of X at x to be the completion of $\mathcal{O}_{X,x}$.

1.2.3   Local decomposition of finite morphisms   Recall that a morphism $f : Y \to X$ of schemes is called **finite**, if for every open affine subset $U = \operatorname{Spec} A \subset X$ the preimage $f^{-1}(U) = \operatorname{Spec} B$ is affine and B is a finite A-algebra, i.e. finitely generated as A-module. Then we can write $Y = \operatorname{Spec} \mathcal{B}$, where $\mathcal{B}$ is a finite $\mathcal{O}_X$-algebra on X, i.e. a coherent sheaf with an additional algebra structure (see [14], II.3 and II.Ex. 5.17).

Let $f : Y = \operatorname{Spec} \mathcal{B} \to X$ be a finite morphism and $x : \operatorname{Spec} k \to X$ a geometric point. Since k is algebraically closed, $Y \times_X \operatorname{Spec} k \cong \operatorname{Spec} k^n$ for some $n \geq 0$. The n idempotents of k^n correspond one to one to the lifts $y_1, \ldots, y_n : \operatorname{Spec} k \to Y$ of x to Y.

If $U = \operatorname{Spec} A \to X$ is an affine étale neighborhood of x, then $V := Y \times_X U = \operatorname{Spec} B$, where B is a finite A-algebra. By Lemma 5.1.2 (i), $V = \operatorname{Spec} B \to Y$ is an affine étale map.

Lemma 1.2.3 *For every sufficiently small connected étale neighborhood* $U = \operatorname{Spec} A$ *of* x, *$V = \operatorname{Spec} B$ is a disjoint union of the form*

$$
V = \coprod_{i=1}^{n} V_i, \qquad V_i = \operatorname{Spec} B_i,
$$

such that $V_i = \operatorname{Spec} B_i \to Y$ *is a connected neighborhood of* y_i. *Moreover,*

$$
\mathcal{O}_{Y,y_i} = \varinjlim B_i \otimes_A A',
$$

where $A' \to k$ runs over $\mathsf{Et}'(A \to k)$.

Proof: Let $\check{A}$ be the henselization of A with respect to $A \to k$ and $\check{B} :=
B \otimes_A \check{A}$. Since $\check{A} \to \check{B}$ is finite and $\check{A}$ henselian, we obtain a decomposition
$\check{B} = \oplus_{i=1}^n \check{B}_i$ into local factors. For a sufficiently small $A' \to k \in \mathsf{Et}'(A \to k)$,
we have $B \otimes_A A' = \oplus_i B_i$ such that $\check{B}_i = B_i \otimes_{A'} \check{A}$. This proves the first
claim. Checking the universal property of henselization (Remark 5.2.2) we
see that $\check{B}_i$ is the henselization of B with respect to the ring morphisms
$B \to k$ corresponding to y_i. Now the second claim follows from the definition
of henselization. $\blacksquare$

We will say that a neighborhood U with the properties formulated in
Lemma 1.2.3 **decomposes** the finite map f. We may also state this as
follows: by choosing an arbitrarily small affine étale neighborhood U of x,
its inverse image on Y is the disjoint union of arbitrarily small affine étale
neighborhoods V_i of y_i.

1.3   Descent

A quasi-coherent sheaf $\mathcal{M}$ on a scheme X is an $\mathcal{O}_X$-module which is, over every affine open subset $U = \operatorname{Spec} A$, represented by an A-module M. More precisely, given an affine open covering $(U_i = \operatorname{Spec} A_i \subset X)_i$ of X, $\mathcal{M}$ corresponds
to a family of A_i-modules M_i together with isomorphisms $M_i \otimes_{A_i} A_{i,j} \xrightarrow{\sim}
M_i \otimes_{A_j} A_{i,j}$ verifying a natural cocycle condition (where $\operatorname{Spec} A_{i,j} = U_i \times_X U_j$).
The theory of descent shows that this is still true if $(U_i = \operatorname{Spec} A_i \to X)_i$ is
an étale covering, as defined in the last section. Moreover, this also works
for finite $\mathcal{O}_X$-algebras $\mathcal{B}$. Since the latter correspond to finite morphisms
$\rho : Y \to X$, descent is the right tool to glue finite morphisms, as described in
1.1.2.

1.3.1   Let X be a scheme and $\mathcal{M}$ a quasi-coherent sheaf of $\mathcal{O}_X$-modules
over X (see [14], II.5). Given an affine étale map $\varphi : U = \operatorname{Spec} A \to X$, the
inverse image $\varphi^* \mathcal{M}$ is a quasi-coherent sheaf on $U = \operatorname{Spec} A$, so it can be
written as

$$\varphi^* \mathcal{M} = M \tag{6}$$

where M is an A-module ([14] II Proposition 5.4). If $\varphi' : U' = \operatorname{Spec} A' \to X$
is another affine étale map and $f : U' \to U$ is an X-morphism, we can
write $(\varphi')^* \mathcal{M} = M'$ and the relation $\varphi' = \varphi \circ f$ induces a natural A'-linear
isomorphism

$$M' \cong M \otimes_A A' \tag{7}$$

Now let $\mathcal{U} = (\varphi_i : U_i = \operatorname{Spec} A_i \to X)_i$ be an affine étale covering and $\mathcal{M}$
a quasi-coherent sheaf on X. Then $\varphi_i^* \mathcal{M} = M_i$, where M_i is an A_i-module.

Using the notation introduced in Section 1.2, the relations $\varphi_i \circ p_{i,j}^{(1)} = \varphi_j \circ p_{i,j}^{(2)}$ induce a family $\alpha_{\mathcal{M}} = (\alpha_{i,j})_{i,j}$ of $A_{i,j}$-linear isomorphisms

$$\alpha_{i,j} : M_i \otimes_{A_i} A_{i,j} \xrightarrow{\sim} M_j \otimes_{A_j} A_{i,j}. \tag{8}$$

For each triple i, j, k, the family $(\alpha_{i,j})$ gives rise to $A_{i,j,k}$-linear isomorphisms

$$
\begin{aligned}
\alpha_{i,j,k}^{(1)} &: M_j \otimes_{A_j} A_{i,j,k} \xrightarrow{\sim} M_k \otimes_{A_k} A_{i,j,k} \\
\alpha_{i,j,k}^{(2)} &: M_i \otimes_{A_i} A_{i,j,k} \xrightarrow{\sim} M_k \otimes_{A_k} A_{i,j,k} \\
\alpha_{i,j,k}^{(3)} &: M_i \otimes_{A_i} A_{i,j,k} \xrightarrow{\sim} M_j \otimes_{A_j} A_{i,j,k}.
\end{aligned}
\tag{9}
$$

For instance, $\alpha_{i,j,k}^{(3)}$ is defined by $\alpha_{i,j} \otimes_{e_{i,j}^{(3)}} A_{i,j,k}$, where $e_{i,j,k}^{(3)} : A_{i,j} \to A_{i,j,k}$ is the natural morphism and where we have identified $(M_i \otimes_{A_i} A_{i,j}) \otimes_{A_{i,j}} A_{i,j,k}$ with $M_i \otimes_{A_i} A_{i,j,k}$ and $(M_j \otimes_{A_j} A_{i,j}) \otimes_{A_{i,j}} A_{i,j,k}$ with $M_j \otimes_{A_j} A_{i,j,k}$. The other two cases are similar. A tedious but formal verification shows that the following cocycle relation holds for every triple i, j, k:

$$\alpha_{i,j,k}^{(1)} \circ \alpha_{i,j,k}^{(3)} = \alpha_{i,j,k}^{(2)}. \tag{10}$$

Conversely, let $(M_i)_i$ be a family of A_i-modules and $(\alpha_{i,j})_{i,j}$ a family of $A_{i,j}$-linear isomorphisms as in (8). The datum $(M_i, \alpha_{i,j})$ is called a **descent datum** on the covering $\mathcal{U}$ if the family of isomorphisms $\alpha_{i,j,k}^{(\mu)}$ derived from the $\alpha_{i,j}$ as in (9) verifies the cocycle relation (10). A morphism of descent data from $(M_i, \alpha_{i,j})$ to $(N_i, \beta_{i,j})$ is given by a family of A_i-linear maps $f_i : M_i \to N_i$ compatible with $\alpha_{i,j}$ and $\beta_{i,j}$ in the obvious way. Another formal verification shows that

$$\mathcal{M} \longmapsto (M_i, \alpha_{i,j}) \tag{11}$$

defines a functor from the category of quasi-coherent sheaves on X to the category of descent data on $\mathcal{U}$.

Theorem 1.3.1 (Descent for quasi-coherent sheaves) *The functor defined by (11) is an equivalence of categories. In particular, for every descent datum $(M_i, \alpha_{i,j})$ on $\mathcal{U}$ there exists a quasi-coherent sheaf $\mathcal{M}$ on X with $\varphi_i^* \mathcal{M} = M_i$.*

Proof: For a proof of this theorem and much more general results, see e.g. [20] Chapter VII or [1] Chapter 6.1. We will explain how to reduce the theorem to a problem on faithfully flat descent of modules.

First assume that $X = \operatorname{Spec} A$ is affine. Let $(M_i, \alpha_{i,j})$ be a descent datum on $\mathcal{U}$. Then $M' := \oplus_i M_i$ has a natural structure of an $A' = \oplus_i A_i$ module. Using Lemma 1.2.2 we obtain canonical A'-linear isomorphisms

$$
\begin{aligned}
M' \otimes_A A' &\cong \bigoplus_{i,j} M_i \otimes_A A_j \cong \bigoplus_{i,j} M_i \otimes_{A_i} A_{i,j} \\
A' \otimes_A M' &\cong \bigoplus_{i,j} M_j \otimes_A A_i \cong \bigoplus_{i,j} M_j \otimes_{A_j} A_{i,j}.
\end{aligned}
\tag{12}
$$

Therefore a descent datum $(M_i, \alpha_{i,j})$ on $\mathcal{U}$ gives rise to a a datum (M', α). where M' is an A'-module and $\alpha : M' \otimes_A A' \xrightarrow{\sim} A' \otimes_A M'$ is an $A'' = A' \otimes_A A'$-linear isomorphism. The cocycle relation (10) corresponds to the relation

$$\alpha^{(3)} \circ \alpha^{(1)} = \alpha^{(2)}, \tag{13}$$

with A'''-linear isomorphisms

$$
\begin{aligned}
\alpha^{(1)} &: A' \otimes_A A' \otimes_A M' \xrightarrow{\sim} A' \otimes_A M' \otimes_A A' \\
\alpha^{(2)} &: A' \otimes_A A' \otimes_A M' \xrightarrow{\sim} M' \otimes_A A' \otimes_A A' \\
\alpha^{(3)} &: A' \otimes_A M' \otimes_A A' \xrightarrow{\sim} M' \otimes_A A' \otimes_A A'
\end{aligned}
\tag{14}
$$

obtained by tensoring α on the left, in the middle and on the right with $\mathrm{Id}_{A'}$.

Remember that we are assuming $X = \operatorname{Spec} A$ and we can therefore identify a quasi-coherent sheaf $\mathcal{M}$ on X with the A-module $M := \Gamma(X, \mathcal{M})$. In this case we have $M' = \oplus_i M_i = M \otimes_A A'$. and α is defined by $a_1 \otimes (m \otimes a_2) \longmapsto (a_1 \otimes m) \otimes a_2$. We can therefore reformulate the statement of Theorem 1.3.1 as follows. Given a faithfully flat ring homomorphism $A \to A'$, the functor

$$M \longmapsto (M' := M \otimes_A A', \alpha) \tag{15}$$

is an equivalence between the category of A-modules and the category of descent data for the morphism $A \to A'$. For a proof of this statement, see e.g. [18] I Remark 2.21.

The general case of Theorem 1.3.1 now follows quite easily. We can cover X be affine Zariski open subsets $V_\mu = \operatorname{Spec} B_\mu$. It is easy to see that for every μ we can restrict our descent datum $(M_i, \alpha_{i,j})$ on $\mathcal{U}$ to a descent datum on the étale covering $\mathcal{U}|_{V_\mu} = (U_i \times_X V_\mu \to V_\mu)$ of $V_\mu = \operatorname{Spec} B_\mu$. Applying Theorem 1.3.1 in the affine case, we obtain quasi-coherent sheaves $\mathcal{M}_\mu$ on V_μ corresponding to B_μ-modules M_μ. Using the fully faithfulness of the functor (11) in the affine case we get isomorphisms $\mathcal{M}_\mu|_{V_\mu \cap V_\nu} \xrightarrow{\sim} \mathcal{M}_\nu|_{V_\mu \cap V_\nu}$, because both sheaves correspond to the restriction of the descent datum $(M_i, \alpha_{i,j})$ to the affine open subset $V_\mu \cap V_\nu$. The uniqueness of these isomorphisms forces them to satisfy the usual cocycle relation. By [14] II Exercise 1.22 we can glue the $\mathcal{M}_\mu$ to a quasi-coherent sheaf $\mathcal{M}$. By construction, $\mathcal{M}$ corresponds to the descent datum $(M_i, \alpha_{i,j})$. A similar argument shows the fully faithfulness of (11) in the general case. ∎

1.3.2 Let X and $\mathcal{U}$ be as before and let $\mathcal{B}$ be a finite $\mathcal{O}_X$-algebra. Regarding $\mathcal{B}$ as a quasi-coherent sheaf on X, we obtain a descent datum $(B_i, \alpha_{i,j})$. where the B_i are finite A_i-algebras and the $\alpha_{i,j}$ are isomorphisms of $A_{i,j}$-algebras. We say that $(B_i, \alpha_{i,j})$ is a descent datum for finite $\mathcal{O}_X$-algebras on $\mathcal{U}$.

Corollary 1.3.2 *The functor* $B \mapsto (B_i, \alpha_{i,j})$ *is an equivalence between the category of finite $\mathcal{O}_X$-algebras and the category of descent data for finite $\mathcal{O}_X$-algebras on $\mathcal{U}$.*

Proof: Let $(B_i, \alpha_{i,j})$ be a descent datum for finite $\mathcal{O}_X$-algebras on $\mathcal{U}$. The algebra structure of the B_i is given by A_i-linear multiplication maps m_i : $B_i \otimes_{A_i} B_i \to B_i$. These maps verify certain identities corresponding to the rules for multiplication in a ring. Using the fact that the $\alpha_{i,j}$ are algebra morphisms, the family $(m_i)_i$ is easily seen to be a morphism of descent data. By Theorem 1.3.1 we obtain a quasi-coherent sheaf B on X together with a morphism $m : B \otimes_{\mathcal{O}_X} B \to B$ of quasi-coherent sheaves. The morphism m verifies the same identities as the m_i and defines thus a structure of $\mathcal{O}_X$-module on B. It remains to show that B is actually a finite $\mathcal{O}_X$-module. In view of the construction of B in the proof of Theorem 1.3.1, this follows from the following fact. Let M be an A-module and $A \to A'$ a faithfully flat ring extension. Then M is finitely generated over A iff $M' := M \otimes_A A'$ is finitely generated over A'.   ∎

1.4   Grothendieck's Existence Theorem

Let R be a complete local ring with maximal ideal $\mathfrak{m}$ and let X be a scheme over R. For $n \geq 0$, let $R_n := R/\mathfrak{m}^{n+1}$ and $X_n := X \times_R R_n$. Note that X_n is a closed subscheme of X. In particular, $\bar{X} := X_0$ is the special fiber of X.

A **formal coherent sheaf** on X is a family $(\mathcal{M}_n)_{n \geq 0}$ of coherent sheaves $\mathcal{M}_n$ on X_n together with a system of isomorphisms

$$\varphi_{n,m} : \mathcal{M}_n \otimes_{R_n} R_m \xrightarrow{\sim} \mathcal{M}_m, \qquad n \geq m \qquad (16)$$

such that $\varphi_{l,m} \circ (\varphi_{n,m} \otimes_{R_m} R_l) = \varphi_{n,l}$ for all $n \geq m \geq l$. For instance, given a coherent sheaf $\mathcal{M}$ on X, we can define the formal coherent sheaf $\mathcal{M} := (\mathcal{M} \otimes_R R_n)_n$, called the **formalization** of $\mathcal{M}$. There is an obvious notion of morphisms between formal coherent sheaves. Formalization is a functor

$$\mathcal{M} \longmapsto \mathcal{M} := (\mathcal{M} \otimes_R R_n)_n \qquad (17)$$

from coherent to formal coherent sheaves. A formal coherent sheaf $\mathcal{M}'$ on X is called **algebraizable**, if there exists a coherent sheaf $\mathcal{M}$ on X with $\mathcal{M}' = \mathcal{M}$.

Theorem 1.4.1 (Grothendieck's Existence Theorem) *Let R be a complete noetherian local ring and X a projective scheme over R. Then (17) is an equivalence of categories. In particular, every formal coherent sheaf on X is algebraizable.*

The proof of this theorem in [7], Chapter 5, uses the cohomology of coherent sheaves. In [14] this cohomological machinery is developed under less general hypotheses. Below we give a proof of Theorem 1.4.1 which follows closely the original lines of [7] but only relies on results proved in [14]. First we state the corollary which we will need to do formal patching.

Let X/R be as in Theorem 1.4.1 and let $\mathcal{B}$ be a finite $\mathcal{O}_X$-algebra (see Section 1.2.3). Regarding $\mathcal{B}$ as a coherent sheaf, we define its formalization $\mathcal{B} := (\mathcal{B}_n)$. Each $\mathcal{B}_n$ is a finite $\mathcal{O}_{X_n}$-algebra and the isomorphisms $\mathcal{B}_n \otimes_{R_n} R_m \xrightarrow{\sim} \mathcal{B}_m$ respect the algebra structure. Therefore we call $\mathcal{B}$ a **formal finite** $\mathcal{O}_X$-**algebra**. The following Corollary is implied by Theorem 1.4.1 in the same way as Corollary 1.3.2 is implied by Theorem 1.3.1.

Corollary 1.4.2 *Assumption as in Theorem 1.4.1. Every formal finite $\mathcal{O}_X$-algebra is uniquely algebraizable.*

1.4.1 Before passing to the proof of Theorem 1.4.1 we fix some notation and recall some general facts. Since $X \subset \mathbb{P}^r_R$ by assumption, we can consider coherent resp. formal coherent sheaves on X as coherent resp. formal coherent sheaves on $\mathbb{P}^r_R$. Therefore, we may assume that $X = \mathbb{P}^r_R$.

$X = \mathbb{P}^r_R$ has a standard affine open covering $(U_i = \operatorname{Spec} A_i)_i$, where $A_i = R[T_0/T_i, \ldots, T_r/T_i]$, $i = 0, \ldots, r$. Let $\mathcal{M}$ be a coherent sheaf on X. For every sequence $i_1, \ldots, i_p$ with $1 \leq i_1 < \ldots < i_p \leq r$ we put $U_{i_1,\ldots,i_p} := U_{i_1} \cap \ldots \cap U_{i_p}$ and $M_{i_1,\ldots,i_p} := \Gamma(U_{i_1,\ldots,i_p}, \mathcal{M})$. In a standard manner we obtain a complex

$$\bigoplus_i M_i \longrightarrow \bigoplus_{i,j} M_{i,j} \longrightarrow \bigoplus_{i,j,k} M_{i,j,k} \longrightarrow \cdots \tag{18}$$

of R-modules (see [14] II.4). Note that $M_i = \Gamma(U_i, \mathcal{M})$ (resp. $M_{i,j} := \Gamma(U_{i,j}, \mathcal{M})$ etc.) are finitely generated A_i-modules (resp. finitely generated $A_{i,j} = A_i[T_k/T_j]$-modules etc.). For $q \geq 0$ we define the cohomology group $H^q(X, \mathcal{M})$ as the q-th cohomology group of the complex (18). By [14] III, Thm. 4.5 this coincides with the definition of cohomology via derived functors. In particular, $H^0(X, \mathcal{M}) = \Gamma(X, \mathcal{M})$ are the global sections of $\mathcal{M}$.

Let $\mathcal{M}' = (\mathcal{M}_n)_n$ be a formal coherent sheaf. Note that we can consider $\mathcal{M}_n$ as a coherent sheaf on X. We will write $M_i^{(n)} := \Gamma(U_i, \mathcal{M}_n)$, $M_{i,j}^{(n)} := \Gamma(U_{i,j}, \mathcal{M}_n)$ etc. For $n \geq m$ the natural morphism $\mathcal{M}_n \to \mathcal{M}_m$ induces morphisms

$$M_i^{(n)} \longrightarrow M_i^{(m)}, \quad M_{i,j}^{(n)} \longrightarrow M_{i,j}^{(m)}, \ldots \tag{19}$$

and hence a compatible system of morphisms $H^q(X, \mathcal{M}_n) \to H^q(X, \mathcal{M}_m)$. We define the cohomology of a formal coherent sheaf by

$$H^q(X, \mathcal{M}') := \varprojlim_n H^q(X, \mathcal{M}_n). \tag{20}$$

The main ingredient of the proof of GET is the Theorem on Formal Functions ([14] III, Theorem 11.1). It states that for any coherent sheaf $\mathcal{M}$ on X and $q \geq 0$ we have

$$H^q(X, \mathcal{M}) = \varprojlim_n H^q(X, \mathcal{M}_n) = H^q(X, \mathcal{M}). \tag{21}$$

1.4.2   Let $\mathcal{M}' = (\mathcal{M}_n)_n$ be a formal coherent sheaf on $X = \mathbb{P}^r_R$. We will say that $\mathcal{M}'$ is generated by a finite number of global sections if there are elements $m_1, \ldots, m_l \in H^0(X, \mathcal{M}')$ whose images generate the A_i-module $M_i^{(n)}$, for all $1 \leq i \leq r$, $n \geq 0$. Given a formal coherent sheaf $\mathcal{M}'$ and $k \in \mathbb{Z}$, we define its k-th twist of $\mathcal{M}'$ to be $\mathcal{M}'(k) := (\mathcal{M}'_n(k))_n$ (see [14] II.5).

Proposition 1.4.3 *Let $\mathcal{M}'$ be a formal coherent sheaf on $X = \mathbb{P}^r_R$. Then for $k \gg 0$, $\mathcal{M}'(k)$ is generated by a finite number of global sections.*

Proof: Let $\hat{R} := \oplus_{n \geq 0} \mathfrak{m}^n / \mathfrak{m}^{n+1}$ be the graded ring associated to R. Consider the sheaf $\hat{\mathcal{M}} := \oplus_{n \geq 0} \mathfrak{m}^n \mathcal{M}_n$ as a quasi-coherent sheaf on $\hat{X} := \mathbb{P}^r_{\hat{R}}$. We claim that $\hat{\mathcal{M}}$ is a coherent sheaf. In fact, $\hat{\mathcal{M}}$ is generated as an $\mathcal{O}_{\hat{X}}$-module by its subsheaf $\mathcal{M}_0$, which is a finitely generated $\mathcal{O}_{X_0}$-module. Note that we have

$$H^q(\hat{X}, \hat{\mathcal{M}}) = \bigoplus_{n \geq 0} H^q(X, \mathfrak{m}^n \mathcal{M}_n). \tag{22}$$

By [17] 10.D, $\hat{R}$ is noetherian. Hence we can apply [14] III, Theorem 5.2, to conclude that for $k \gg 0$ the coherent sheaf $\hat{\mathcal{M}}(k)$ is generated by a finite number of global sections and $H^q(\hat{X}, \hat{\mathcal{M}}(k)) = 0$ for all $q > 0$. In particular, $\mathcal{M}_0(k)$ is generated by a finite number of global sections and $H^1(X, \mathfrak{m}^n \mathcal{M}_n(k)) = 0$ for all $n \geq 0$ (use (22)).
     The short exact sequence $0 \to \mathfrak{m}^{n+1} \mathcal{M}_{n+1}(k) \to \mathcal{M}_{n+1}(k) \to \mathcal{M}_n(k) \to 0$ yields an exact sequence

$$H^0(X, \mathcal{M}_{n+1}(k)) \longrightarrow H^0(X, \mathcal{M}_n(k)) \longrightarrow H^1(X, \mathfrak{m}^{n+1} \mathcal{M}_{n+1}(k)) = 0, \tag{23}$$

showing that the first morphism is surjective. Choose a set $\bar{m}_1, \ldots, \bar{m}_l \in H^0(X, \mathcal{M}_0(k))$ of global generators of $\mathcal{M}_0(k)$. Using inductively that the first map in (23) is surjective, we can lift them to global sections $m_1, \ldots, m_l \in H^0(X, \mathcal{M}'(k))$. By Nakayama's Lemma, their images in $H^0(X, \mathcal{M}_n(k))$ generate $\mathcal{M}_n$ for all $n \geq 0$. This is exactly what we wanted to prove.   ∎

1.4.3   Now we are going to prove GET. Let $\mathcal{M}$, $\mathcal{N}$ be coherent sheaves on X. Then there is a coherent sheaf $\mathcal{H}om(\mathcal{M}, \mathcal{N})$ such that $\mathrm{Hom}(\mathcal{M}, \mathcal{N}) =$

$H^0(\mathcal{H}om(\mathcal{M},\mathcal{N}))$ ([14] II.5). Formalizing this construction we obtain a formal sheaf $\mathcal{H}om(\mathcal{M},\mathcal{N})$ such that $\text{Hom}(\mathcal{M},\mathcal{N}) = H^0(\mathcal{H}om(\mathcal{M},\mathcal{N}))$. The Theorem on Formal Functions (21) implies

$$\text{Hom}(\mathcal{M},\mathcal{N}) = \text{Hom}(\mathcal{M},\mathcal{N}). \qquad (24)$$

This proves the fully faithfulness of the formalization functor $\mathcal{M} \mapsto \mathcal{M}$.

It remains to show that for any formal coherent sheaf $\mathcal{M}'$ on X we can find a coherent sheaf $\mathcal{M}$ with $\mathcal{M}' = \mathcal{M}$. By Proposition 1.4.3 we can find k such that $\mathcal{M}'(k)$ is generated by a finite number of global sections. Therefore there exists a natural number l and a surjective morphism $\mathcal{O}_X^l \to \mathcal{M}'(k)$ of formal coherent sheaves. Twisting with $-k$ we obtain a surjective morphism $f' : \mathcal{O}_X(-k)^l \to \mathcal{M}'$. Applying this procedure once more to the formal coherent sheaf $\text{Ker}(f')$, we obtain an exact sequence

$$\mathcal{O}_X(-k')^{l'} \xrightarrow{g'} \mathcal{O}_X(-k)^l \xrightarrow{f'} \mathcal{M}' \longrightarrow 0. \qquad (25)$$

The first two formal sheaves are obviously algebraizable. Hence the fully faithfulness of the formalization functor guarantees the existence of a morphism $g : \mathcal{O}_X(-k')^{l'} \to \mathcal{O}_X(-k)^l$ of coherent sheaves such that $g' = g$. Let $\mathcal{M}$ be the cokernel of g. Then $\mathcal{M} \otimes_R R_n = \mathcal{M}_n$ because both sides are the cokernel of g_n. Hence $\mathcal{M}' = \mathcal{M}$. ∎

2   Tame admissible covers

Tame admissible covers are finite morphism between relative nodal curves with a particular local ramification behavior. Our goal is to show that locally in the étale topology all tame admissible covers have a certain standard shape. This is done in Section 2.3. Section 2.1 and 2.2 contain some preliminary results.

Section 2.1 gives a detailed study of the strict complete local ring of a curve at an ordinary double point. The deformation theory of a curve in a neighborhood of an ordinary double point depends on the way one can choose and lift the so called *formal coordinate systems*. Since this will turn out to be crucial for the deformation theory of tame admissible covers, we do it very carefully. Note however that under the assumption of a reduced base ring (which suffices for most applications), proofs would become substantially simpler.

In Section 2.2 we use the results of 2.1 to study étale neighborhoods of ordinary double points. Technically, this means to compare the strict complete local ring of such a point to the strict henselian local ring. We also give the analogous statements for smooth points, which are much easier to prove.

2.1   Formal double points

Let A and R be complete noetherian local rings. Denote their maximal ideals by $\mathfrak{m} \lhd R$ and $\mathfrak{M} \lhd A$ and the residue field $R/\mathfrak{m}$ by k. Let $R \to A$ be a faithfully flat local ring extension.

The ring A is called a **formal double point** over R if there exists a pair u, v of elements of A such that

(i) $t := uv \in \mathfrak{m}$,

(ii) u, v induce an isomorphism $A \cong R[[u, v | uv = t]]$ of R-algebras.

In this case, the pair (u, v) is called a **(formal) coordinate system** for A/R. Any element $f \in A$ can be written uniquely in the form

$$f = a_0 + \sum_{i>0} a_i' u^i + \sum_{i>0} a_i'' v^i \tag{26}$$

with $a_0, a_i', a_i'' \in R$. We will call (26) the (u, v)-**expansion** of f.

Proposition 2.1.1 *Let A, R be as in the first paragraph of this section. Assume that $\bar{A} := A/\mathfrak{m}A$ is a formal double point over k. Then:*

(i) *Every coordinate system $(\bar{u}, \bar{v})$ of $\bar{A}/k$ lifts to a coordinate system (u, v) of A/R. In particular, A/R is a formal double point.*

(ii) *Every pair $r, s \in A$ with $rs \in \mathfrak{m}$ such that*

$$\mathfrak{M} = <r, s, \mathfrak{m}>$$

is a formal coordinate system.

Proof: Let $(\bar{u}, \bar{v})$ be a coordinate system for $\bar{A}$. Note that for every pair $u, v \in A$ lifting $\bar{u}, \bar{v}$ the induced ring homomorphism $R[[u, v]] \to A$ is surjective, since it is surjective mod $\mathfrak{m}$. We have to find lifts u, v such that $uv \in R$. To do this, we will construct inductively a sequence of lifts $u_n, v_n \in A$ of $\bar{u}, \bar{v}$ and a sequence of elements $t_n \in \mathfrak{m}$ (for $n \geq 0$) such that $u_n \equiv u_{n+1}, v_n \equiv v_{n+1}$ mod $\mathfrak{m}^{n+1}A$, $t_n \equiv t_{n+1}$ mod $\mathfrak{m}^{n+1}$ and $u_n v_n \equiv t_n$ mod $\mathfrak{m}^{n+1}$.

To start, take any pair u_0, v_0 lifting $\bar{u}, \bar{v}$ and let $t_0 := 0$. Suppose we have already constructed u_n, v_n, t_n for some $n \geq 0$. Then we can write

$$\begin{aligned}
u_n v_n &= t_n + \sum_{i,j \geq 0} a_{i,j} u_n^i v_n^j \\
&\equiv t_n + a_{0,0} + \sum_{i>0} a_{i,0} u_n^i + \sum_{j>0} a_{0,j} v_n^j \mod \mathfrak{m}^{n+2}A
\end{aligned} \tag{27}$$

with $a_{i,j} \in \mathfrak{m}^{n+1}$. For the first equality we have used the fact that an element of $\mathfrak{m}^{n+1}A$ can be written as a power series in u_n, v_n with coefficients in $\mathfrak{m}^{n+1}$.

For the congruence we have used $u_n v_n \in \mathfrak{m}A$. Let $u_{n+1} := u_n - \sum_{j>0} a_{0,j}\, v_n^{j-1}$
and $v_{n+1} := v_n - \sum_{i>0} a_{i,0}\, u_n^{i-1}$. Using (27) we obtain

$$
\begin{aligned}
u_{n+1}v_{n+1} &\equiv u_n v_n - \sum_{i>0} a_{i,0}\, u_n^i - \sum_{j>0} a_{0,j}\, v_n^j \qquad &&\mathrm{mod}\ \mathfrak{m}^{n+2}A \\
&\equiv t_n + a_{0,0} &&\mathrm{mod}\ \mathfrak{m}^{n+2}A.
\end{aligned}
\tag{28}
$$

Hence the induction step is done if we define $t_{n+1} := t_n + a_{0,0}$. Let $u := \lim u_n$, $v := \lim v_n$ and $t := \lim t_n$. Then u, v lift $\bar{u}, \bar{v}$ and $uv = t \in \mathfrak{m}$.

The following argument is taken from [6], Lemma 2.2. Consider the short exact sequence of R-modules

$$
0 \longrightarrow I \longrightarrow R[[u, v | uv = t]] \longrightarrow A \longrightarrow 0.
\tag{29}
$$

The homomorphism on the right is the one induced from the above choice of $u, v \in A$. Since A is R-flat, (29) remains exact after reduction modulo $\mathfrak{m}$ (see [16] XVI, Lem. 3.3). But $k[[\bar{u}, \bar{v} | \bar{u}\bar{v} = 0]] \xrightarrow{\sim} A/\mathfrak{m}A$ is an isomorphism by hypothesis, hence $I = \mathfrak{m}I$. Since $R[[u, v | uv = t]]$ is noetherian, the ideal I is finitely generated and Nakayama's lemma implies $I = 0$. This proves (i).

Let r, s be as in Statement (ii) of the proposition and let $(\bar{u}, \bar{v})$ be some coordinate system for $\bar{A}/k$. Denote by $(\bar{r}, \bar{s})$ the reduction of (r, s) to $\bar{A}$. Write

$$
\begin{aligned}
\bar{r} &= a_0 + \sum_{i>0} a_i'\, \bar{u}^i + \sum_{i>0} a_i''\, \bar{v}^i, \\
\bar{s} &= b_0 + \sum_{i>0} b_i'\, \bar{u}^i + \sum_{i>0} b_i''\, \bar{v}^i,
\end{aligned}
\tag{30}
$$

with coefficients in k. Using $\bar{r}\bar{s} = \bar{u}\bar{v} = 0$ one finds step by step that the following holds: first, $a_0 b_0 = 0$, next $a_0 = b_0 = 0$ and finally that either $a_i' = b_i'' = 0$ or $a_i'' = b_i' = 0$ for all $i > 0$ (we may assume the latter). By assumption the maximal ideal of $\bar{A}$ is generated by $\bar{r}, \bar{s}$. This can only happen if $a_1', b_1'' \neq 0$. Therefore the power series in (30) are in one variable and invertible. This proves that $(\bar{r}, \bar{s})$ is a coordinate system for $\bar{A}/k$. The argument used at the end of the proof of (i) shows that (r, s) is a coordinate system for A/R. ∎

Proposition 2.1.2 *Let A/R be a formal double point. Then the set of ideals $\{uA, vA\}$ and the ideal tR (with $t := uv$) are the same for any choice of a coordinate system (u, v).*

Proof: First, let us assume that there exists a coordinate system (u, v) of A/R with $uv = 0$. Under this assumption, we will prove that for every coordinate system (r, s) of A/R, we have $rs = 0$ and either $rA = uA$ and $sA = vA$ or

$rA = vA$ and $sA = uA$. Write

$$r = a_0 + \sum_{i>0} a_i' u^i + \sum_{i>0} a_i'' v^i,$$

$$s = b_0 + \sum_{i>0} b_i' u^i + \sum_{i>0} b_i'' v^i. \tag{31}$$

Since (r, s) is a coordinate system, the (u, v)-expansion of rs consists only of a constant term. Computing the (u, v)-expansion of rs using (31) and $uv = 0$, one obtains

$$a_0 b_n' + \sum_{i=1}^{n-1} a_i' b_{n-i}' + a_n' b_0 = 0,$$

$$a_0 b_n'' + \sum_{i=1}^{n-1} a_i'' b_{n-i}'' + a_n'' b_0 = 0, \tag{32}$$

for all $n > 0$. Reducing (31) mod $\mathfrak{m}$ and applying the arguments from the proof of Proposition 2.1.1 (ii) (following (30)) we can assume $a_1', b_1'' \not\equiv 0$ mod $\mathfrak{m}$ and $a_1'', b_1' \equiv 0$ mod $\mathfrak{m}$. For $n = 1$, (32) states

$$a_0 b_1' + a_1' b_0 = 0, \qquad a_0 b_1'' + a_1'' b_0 = 0 \tag{33}$$

Rewrite this as $b_0 = -(a_1')^{-1} a_0 b_1'$ and $a_0 = -(b_1'')^{-1} a_1'' b_0$. Plugging the second equation into the first yields $b_0(1 - (a_1' b_1'')^{-1} a_1'' b_1') = 0$. But the second factor is congruent to 1 mod $\mathfrak{m}$, hence $b_0 = 0$. By symmetry we obtain $a_0 = 0$. In particular, this proves $rs = 0$.

Next we prove by induction that $a_i'', b_i' = 0$ for all $i > 0$. Assume that this is true for all $i < N$ for some $N > 0$. Then (32) with $n := N + 1$ states $a_1' b_N' = 0$ and $a_N'' b_1'' = 0$, therefore $b_N' = a_N'' = 0$. We have shown that r (resp. s) is a power series in u (resp. v) starting with an invertible coefficient for the first power. This proves $rA = uA$ and $sA = vA$ and hence the Proposition in this special case.

The general case follows easily. Let (u, v) be a coordinate system and $t := uv$. The ideal $I := tR$ is the minimal one such that $\bar{u}\bar{v} = 0$ for *every* coordinate system $(\bar{u}, \bar{v})$ of A/IA over R/I. The ideals uA and vA are the inverse images of uA/tA and vA/tA. ∎

Let A/R be a formal double point, (u, v) a coordinate system and n a natural number. Define

$$P_{u,n} := \mathrm{Ann}_A(u^n A) = \{a \in A \mid au^n = 0\} \lhd A,$$

$$P_{v,n} := \mathrm{Ann}_A(v^n A) = \{a \in A \mid av^n = 0\} \lhd A, \tag{34}$$

and $P_u := P_{u,1}$, $P_v := P_{v,1}$. Using the uniqueness of the (u, v)-expansion, a straightforward verification shows:

$$P_{u,n} = \{ \sum_{i=1}^{\infty} a_i v^i \mid a_i t^i = 0 \ (i = 1, \ldots, n-1), \ a_i t^n = 0 \ (i \geq n) \} \subset vA,$$

$$P_{v,n} = \{ \sum_{i=1}^{\infty} a_i u^i \mid a_i t^i = 0 \ (i = 1, \ldots, n-1), \ a_i t^n = 0 \ (i \geq n) \} \subset uA, \tag{35}$$

where $t := uv \in \mathfrak{m}$. From this we see immediately that

$$P_{u,n} \cap P_{v,n} = P_{u,n} \cdot P_{v,n} = (0) \tag{36}$$

and (inside A)

$$R \cap (P_{u,n} + P_{v,n}) = (0). \tag{37}$$

Proposition 2.1.3 *(compare with [19], §3.7 and §3.8) Let (u,v) and (u',v') be coordinate systems of A/R and $n \geq 1$ such that $uA = u'A$. There are unique units $a, b \in A^{\times}$ with*

$$(u')^n = au^n, \ (v')^n = bv^n, \ ab \in R^{\times}$$

Proof: Put $t := uv$ and $t' := u'v'$. Consider the case $n = 1$ first. By Proposition 2.1.2 there are units $a', b' \in A^{\times}$ with $u' = a'u$, $v' = b'v$. Let

$$a'b' = c_0 + \sum_{i=1}^{\infty} c_i' u^i + \sum_{i=1}^{\infty} c_i'' v^i, \quad c_0, c_i', c_i'' \in R. \tag{38}$$

be the (u,v)-expansion of $a'b'$. Using its uniqueness and $t' = u'v' = (a'b')t \in R$ we conclude $c_i' t = c_i'' t = 0$ for every $i \geq 1$. Therefore (35) tells us that the sum in (38) is of the form

$$a'b' = c_0 + c_2 + c_1, \quad c_0 \in R, \ c_1 \in P_u, \ c_2 \in P_v \tag{39}$$

which is unique by (36) and (37).

Every pair (a, b) of units of A with $u' = au$, $v' = bv$ is of the form

$$a = a' + \lambda, \ b = b' + \mu, \quad \lambda \in P_u, \ \mu \in P_v. \tag{40}$$

By (36) we have $\lambda\mu = 0$, so using (39) and (40), we get the decomposition

$$ab = a'b' + b'\lambda + a'\mu = c_0 + (c_1 + b'\lambda) + (c_2 + a'\lambda) \tag{41}$$

in $R + P_u + P_v$. Since it is unique, $ab \in R$ is equivalent to

$$\lambda = -(b')^{-1}c_1, \quad \mu = -(a')^{-1}c_2. \tag{42}$$

This proofs at the same time the existence and the uniqueness in the case $n = 1$.

For general n, deduce the existence of $a, b \in A^{\times}$ with $(u')^n = au^n$, $(v')^n = bv^n$ and $ab \in R$ by using the case $n = 1$ and taking n-th powers. Any other pair $a', b' \in A^{\times}$ with the same properties can be written as $a' = a + \lambda$, $b' = b + \mu$, with $\lambda \in P_{u,n}$ and $\mu \in P_{v,n}$. As above, one gets the unique decomposition $a'b' = ab + b\lambda + a\mu \in R + P_{u,n} + P_{v,n}$, and one can deduce $\lambda = \mu = 0$, which proves the uniqueness of a, b. ∎

2.2   Marked nodal curves

2.2.1   Let R be a noetherian ring. A **curve** over R is a scheme X which is flat and of finite presentation over R such that all geometric fibers of X/R are reduced curves. A **nodal curve** is a curve X/R such that all geometric fibers have at most ordinary double points as singularities. We write $X^{\mathrm{sm}} \subset X$ for the smooth locus of the morphism $X \to \operatorname{Spec} R$. A **mark** on a nodal curve X/S is a closed subscheme $D \subset X^{\mathrm{sm}}$ such that the natural morphism $D \to \operatorname{Spec} R$ is finite étale. Hence $D = \operatorname{Spec} R'$ for a finite étale R-algebra R'. We call the pair $(X/R, D)$ a marked nodal curve.

Remark 2.2.1 Being a marked nodal curve is local in the étale topology. More precisely, let X be a scheme over a noetherian ring R, $D \subset X$ a closed subscheme and $(U_i \to X)_i$ an étale covering. Then $(X/R, D)$ is a marked nodal curve if and only if $(U_i/R, D \times_X U_i)$ is a marked nodal curve, for all i.

2.2.2   Let X/R be a nodal curve, k a field and $x : \operatorname{Spec} k \to X$ a geometric point whose image is a closed point of X. Let $\bar{X} \subset X$ be the fiber of the morphism $X \to \operatorname{Spec} R$ containing the image of x. The composition of x with the morphism $X \to \operatorname{Spec} R$ corresponds to a ring morphism $R \to k$. Let $\check{R} := \mathcal{O}_{\operatorname{Spec} R, s}$ (resp. $\hat{R} := \mathcal{O}_{\operatorname{Spec} R, s}$) be the strict henselization (resp. strict completion) of R with respect to k.

In this subsection we assume in addition that the image of x is a singular point of $\bar{X}$. Since X/R is nodal we have $\mathcal{O}_{\bar{X}, x} = k[[\bar{u}, \bar{v}|\bar{u}\bar{v} = 0]]$. We will say that x is a **geometric double point** of X/R.

Proposition 2.2.2 *Let X/R be as above and $x : \operatorname{Spec} k \to X$ a geometric double point of X/R. Then the complete local ring $\mathcal{O}_{X,x}$ is a formal double point over $\hat{R}$ (see Section 2.1).*

Proof: By the definition of a curve, there is a Zariski open subset $U = \operatorname{Spec} A \subset X$ containing the image of x such that A is flat and of finite presentation over R. By Proposition 5.2.3 (i) and (ii) the complete local ring $\mathcal{O}_{X,x}$ is a local flat $\hat{R}$-algebra. Let $\mathfrak{m} := \operatorname{Ker}(R \to k)$. Since $\bar{U} := \operatorname{Spec}(A/\mathfrak{m}A)$ is an open subset of $\bar{X}$ containing the image of x, Proposition 5.2.3 (iii) shows that $\mathcal{O}_{\bar{X},x} = \mathcal{O}_{X,x}/\mathfrak{m}\mathcal{O}_{X,x}$. Hence the Proposition follows from the definition of geometric double points and Proposition 2.1.1 (i).   ∎

If $x : \operatorname{Spec} k \to X$ is a rational double point, Proposition 2.2.2 shows that there exists a pair u, v of elements of $\mathcal{O}_{X,x}$ such that $t := uv \in \hat{R}$ and $\mathcal{O}_{X,x} = \hat{R}[[u, v|uv = t]]$. Such a pair (u, v) will be called a **formal coordinate system** for X/R at x. Note that the ring extensions $\check{R} \to \hat{R}$ and $\mathcal{O}_{X,x} \to \mathcal{O}_{X,x}$ are faithfully flat, in particular injective (Proposition 5.2.3 (ii)). We will call a pair u, v of elements of $\mathcal{O}_{X,x}$ a **coordinate system** for

X/R at x if $t := uv \in \mathring{R}$ and (u,v) (regarded as a pair of elements of $\mathcal{O}_{X,\hat{x}}$) is a formal coordinate system.

Proposition 2.2.3 *Let X/R be as before and $x :$ Spec $k \to X$ a geometric double point. There exists a coordinate system (u,v) for X/R at x. Let (u,v) be any coordinate system for X/R at x and $R' \in \mathsf{E}\mathfrak{t}(R \to k)$ such that $t := uv \in R'$. Then for every sufficiently small affine étale neighborhood $U =$ Spec $A \to X$ of x, the natural morphism $R'[u,v|uv = t] \to A$ is étale.*

Proof: The first assertion is a consequence of Proposition 2.2.2 and Artin's Approximation Theorem. We will not give the details of the argument, because a much more general statement is proved in [3] XV, Corollaire 1.3.2. For the second assertion, let (u,v) be any coordinate system for X/R at x. By definition, $t := uv \in \mathring{R}$, hence we can find $R' \in \mathsf{E}\mathfrak{t}'(R \to k)$ with $t \in R'$. Let $U =$ Spec $A \to X$ be a neighborhood of x. Replacing A by $A \otimes_R R'$ we may assume that A is an R'-algebra. By the definition of $\mathcal{O}_{X,x}$ we may assume that $u,v \in A$. We obtain a natural morphism $R'[u,v|uv = t] \to A$. Since $U =$ Spec $A \to X$ is a neighborhood of x, A is equipped with a natural morphism $A \to k$; the composition $R'[u,v|uv = t] \to k$ sends u,v,t to 0. Taking the completion of the rings $R'[u,v|uv = t]$ and A with respect to the morphism to k, we obtain an isomorphism $\hat{R}[[u,v|uv = t]] \xrightarrow{\sim} \mathcal{O}_{X,\hat{x}}$. Hence, by Proposition 5.2.3 (v) the morphism $R'[u,v|uv = t] \to A$ is étale in a neighborhood of the maximal ideal $\mathrm{Ker}(A \to k)$. Replacing $U =$ Spec A by a Zariski open neighborhood of this point completes the proof of the Proposition. ∎

Given a coordinate system (u,v) of a geometric double point on X/R, a pair $(R', U =$ Spec $A \to X)$ as in Proposition 2.2.3 will be called a **coordinate neighborhood** for (u,v). Frequently we will omit the ring R' from our notation.

2.2.3 Let R, X and $x :$ Spec $k \to X$ be as in the first paragraph of 2.2.2. Now we assume that x is a geometric smooth point on X/R. By this we mean that the image of x is a smooth point of the fibre $\bar{X} \subset X$ on which it lies. In addition, let $D \subset X$ be a mark on the nodal curve X/R. For an affine étale neighborhood $U =$ Spec $A \to X$ of x, let $D_U := D \times_X U$. Then (U, D_U) is again a marked nodal curve. Since $U =$ Spec A is affine, $D_U =$ Spec(A/I) for an ideal $I \lhd A$ ([14] II, Corollary 5.10). By definition, $R \to A/I$ is finite étale. There are two cases to consider. First, if the image of x lies on $X - D$, then for any sufficiently small neighborhood $U =$ Spec $A \to X$ of x we have $I = A$ and $D_U = \emptyset$. On the other hand, if x lies on D, Proposition 5.2.3 (iii) and (v) imply

$$\mathring{R} = \mathcal{O}_{D,x} = \mathcal{O}_{X,x}/I\mathcal{O}_{X,x}, \qquad \hat{R} = \mathcal{O}_{D,x} = \mathcal{O}_{X,\hat{x}}/I\mathcal{O}_{X,\hat{x}}. \qquad (43)$$

Proposition 2.2.4 *Let $x : \operatorname{Spec} k \to X$ be a geometric smooth point on the marked nodal curve $(X/R, D)$. There exists an element $z \in \mathcal{O}_{X,x}$ such that for every sufficiently small affine étale neighborhood $U = \operatorname{Spec} A \to X$ of x the natural morphism $R[z] \to A$ is étale. Moreover, if x lies on D, we can choose z such that D_U is defined by the equation $z = 0$.*

Proof: We will assume that x lies on D. Since x is a smooth point of $\bar{X}$ we have $\mathcal{O}_{\bar{X},x} = k[[\bar{z}]]$. Similarly to what we did in the proof of Proposition 2.1.1 (i) and Proposition 2.2.2 one shows that $\mathcal{O}_{X,x} = \hat{R}[[z]]$. Here z is any lift of $\bar{z}$ to $\mathcal{O}_{X,x}$. Let $U = \operatorname{Spec} A \to X$ be a neighborhood of x and $I \lhd A$ such that $D \times_X U = \operatorname{Spec}(A/I)$. Let $\hat{I} := I\mathcal{O}_{X,x}$. By (43) we have $\hat{R}[[z]]/\hat{I} \cong \hat{R}$. Therefore $z - a \in \hat{I}$ for some $a \in \hat{\mathfrak{m}} \lhd \hat{R}$. We may replace z by $z + a$, hence $z \in \hat{I}$. Now Nakayama's Lemma implies $\hat{I} = z\mathcal{O}_{X,x}$. Using [17] 24.E (i) we conclude that $I\mathcal{O}_{X,x} = z'\mathcal{O}_{X,x}$. Actually, we may assume that $z = z' \in \mathcal{O}_{X,x}$. After shrinking $U = \operatorname{Spec} A$ we may assume that $z \in A$ and $I = zA$. Completing the natural morphism $R[z] \to A$ with respect to k we obtain an isomorphism $\hat{R}[[z]] \xrightarrow{\sim} \mathcal{O}_{X,x}$. Hence, after shrinking $U = \operatorname{Spec} A$ a little bit more, $R[z] \to A$ will be étale. ∎

If x lies on D, an element $z \in \mathcal{O}_{X,x}$ with $\mathcal{O}_{X,x} = \hat{R}[[z]]$ and $\hat{I} = z\mathcal{O}_{X,x}$ will be called a **formal coordinate** for D at x. If z is moreover an element of $\mathcal{O}_{X,x}$, it will be called a **coordinate** for D at x.

2.3   Tame admissible covers

2.3.1   Let $(X/R, D)$ be a marked nodal curve over a noetherian ring R (see Section 2.2) and $\rho : Y \to X$ a finite morphism of schemes. Moreover, let $y : \operatorname{Spec} k \to Y$ be a closed geometric point and $x := \rho \circ y$. Since ρ is finite. it induces finite local ring extensions

$$\mathcal{O}_{X,x} \longrightarrow \mathcal{O}_{Y,y}, \quad \mathcal{O}_{X,x} \longrightarrow \mathcal{O}_{Y,y}. \tag{44}$$

As in Section 2.2, $\tilde{R}$ (resp. $\hat{R}$) denotes the strict henselization (resp. the strict completion) of R with respect to k. The first (resp. second) arrow in (44) is a morphism of $\tilde{R}$-algebras (resp. $\hat{R}$-algebras).

Definition 2.3.1 *(see [13] §4) Let ρ, x and y be as above. We say that $\rho : Y \to X$ is **tame admissible** at y, if the following holds.*

(i) *Assume that x is a geometric double point. Then $\mathcal{O}_{Y,y}$ is a formal double point over $\hat{R}$. Moreover, there exist a formal coordinate system (u, v) of $\mathcal{O}_{X,x}$, a formal coordinate system (r, s) of $\mathcal{O}_{Y,y}$ and an integer n prime to the characteristic of k such that (44) sends u to r^n and v to s^n.*

(ii) *Assume that x is a smooth point lying on D. Then there exist a formal coordinate $z \in \mathcal{O}_{X,x}$ for D, an element $w \in \mathcal{O}_{Y,y}$ and an integer n prime* to the characteristic of k such that $\mathcal{O}_{Y,y} = \hat{R}[[w]]$ and (44) sends z to w^n.

(iii) *If x is a smooth point not lying on D then $\mathcal{O}_{Y,y} \cong \mathcal{O}_{X,x}$.*

*The morphism ρ is called a **tame admissible cover** of $(X/R, D)$ if it is tame admissible at every geometric point y of Y.*

Part (ii) and (iii) of Definition 2.3.1 are usually stated as follows: the morphism ρ is étale over $X^{\text{sm}} - D$ and tamely ramified along D. Let

$$\text{Rer}_R^D(X) := \{\, \rho : Y \to X \,\}$$

be the category whose objects are tame admissible covers of $(X/R, D)$ and whose morphisms between objects $Y \to X$ and $Z \to X$ are morphisms of schemes $Y \to Z$ compatible with the maps to X.

2.3.2 Consider the following situation. Let $(X/R, D)$ be a marked nodal curve and $\rho : Y \to X$ a finite morphism. Let $y : \operatorname{Spec} k \to Y$ be a geometric closed point such that $x := \rho \circ y$ is a geometric double point of X and such that ρ is admissible at y. We want to study ρ in a neighborhood of y. Let $U = \operatorname{Spec} A \to X$ be a neighborhood of x which decomposes ρ (see Section 1.2.3). Then there is a unique component $V = \operatorname{Spec} B$ of $U \times_X Y$ which is a neighborhood of y. We will refer to V as the local inverse image of U at y.

Proposition 2.3.2 *Let (u, v) be a coordinate system at x. If $U = \operatorname{Spec} A - X$ is a sufficiently small coordinate neighborhood for (u, v) then*

$$B = A[r, s \mid r^n = u, s^n = v, rs = \tau],$$

where $V = \operatorname{Spec} B$ is the local inverse image of U at y, n an integer prime to the characteristic of k and $\tau \in \tilde{R} \cap A$. In particular, Y/R is a nodal curve in a neighborhood of y and (r, s) is a coordinate system at y.

Proof: Let $\tilde{A} := \mathcal{O}_{X,x}$, $\tilde{B} := \mathcal{O}_{Y,y}$, $\hat{A} := \mathcal{O}_{X,x}$ and $\hat{B} := \mathcal{O}_{Y,y}$. By hypothesis we have $u, v \in \tilde{A}$, $t := uv \in \tilde{R}$ and $\hat{A} = \hat{R}[[u, v | uv = t]]$. By Definition 2.3.1 there exist formal coordinate systems (u', v') for $\hat{A}$ and (r', s') for $\hat{B}$ such that $(r')^n = u'$ and $(s')^n = v'$. From Proposition 2.1.2 we know that there exist unique elements $a, b \in \hat{A}^\times$ with $u = au'$, $v = bv'$ and $ab \in \hat{R}^\times$. By Hensel's Lemma we can choose $c, d \in \hat{A}^\times$ with $c^n = a$ and $d^n = b$. Then $(cd)^n = ab \in \hat{R}^\times$, therefore we have $cd \in \hat{R}^\times$, by Hensel's Lemma. Let $r := cr'$ and $s := ds'$. Then $r^n = u$, $s^n = v$ and $\tau := rs \in \hat{R}$. But $r^n, s^n \in \tilde{B}$ and $\tau^n \in \tilde{R}$, hence by Proposition 5.2.3 (vi) we have $r, s \in \tilde{B}$ and $\tau \in \tilde{R}$.

Let $\tilde{B}_1 := \tilde{A}[r, s | r^n = u, s^n = v, rs = \tau]$. Since $\tilde{A} \to \tilde{B}_1$ is a finite local extension, $\tilde{B}_1$ is henselian. There is a natural local morphism $\tilde{B}_1 \to \tilde{B}$. Taking completions at both sides we obtain an isomorphism, because the completion of $\check{B}_1$ is easily seen to be isomorphic to $\hat{B} = \hat{R}[[r, s \mid rs = \tau]]$. With Proposition 5.2.3 (v) we conclude $\hat{B} = \check{B}_1$. Therefore, if $U = \operatorname{Spec} A$ is sufficiently small, $V = \operatorname{Spec} B$ is as stated in the Proposition. The other statements follow easily.   ∎

We need a slight strengthening of Proposition 2.3.2. Let $\bar{Y} \subset Y$ be the fiber of Y/R on which y lies. Then $\mathcal{O}_{\bar{Y}, y} = \mathcal{O}_{Y, y}/\tilde{m}\mathcal{O}_{Y, y}$, where $\tilde{m}$ is the maximal ideal of $\hat{R}$ (compare with the proof of Proposition 2.2.2). Let (u, v) be a coordinate system for X/R at x. Write $\bar{u}, \bar{v}$ for the image of u, v in $\mathcal{O}_{\bar{X}, x} = \mathcal{O}_{X, x}/\tilde{m}\mathcal{O}_{X, x}$.

Proposition 2.3.3 *Notation as above. Let* $(\bar{r}, \bar{s})$ *be a coordinate system for* $\bar{Y}$ *at* y *such that* $\bar{r}^n = \bar{u}$ *and* $\bar{s}^n = \bar{v}$. *Then there is a unique coordinate system* (r, s) *for* Y *at* y *lifting* $(\bar{r}, \bar{s})$ *such that* $r^n = u$ *and* $s^n = v$.

Proof: By Proposition 2.3.2 we can choose a coordinate system (r', s') for Y at y with $(r')^n = u$ and $(s')^n = v$. Then the coordinate system (r, s) we are looking for is of the form $r = ar'$, $s = bs'$ with $a, b \in \mathcal{O}_{Y, y}^\times$ and $ab \in \hat{R}^\times$. Then $r^n = u$ and $s^n = v$ is equivalent to $a^n(r')^n = (r')^n$ and $b^n(s')^n = (s')^n$. By Proposition 2.1.3 this is equivalent to $a^n = 1$ and $b^n = 1$. Applying Proposition 2.1.3 to $\mathcal{O}_{\bar{Y}, y}$ we find that (r, s) is a lift of $(\bar{r}, \bar{s})$ if and only if the reductions $\bar{a}, \bar{b}$ of a, b are uniquely determined n-th roots of unity. Now the Proposition follows from Hensel's Lemma.   ∎

2.3.3   We continue with the notation fixed in the first paragraph of 2.3.3, but this time under the assumption that x is a smooth point of X/R lying on $D \subset X$. In analogy to Proposition 2.3.2 and Proposition 2.3.3 we can show the following.

Proposition 2.3.4 *Let* z *be a coordinate for* D *at* x. *If* $U = \operatorname{Spec} A \to X$ *is a sufficiently small coordinate neighborhood for* z *and* $V = \operatorname{Spec} B$ *its local inverse image at* y, *then*

$$B = A[w \mid w^n = z].$$

Here n *is an integer prime to the characteristic of* k. *As an element of* $\mathcal{O}_{Y, y}$, w *is uniquely determined by* z *and the image of* w *in* $\mathcal{O}_{\bar{Y}, y} = \mathcal{O}_{Y, y}/\tilde{m}\mathcal{O}_{Y, y}$.

3   Deformation theory

This section contains the main deformation result. In 3.1 we give the necessary notation and the precise statement (Theorem 3.1.1). Moreover, we sketch how this theorem can be generalized to the case of a non algebraically closed residue field, using Galois descent.

3.1   Statement of the main result

3.1.1   Let R be a noetherian *complete* local ring with algebraically closed residue field $k = R/\mathfrak{m}$. Let $(X/R, D)$ be a marked nodal curve where X is projective over R. We write $\bar{X} := X \times_R k$ for the special fiber and $\bar{D} := D \times_R k \subset \bar{X}$.

Let $\bar{\rho} : \bar{Y} \to \bar{X} \in \mathsf{Rev}_k^{\bar{D}}(\bar{X})$ be a tame admissible cover. A **deformation** of $\bar{\rho}$ to R is a pair (ρ, λ), where $\rho : Y \to X \in \mathsf{Rev}_R^D(X)$ is a tame admissible cover and $\lambda : Y \times_R k \xrightarrow{\sim} \bar{Y}$ is an isomorphism in the category $\mathsf{Rev}_k^{\bar{D}}(\bar{X})$. An isomorphism from one deformation (ρ_1, λ_1) to another (ρ_2, λ_2) is an isomorphism $f : Y_1 \xrightarrow{\sim} Y_2$ with $\rho_2 \circ f = \rho_1$ and $\lambda_2 \circ (f \times \mathrm{Id}_k) = \lambda_1$. Let

$$\mathrm{Def}_{\bar{\rho}}(R) = \{\,(\rho, \lambda)\,\}/_{\cong} \tag{45}$$

be the set of isomorphism classes of deformations of $\bar{\rho}$ to R. Most of the time we will identify the special fiber of a deformation with $\bar{Y}$, i.e. we will assume $\lambda = \mathrm{Id}_{\bar{Y}}$ and write $\rho \in \mathrm{Def}_{\bar{\rho}}(R)$.

3.1.2   Before stating the main theorem we have to fix some notations. The special fiber $\bar{X}$ has a finite number of ordinary double points, which we denote by $x_1, \ldots, x_r \in \bar{X}$ (they can be identified with the corresponding geometric points $x_i : \mathrm{Spec}\, k \to X$). Let $I' := \{1, \ldots, r\}$. By Proposition 2.2.2 we can choose coordinate systems (u_i, v_i) for X at x_i, $i \in I'$. Recall that u_i, v_i are elements of the strict henselian local ring $\mathcal{O}_{X, x_i}$ such that $t_i := u_i v_i$ is an element of R and $\mathcal{O}_{X, x_i} = R[[u_i, v_i | u_i v_i = t_i]]$. We write $\bar{u}_i, \bar{v}_i$ for the image of u_i, v_i in $\mathcal{O}_{\bar{X}, x_i} = \mathcal{O}_{X, x_i}/\mathfrak{m}\mathcal{O}_{X, x_i}$. Then $(\bar{u}_i, \bar{v}_i)$ is a coordinate system for $\bar{X}$ at x_i.

Fix $i \in I'$ for a moment and let $\bar{\rho}^{-1}(x_i) = \{y_j \in \bar{Y} \mid j \in J_i'\}$ be the fiber of $\bar{\rho}$ over x_i, indexed by a finite set J_i'. By Proposition 2.3.2 there exist coordinate systems $(\bar{r}_j, \bar{s}_j)$ for $\bar{Y}$ at y_j such that $\bar{r}_j^{n_j} = \bar{u}_i$ and $\bar{s}_j^{n_j} = \bar{v}_j$ for integers n_j prime to the characteristic of k, for all $j \in J_i'$. Let J' be the disjoint union of the sets J_i' for $i \in I'$ and let $\kappa : J' \to I'$ be the map sending $j \in J_i'$ to i.

Let $\rho \in \mathrm{Def}_{\bar{\rho}}(R)$ be a deformation of $\bar{\rho}$ to R. For $j \in J'$ and $i := \kappa(j)$, Proposition 2.3.3 shows that there is a unique lift (r_j, s_j) of $(\bar{r}_j, \bar{s}_j)$ to a coordinate system of Y at y_j such that $r_j^{n_j} = u_i$ and $s_j^{n_j} = v_i$. Note that $\tau_j := r_j s_j$ is an element of R with $\tau_j^{n_j} = t_i$. This defines a map

$$\begin{array}{ccc} \mathrm{Def}_{\bar{\rho}}(R) & \longrightarrow & T(R) \\ (\rho, \lambda) & \longmapsto & \cdot \ (\tau_j := s_j r_j)_{j \in J'}. \end{array} \tag{46}$$

where $T(R) := \{(\tau_j)_{j \in J'} \mid \tau_j \in R, \ \tau_j^{n_j} = t_{\kappa(j)}\}$. An element $(\tau_j)_j$ of $T(R)$ will be called a **deformation datum**. Note that the map (46) depends on the choice of the coordinate systems (u_i, v_i) and $(\bar{r}_j, \bar{s}_j)$. Using the assumptions and notations of 3.1.1 and 3.1.2, we can now state the main theorem.

Theorem 3.1.1 *The map (46) is bijective. Moreover, if two deformations of $\bar{\rho}$ are isomorphic, then this isomorphism is unique.*

The proof of this theorem will be given in Section 3.2, following the outline given in Section 1.1.

3.1.3   In the statement of Theorem 3.1.1 we have assumed that the residue field k of R is algebraically closed. This is more than we really need. A careful inspection of the proof of Theorem 3.1.1 shows that we only have to assume that the singularities of $\bar{X}$ and the points on $\bar{D} = D \times_R k$ are k-rational and that k contains enough roots of unity. For instance, this holds if k is separably closed. But one can prove much more.

Suppose R is a complete noetherian local ring with arbitrary residue field k. Let k^s be a separable closure of k and let $\hat{R}$ be the strict completion of R with respect to k^s. Let $(X/R, D)$ be a projective marked nodal curve. Write $\bar{X} := X \times_R k$ for the special fiber and $\hat{X} := X \times_R \hat{R}$. Let $\bar{\rho} : \bar{Y} \to (\bar{X}, \bar{D})$ be a tame admissible cover and $\bar{\rho}^s$ its base change to k^s. Base change from R to $\hat{R}$ induces a map

$$\mathrm{Def}_{\bar{\rho}}(R) \; \longrightarrow \; \mathrm{Def}_{\bar{\rho}^s}(\hat{R}). \tag{47}$$

Let $G := \mathrm{Gal}(k^s/k)$ be the Galois group of k. Note that the action of G on k^s extends naturally to an action on $\hat{R}$.

Remark 3.1.2 There is a natural action of G on $\mathrm{Def}_{\bar{\rho}^s}(\hat{R})$. The map (47) induces a bijection $\mathrm{Def}_{\bar{\rho}}(R) \cong \mathrm{Def}_{\bar{\rho}^s}(\hat{R})^G$. In particular, if $\bar{\rho}$ is unramified over the singular points then there is a unique deformation ρ of $\bar{\rho}$ to R.

The first two statement of this remark are a variant of Weil's Descent Criterion and can be proved using étale descent. The third statement follows from the first two and from Theorem 3.1.1. Note that Theorem 3.1.1 can be applied to $\hat{X}/\hat{R}$. Hence we have a bijection $\mathrm{Def}_{\bar{\rho}^s}(\hat{R}) \cong T(\hat{R})$, where $T(\hat{R})$ is defined as in (46) in terms of coordinate systems of $\hat{X}$ at the singular points. The induced G-action on $T(\hat{R})$ can be determined from the natural action of G on the strict complete local rings of the singular points of Y. In particular, if X/R is smooth or if $\bar{\rho}$ is unramified over the singular points, $T(\hat{R})$ has exactly one element.

As remarked in the introduction, the results of [9] about deformation of mock covers with tame ramification can be reformulated in terms of tame admissible covers which are unramified over the singular points. Therefore, by Remark 3.1.2, Theorem 3.1.1 implies the results of [9]. However, the more general results of [10] on mock covers with wild ramification do not follow from Theorem 3.1.1.

3.1.4   Suppose that X is a smooth projective curve over a complete noetherian local ring R, and let $D \subset X$ be a mark. We will call tame admissible covers of (X, D) **tamely ramified covers**.

Corollary 3.1.3 *The functor*

$$\mathrm{Rev}_R^D(X) \xrightarrow{\sim} \mathrm{Rev}_k^{\bar{D}}(\bar{X})$$

reducing tamely ramified covers of X to the special fiber $\bar{X} := X \times_R k$ is an equivalence of categories.

Proof: In the case of smooth curves, Theorem 3.1.1 and Remark 3.1.2 show that tamely ramified covers lift uniquely. To prove Corollary 3.1.3, it remains to show the following. Let $\rho : Y \to (X, D)$ and $\eta : Z \to (X, D)$ be tamely ramified covers. Denote their reductions to $\bar{X}$ by $\bar{\rho}$ and $\bar{\eta}$. Let $\bar{f} : \bar{Z} \to \bar{Y}$ be an $\bar{X}$-morphism. Then $\bar{f}$ lifts uniquely to an X-morphism $f : Z \to Y$.

Using the result of Section 2.3, it is easy to see that $C := \rho^{-1}(D) \subset Y$ is a mark and $\bar{f}$ a tame admissible cover of $(\bar{Y}, \bar{C})$. We already know that $\bar{f}$ lifts to a tame admissible cover $f : Z' \to Y$. It follows that $\rho \circ f : Z' \to X$ is a tame admissible cover. Moreover, $\rho \circ f$ is a lift of $\bar{\eta}$. But since lifting is unique we may assume that $Z' = Z$ and $\rho \circ f = \eta$.   ∎

In Section 4 we will use Corollary 3.1.3 to construct a specialization morphism for tame fundamental groups.

3.2   Proof of the main theorem

In this Section we give a proof of Theorem 3.1.1 following the outline given in Section 1.1. In 3.2.1 we choose an étale covering of the curve X and fix some notation. In 3.2.2 we prove the theorem in the special case of an artinian base ring. In Section 3.2.3 we prove the general case by successively lifting $\bar{\rho}$ to the artinian rings $R_n := R/\mathfrak{m}^{n+1}$ and applying Grothendieck's Existence Theorem.

3.2.1   Notations and assumptions are as in Section 3.1.1 and 3.1.2. In addition, we assume that R is artinian. This means that $\mathfrak{m}^N = 0$ for some integer $N > 0$. The scheme $\mathrm{Spec}\, R$ consists of a single point and the closed embedding $\bar{X} \subset X$ induces a homeomorphism of the underlying topological spaces. Recall from 3.1.2 that x_i, $i \in I'$, are the singular points of $\bar{X}$, that y_j, $j \in J'$, are the singular points of $\bar{Y}$ and that the map $\kappa : J' \to I'$ is defined by $\bar{\rho}(y_j) = x_{\kappa(j)}$. We have chosen coordinate systems (u_i, v_i) (resp. $(\bar{r}_j, \bar{s}_j)$) for X at x_i, $i \in I'$, (resp. for $\bar{Y}$ at y_j, $j \in J'$,) such that $\bar{r}_j^{n_j} = \bar{u}_{\kappa(j)}$ and $\bar{s}_j^{n_j} = \bar{v}_{\kappa(j)}$.

The closed subscheme $\bar{D} \subset \bar{X}$ consists of a finite set of smooth points $x_{r+1}, \ldots, x_s \in \bar{X}$. Let $I'' := \{r+1, \ldots, s\}$. For $i \in I''$ let $\bar{\rho}^{-1}(x_i) = \{y_j \in \bar{Y} \mid j \in J_i''\}$ be the fiber over x_i, indexed by J_i''. Let $J'' := \cup_{i \in I''} J_i''$, $I := I' \cup I''$.

$J := J' \cup J''$ and extend κ to a map $\kappa : J \to I$ with $\kappa(j) = i$ for $j \in J_i''$. For all $i \in I''$, choose a coordinate $z_i \in \mathcal{O}_{X,x_i}$ for D at x_i (Proposition 2.2.4). For all $j \in J_i''$, choose an element $\bar{w}_i \in \mathcal{O}_{\bar{Y},y_j}$ such that $\mathcal{O}_{\bar{Y},y_j} = \mathcal{O}_{\bar{X},x_i}[\bar{w}_j | \bar{w}_j^{n_j} = \bar{z}_i]$, as in Proposition 2.3.4.

For $i \in I$, let $U_i = \operatorname{Spec} A_i \to X$ be an affine étale neighborhood of x_i. We may assume that U_i is a coordinate neighborhood, for the coordinate system (u_i, v_i) if $i \in I'$ and for the coordinate z_i of D if $i \in I''$. Hence, we have étale ring morphisms $R[u_i, v_i \mid u_i v_i = t_i] \to A_i$ for $i \in I'$ and $R[z_i] \to A_i$ for $i \in I''$. Let $U_0 := X - \{x_i \mid i \in I\} \subset X$ and $I_0 := I \cup \{0\}$. We may assume that $U_0 = \operatorname{Spec} A_0$ is affine. Then $\mathcal{U} := (U_i)_{i \in I_0}$ is an affine étale covering of X. In the sequel we will keep the open subset U_0 fixed and continue to shrink the neighborhoods U_i, $i \in I$, as necessary. Then $\mathcal{U}$ will always remain a covering of X. Hence we may assume that for all $i \in I$ the image of U_i on X is contained in $U_0 \cup \{x_i\}$. This implies that for $i \neq j$ the image of $U_{i,j} := U_i \times_X U_j$ on X is contained in U_0.

Let $\bar{\mathcal{U}} := (\bar{U}_i := U_i \times_X \bar{X})_{i \in I_0}$ be the restriction of $\mathcal{U}$ to $\bar{X}$. For $i \in I'$ (resp. $i \in I''$), $\bar{U}_i := \operatorname{Spec} \bar{A}_i \to \bar{X}$ is a coordinate neighborhood for $(\bar{u}_i, \bar{v}_i)$ (resp. for $\bar{z}_i$), where $\bar{A}_i = A_i / \mathfrak{m} A_i$. Since $\bar{\rho} : \bar{Y} \to \bar{X}$ is a finite map, we can write $\bar{Y} = \operatorname{Spec} \bar{B}$ for a finite $\mathcal{O}_{\bar{X}}$-algebra $\bar{B}$. Let $\bar{\alpha} = (\bar{C}_i, \alpha_{i,j})$ be the descent datum for $\bar{B}$ on $\bar{\mathcal{U}}$ (Section 1.3). In particular, $\bar{C}_i$ is a finite $\bar{A}_i$-algebra such that $\bar{U}_i \times_{\bar{X}} \bar{Y} = \operatorname{Spec} \bar{C}_i$. Note that the neighborhoods $\bar{U}_i$ of x_i can be made arbitrarily small by choosing U_i small. We may therefore assume that we have $\bar{C}_i = \oplus_{\kappa(j)=i} \bar{B}_j$, where

$$\bar{B}_j = \bar{A}_i[\bar{r}_j, \bar{s}_j \mid \bar{r}_j^{n_j} = \bar{u}_i, \ \bar{s}_j^{n_j} = \bar{v}_i, \ \bar{r}_j \bar{s}_j = 0] \tag{48}$$

for $j \in J_i'$ and

$$\bar{B}_j = \bar{A}_i[\bar{w}_j \mid \bar{w}_j^{n_j} = \bar{z}_i] \tag{49}$$

for $j \in J_i''$ (see Lemma 1.2.3, Proposition 2.3.2 and Proposition 2.3.4).

3.2.2   Let $\rho : Y \to X$ be a deformation of $\bar{\rho}$ to R. We can write $Y = \operatorname{Spec} B$ for a finite $\mathcal{O}_{\bar{X}}$-algebra B with $\bar{B} = B \otimes_R k$. Let $\alpha = (C_i, \alpha_{i,j})$ be the descent datum for B on $\mathcal{U}$. It follows that $\bar{\alpha} = \alpha \otimes_R k$. By the Propositions 2.3.3 and 2.3.4 there are unique lifts (r_j, s_j) of $(\bar{r}_j, \bar{s}_j)$ (resp. w_j of $\bar{w}_j$) with $r_j^{n_j} = u_{\kappa(j)}$ and $s_j^{n_j} = v_{\kappa(j)}$, $j \in J'$ (resp. $w_j^{n_j} = z_{\kappa(j)}$, $j \in J''$). Moreover, we may assume that $C_i = \oplus_{\kappa(j)=i} B_j$, where

$$B_j = A_i[r_j, s_j \mid r_j^{n_j} = u_i, \ s_j^{n_j} = v_i, \ r_j s_j = \tau_j], \qquad \tau_j := r_j s_j \in R \tag{50}$$

for $j \in J_i'$,

$$B_j = A_i[w_j \mid w_j^{n_j} = z_i] \tag{51}$$

for $j \in J_i''$ and $\bar{B}_j = B_j \otimes_R k$ for all $j \in J$. Since ρ is étale over U_0, C_0 is an étale A_0-algebra. For the same reason and our choice of the covering $\mathcal{U}$, $C_i \otimes_{A_i} A_{i,j}$ and $C_j \otimes_{A_j} A_{i,j}$ are étale $A_{i,j}$-algebras, for $i, j \in I$, $i \neq j$.

Now let $\rho' : Y' \to X$ be another deformation of $\bar{\rho}$ to R inducing the same deformation data $(\tau_j)_j$ as ρ. Let $Y' = \operatorname{Spec} B'$ and $\alpha' = (C_i', \alpha_{i,j}')$ be the corresponding descent datum for B' on $\mathcal{U}$. Since C_0 and C_0' are étale A_0-algebras with $C_0 \otimes_R k = \bar{C}_0 = C_0' \otimes_R k$, there is a unique A_0-linear isomorphism $f_0 : C_0 \xrightarrow{\sim} C_0'$ with $f \otimes_R k = \operatorname{Id}_{\bar{C}_0}$ (Lemma 5.1.1). After shrinking the neighborhoods U_i, $i \in I$, we find A_i-linear isomorphisms $f_i : C_i \to C_i'$ with $f_i \otimes_R k = \operatorname{Id}_{\bar{C}_i}$, because (48) and (49) depend, up to unique isomorphism, only on the deformation data $(\tau_j)_j$. Moreover, the f_i are uniquely determined by their values on the coordinates r_j, s_j and w_j. It follows from the Propositions 2.3.3 and 2.3.4 that the f_i are unique. We claim that the family (f_i) is an isomorphism of descent data between α and α'. This follows from the fact that $C_i \otimes_{A_i} A_{i,j}$, $C_j \otimes_{A_j} A_{i,j}$, $C_i' \otimes_{A_i} A_{i,j}$ and $C_j' \otimes_{A_j} A_{i,j}$ are étale $A_{i,j}$-algebras and from Lemma 5.1.1. By Corollary 1.3.2 the family (f_i) descends to a unique isomorphism $f : Y' \to Y$ between the deformations ρ' and ρ. This proves the second statement of Theorem 3.1.1 and the injectivity of the map (46) in our special case.

To prove the surjectivity of the map (46), let $(\tau_j)_j \in T(R)$. For $i \in I$, define a finite A_i-algebra $C_i = \oplus_{\kappa(j)=i} B_j$ by the expression given in (50) and (51). By Lemma 5.1.1 we can lift $\bar{C}_0$ to a finite étale A_0-algebra C_0. Using our assumptions on the covering $\mathcal{U}$, it is easy to see that for all $i \neq j$, $i, j \in I_0$, $C_i \otimes_{A_i} A_{i,j}$ and $C_j \otimes_{A_j} A_{i,j}$ are étale $A_{i,j}$-algebras. Therefore we can apply Lemma 5.1.1 once more to construct $A_{i,j}$-linear isomorphisms $\alpha_{i,j} : C_i \otimes_{A_i} A_{i,j} \xrightarrow{\sim} C_j \otimes_{A_j} A_{i,j}$ with $\alpha_{i,j} \otimes_R k = \bar{\alpha}_{i,j}$. Let $\alpha := (C_i, \alpha_{i,j})$. Since $\alpha \otimes_R k = \bar{\alpha}$, the uniqueness statement of Lemma 5.1.1 forces α to be a descent datum. By Corollary 1.3.2, α determines a finite $\mathcal{O}_X$-algebra $\mathcal{B}$, hence a finite morphism $\rho : Y = \operatorname{Spec} \mathcal{B} \to X$ lifting $\bar{\rho}$. By construction, ρ is a tame admissible cover. This completes the proof of Theorem 3.1.1 in the special case.

3.2.3   We are now going to prove Theorem 3.1.1 in the general case. For $n \geq 0$, let $R_n := R/\mathfrak{m}^{n+1}$ and $X_n := X \times_R R_n$. In particular, $X_0 = \bar{X}$. Let $\rho : Y \to X$ be a deformation of $\bar{\rho}$ to R with deformation datum $(\tau_j)_j$. Think of ρ as a finite $\mathcal{O}_X$-algebra $\mathcal{B}$ with $Y = \operatorname{Spec} \mathcal{B}$. The finite $\mathcal{O}_{X_n}$-algebra $\mathcal{B}_n := \mathcal{B} \otimes_R R_n$ corresponds to the deformation $\rho \times_R R_n$ of $\bar{\rho}$ to R_n with deformation datum $(\tau_j^{(n)})_j$. Here $\tau_j^{(n)}$ denotes the image of τ_j in R_n. Using the special case of Theorem 3.1.1 and Grothendieck's Existence Theorem (Corollary 1.4.2), we see that ρ is determined by the deformation datum $(\tau_j)_j$ up to unique isomorphism.

Conversely, let $(\tau_j)_j \in T(R)$ be a deformation datum. Applying Theorem 3.1.1 in the case of an artinian base ring, we obtain a compatible system of lifts $\rho_n : Y_n \to X_n$ with deformation datum $(\tau_j^{(n)})_j$. Write $Y_n = \operatorname{Spec} \mathcal{B}_n$, then $\mathcal{B}' =$

$(\mathcal{B}_n)_n$ is a formal finite $\mathcal{O}_X$-algebra. By Grothendieck's Existence Theorem (Corollary 1.4.2) we obtain a finite $\mathcal{O}_X$-algebra $\mathcal{B}$ with $\mathcal{B}_n = \mathcal{B} \otimes_R R_n$. Let $Y := \operatorname{Spec} \mathcal{B}$. All we have to show is that $\rho : Y \to X$ is a tame admissible cover.

Let $x \in \bar{X}$ be a singular point of the special fibre and $y \in \bar{\rho}^{-1}(x)$. Choose a formal coordinate system (u, v) for X at x and write $\hat{A} := \mathcal{O}_{X,x} = R[[u, v | uv = t]]$. Since ρ is finite, $\hat{B} := \mathcal{O}_{Y,y}$ is a finite local $\hat{A}$-algebra. It follows from Proposition 5.2.3 (iii) that $\mathcal{O}_{Y_n,y} = \hat{B}/\mathfrak{m}^{n+1}\hat{B}$. By the construction of Y_n, there exist elements $r_n, s_n \in \hat{B}/\mathfrak{m}^{n+1}\hat{B}$ with $r_n^e = u_n$, $s_n^e = v_n$, $\tau_n := r_n s_n \in R_n$ and $\hat{B}/\mathfrak{m}^{n+1}\hat{B} = R_n[[r_n, s_n \mid r_n s_n = \tau_n]]$. Moreover, we may assume that (r_{n+1}, s_{n+1}) lifts (r_n, s_n). Since $\hat{B}$ is complete, we find elements $r, s \in \hat{B}$ with image r_n, s_n in $\hat{B}/\mathfrak{m}^{n+1}$ such that $r^e = u$, $s^e = v$ and $\tau := rs \in R$. By the local criterium of flatness ([17] 20.C Theorem 49), $\hat{B}$ is a flat R-algebra. Hence, by Proposition 2.1.1, we have $\hat{B} = R[[r, s | rs = \tau]]$. We have shown that ρ is tame admissible at y (Definition 2.3.1 (i)). By similar arguments one shows that the same is true if x is a smooth point of $\bar{X}$. This completes the proof of Theorem 3.1.1.   ∎

4   Tame fundamental groups of smooth curves

Let X be a smooth projective curve over a complete local ring R with algebraically closed residue field, and let $D \subset X$ be a mark. Corollary 3.1.3 states that tamely ramified covers of the special fiber $(\bar{X}, \bar{D})$ lift uniquely to tamely ramified covers of (X, D). This result was first obtained by A. Grothendieck and used to prove his famous theorem stating that the prime to p part of the tame fundamental group of a smooth projective curve over an algebraically closed field of characteristic p is the same as it would be in characteristic 0.

So far, [8] is the only complete reference for this result. For the special case of étale fundamental groups there is the more accessible account of [20]. In both expositions Grothendieck's Theorem is deduced from facts about fundamental groups of rather general schemes. We will see in this section that the case of tame covers of smooth projective curves can be handled with much less machinery. We are roughly going to prove the following. Let R be a mixed characteristic discrete valuation ring with algebraically closed residue field of characteristic $p > 0$ and X a connected smooth projective curve over R with a mark $D \subset X$. Then there is a surjective specialization morphism from the tame fundamental group of the generic geometric fiber $(X_{\bar{K}}, D_{\bar{K}})$ of X to the tame fundamental group of the special fiber $(\bar{X}, \bar{D})$. Moreover, this specialization morphism induces an isomorphism on the prime to p parts of the tame fundamental groups.

4.1 The tame fundamental group as a Galois group

Let k be an algebraically closed field and X a smooth connected projective curve over k. In this case, a mark $D \subset X$ of X/k is a finite set $D = \{x_1, \ldots, x_r\}$ of closed points of X. Tamely ramified covers of (X, D) correspond to finite tamely ramified extensions of the function field $k(X)$ of X. Thus we can define the tame fundamental group $\pi_1^D(X)$ as the Galois group of the maximal algebraic extension of $k(X)$, tamely ramified over $x_1, \ldots, x_r$. This is classical valuation theory for function fields in one variable. Moreover, if k is of characteristic 0, $\pi_1^D(X)$ is the profinite completion of the topological fundamental group $\pi_1^{\mathrm{top}}(X_{\mathbb{C}} - \{x_1, \ldots, x_r\})$. This follows from the Riemann Existence Theorem. In this subsection, we list all the facts we need about tame fundamental groups of smooth projective curves over an algebraically closed field.

4.1.1   If $a \in X$ is a closed point, the local ring $\mathcal{O}_{X,a}$ is a discrete valuation ring with quotient field $k(X)$. This induces a bijection between closed points of X and discrete valuations of $k(X)$ which are trivial on k. If $z \in k(X)$ is a local parameter for a, the complete local ring is of the form $\mathcal{O}_{X,a} = k[[z]]$.

If $\rho : Y \to (X, D) \in \mathrm{Rev}_k^D(X)$ is a tamely ramified cover (see Definition 2.3.1) and Y is connected, the function field $k(Y)$ is a finite extension of $k(X)$, tamely ramified over the valuations corresponding to the points $x_1, \ldots, x_r$. In fact, let $y \in \rho^{-1}(x_i)$ and consider the natural extension $\mathcal{O}_{X,x_i} \to \mathcal{O}_{Y,y}$ of valuation rings; passing to the complete local rings, we obtain $\mathcal{O}_{Y,y} = k[[w]] = \mathcal{O}_{X,x_i}[w | w^n = z]$ for a suitable choice of local parameters z and w and with n prime to the characteristic of k (see Definition 2.3.1 (ii)).

Conversely, let $L/k(X)$ be a finite field extension which is tamely ramified over $x_1, \ldots, x_r$ and unramified everywhere else. Then $L = k(Y)$ for a smooth connected projective curve Y over k. Moreover, the birational map induced by the inclusion $k(X) \hookrightarrow L = k(Y)$ extends uniquely to a tamely ramified cover $\rho : Y \to (X, D)$.

This correspondence between function fields and connected curves carries over to non connected curves. Let $\rho : Y \to (X, D)$ be any tamely ramified cover and let $Y_1, \ldots, Y_s$ be the connected components of Y. The function ring of Y is defined as the finite $k(X)$-algebra $k(Y) := k(Y_1) \oplus \ldots \oplus k(Y_s)$. The extension $k(Y)/k(X)$ is tamely ramified over $x_1, \ldots, x_r$ and unramified everywhere else. We obtain an equivalence between $\mathrm{Rev}_k^D(X)$ and the category of finite $k(X)$-algebras, tamely ramified over $x_1, \ldots, x_r$ (from now on we will tacitly understand that 'tamely ramified over $x_1, \ldots, x_r$' implies 'unramified everywhere else').

4.1.2   Choose an algebraic closure $k(X) \hookrightarrow \Omega$ and let $\Omega \subset \Omega$ be the maximal subextension tamely ramified over $x_1, \ldots, x_r$. Then $\Omega/k(X)$ is a Galois extension. Choose a closed point $a \in X - \{x_1, \ldots, x_r\}$ and a discrete valuation $\tilde{a}$ of Ω extending the valuation of $k(X)$ corresponding to a. Define the **tame fundamental group** of (X, D) with **base point** a as the Galois group of Ω over $k(X)$:

$$\pi_1^D(X, a) := \text{Gal}(\Omega/k(X)). \tag{52}$$

It may seem strange that $\pi_1^D(X, a)$ does not depend on a. But the choice of a and $\tilde{a}$ determines the way in which $\pi_1^D(X, a)$ classifies all tamely ramified covers of (X, D). For our purposes it suffices to consider only Galois covers of (X, D).

Let G be a finite group. A G-**Galois cover** of (X, D, a) is a tamely ramified cover $\rho : Y \to (X, D) \in \text{Rev}_k^D(X)$ together with an isomorphism $G \xrightarrow{\sim} \text{Aut}_X(Y)$ such that Y is connected and G (as automorphism group of the cover ρ) acts transitively and without fixed points on the fiber $\rho^{-1}(a)$. A **pointed G-Galois cover** of (X, D, a) is a G-Galois cover $\rho : Y \to (X, D)$ together with a choice of an element $b \in \rho^{-1}(a)$.

Let $\rho : Y \to (X, D)$ be a G-Galois cover. Then $k(Y)/k(X)$ is a finite field extension and G acts as a group of $k(X)$-automorphisms on $k(Y)$. We can identify $\rho^{-1}(a)$ with the set of valuations of $k(Y)$ extending a. Since G acts transitively and fixed point free on this set, $k(Y)/k(X)$ is a Galois extension with Galois group G. Moreover, if we choose an element $b \in \rho^{-1}(a)$ there is a unique embedding $\lambda_b : k(Y) \hookrightarrow \Omega$ over $k(X)$ such that $\lambda_b^{-1}(\tilde{a}) = b$ (as valuation of $k(Y)$). The restriction map on the Galois groups induced by λ_b is a surjective continuous homomorphism $\pi_1^D(X, a) \twoheadrightarrow G$.

Conversely, any surjective and continuous homomorphism $\pi_1^D(X, a) \twoheadrightarrow G$ corresponds to a subfield $L \subset \Omega$ which is Galois over $k(X)$ with Galois group G. If Y is the smooth projective model of L, i.e. $L = k(Y)$, we obtain a G-Galois cover $\rho : Y \to (X, D)$. Note that G acts without fixed points on $\rho^{-1}(a)$ because a is unramified in $L = k(Y)$. Moreover, there is a unique distinguished element $b \in \rho^{-1}(a)$ corresponding to the restriction of $\tilde{a}$ on $L = k(Y)$. We have proved the following:

Proposition 4.1.1 *The choice of $\tilde{a}$ induces a bijection between isomorphism classes of pointed G-Galois covers of (X, D, a) and surjective continuous morphisms $\pi_1^D(X, a) \twoheadrightarrow G$.*

Let $g, r \geq 0$. Let $\Gamma_{g,r}$ be the profinite group with $2g + r$ generators $a_1, b_1, \ldots, a_g, b_g, c_1, \ldots, c_r$ and the single relation

$$[a_1, b_1] \cdot \ldots \cdot [a_g, b_g] \cdot c_1 \cdot \ldots \cdot c_r = 1. \tag{53}$$

Remark 4.1.2 If the ground field k is of characteristic 0 and the curve X of genus g, we obtain an isomorphism

$$\Gamma_{g,r} \xrightarrow{\sim} \pi_1^D(X, a).$$

To prove this fact, one may assume that k is a subfield of $\mathbb{C}$. One can show that the natural homomorphism $\pi_1^D(X_\mathbb{C}, a) \to \pi_1^D(X, a)$ is actually an isomorphism, because any G-Galois cover of $(X_\mathbb{C}, D_\mathbb{C})$ is already defined over k. By the Riemann Existence Theorem, $\pi_1^D(X_\mathbb{C}, a)$ is isomorphic to the profinite completion of the topological fundamental group $\pi_1^{\text{top}}(X_\mathbb{C} - \{x_1, \ldots, x_r\}, a)$, which is $\Gamma_{g,r}$. See e.g. [24] for more details.

4.2 Tame covers over complete discrete valuation rings

In this section R will be a complete discrete valuation ring with algebraically closed residue field k. We denote by K the quotient field of R and by $\bar{K}$ an algebraic closure of K. We work out several details about smooth projective curves and tamely ramified covers over R which are used in the proof of Grothendieck's Theorem.

4.2.1 Purity of branch locus

First we need some preliminaries about regular local rings of dimension 2. See [17], Chapter 17, or [2] VIII §3 for more details.

Let A be a noetherian local domain; denote its residue field by k, its maximal ideal by $\mathfrak{m}$ and its quotient field by K. The ring A is regular of dimension d iff its completion $\hat{A}$ is.

From now on, assume A to be regular of dimension 2. Let S_A be the set of discrete valuations v of K such that the valuation ring $\mathcal{o}_v$ contains A. If $v \in S_A$ then $\mathfrak{p} := A \cap \mathfrak{m}_v$ is a prime ideal of height 1. The localization A is a discrete valuation ring, hence $\mathcal{o}_v = A$. Regular local rings are normal, therefore equal to the intersection of their localizations at primes of height one. We see that

$$A = \bigcap_{v \in S_A} \mathcal{o}_v. \tag{54}$$

Let $v \in S_A$ and $\mathfrak{p} := A \cap \mathfrak{m}_v$ be as above and choose $a \in \mathfrak{p} - \mathfrak{m}^2$. By [2] VIII, 5.3 Proposition 2 and Corollaire 1 we conclude that $\bar{A} := A/aA$ is regular and that (a) is a prime ideal. Since $\dim A = 2$ we have in fact $\mathfrak{p} = (a)$, and $\bar{A}$ is a discrete valuation ring. Moreover, we have $v'(a) = 0$ for all $v' \in S_A - \{v\}$. We will call an element $a \in A$ with the above properties a **parameter** of A at v.

The following lemma is a special case of the Purity Theorem of Nagata and Zariski, also called 'Purity of branch locus'. Its proof is an adaption of a proof given in the Appendix of [4]. For much more general versions of Purity, see e.g. [8] X.3.

Lemma 4.2.1 *Let A be a noetherian regular local domain of dimension 2, K the quotient field of A and L/K a finite extension. Then the integral closure*

B of A in L is a finite A-module of rank $[L : K]$. Moreover, B is étale over A iff L/K is unramified over S_A (the set of valuations dominating A).

Proof: Let S_B be the set of discrete valuations w of L lying over some valuation $v \in S_A$. Denote the corresponding valuation ring by $\mathcal{O}_w$ and put

$$B := \bigcap_{w \in S_B} \mathcal{O}_w \quad (\subset L) \tag{55}$$

From (54) we can see that $A \subset B$. As intersection of integrally closed domains of L, B is itself an integrally closed domain of L. Hence, if we can show that B is a finite A-module then it will follow that B is the integral closure of A in L.

Let $v \in S_A$ be any valuation of K dominating A and choose a parameter a for A at v. Since the localization $A_{(a)}$ is a discrete valuation ring, $B_{(a)} := \cap_{w|v_a} \mathcal{O}_w$ is the integral closure of $A_{(a)}$ in L. By standard valuation theory $B_{(a)}$ is a free $A_{(a)}$-module of rank $[L : K]$. Therefore $B_{(a)}/aB_{(a)}$ is a free $A_{(a)}/aA_{(a)}$-module of rank $[L : K]$. Let $\bar{A} := A/aA$ and $\bar{B} := B/aB$. Using the fact that $v(a) = 0$ for all $v \in S_A - \{v_a\}$ it is easy to see that the $\bar{A}$-algebra $\bar{B} := B/aB$ injects into $B_{(a)}/aB_{(a)}$. Since $\bar{A}$ is a discrete valuation ring, it follows that $\bar{B}$ is a free $\bar{A}$-module of rank $\leq [L : K]$. Nakayama's Lemma implies that B can be generated (as an A-module) by $[L : K]$ elements. But L is the quotient field of B, therefore B is free over A of rank exactly $[L : K]$.

Let $b_1, \ldots, b_n$ be an A-basis of B and δ be the discriminant of B over A with respect to $b_1, \ldots, b_n$. Now L/K is unramified over S_A iff $v(\delta) = 0$ for all $v \in S_A$ iff $\delta \in A^*$ iff B is étale over A (Lemma 5.1.2 (iv)). This completes the proof of the lemma. ∎

If L/K is tamely ramified in only one place of S_A, we can use Abhyankar's Lemma to prove another version of purity. Its formulation is simpler if we work with strict complete local rings.

Lemma 4.2.2 *In the situation of Lemma 4.2.1, assume that A is complete with algebraically closed residue field. Assume moreover that L/K is tamely ramified over some valuation $v_0 \in S_A$ and unramified over $S_A - \{v_0\}$. Then L/K is purely ramified at v_0 and the integral closure of A in L is of the form*

$$B = A[b \mid b^e = a].$$

Here e is the ramification index of v_0 in L and $a \in A$ is a parameter of A at v_0.

Proof: For any natural number e prime to the residue characteristic of v_0 we put $K_e := K[b \mid b^e = a]$; this is a finite extension of K of degree e, purely ramified in v_0 and unramified over $S_A - \{v_0\}$. Moreover, $A_e := A[b \mid b^e = a]$ is the integral closure of A in K_e; it is a complete regular local domain of dimension two (see [2] VIII, 5.4).

Let $L_e := L \cdot K_e$ (inside some algebraic closure of K). By Abhyankar's Lemma (see e.g. [20], Appendix to Chapter IX) and the hypothesis we can choose ϵ such that L_e/K_e is unramified at the unique extension of v_0 to K_e. Hence it is unramified at all discrete valuations dominating A_e. Now Lemma 4.2.1 applies, showing that the integral closure B_e of A_e in L_e is finite étale over A_e. But A_e is strictly henselian, therefore we have $B_e = A_e$ (Proposition 5.2.3 (v)) and $L \subset K_e$. In particular, v_0 is purely ramified in L. Hence, if we choose ϵ to be its ramification index, then $L = K_e$ and $B = A_e$. ∎

4.2.2 Specialization

Let X be a connected smooth projective curve over R. It is a regular scheme of dimension 2. We will use the notation $\bar{X} := X \times_R k$ for the special fiber, $X_K := X \times_R K$ for the generic fiber and $X_{\bar{K}} := X \times_R \bar{K}$ for the geometric generic fiber. Note that X is an integral scheme, because it is both regular and connected (use [14] II Proposition 3.1 and the fact that a regular connected scheme is irreducible). We will denote its function field by $K(X)$.

Let x be a closed point of the generic fiber X_K (which we consider as an open subset of X). The point x corresponds to a discrete valuation of $K(X)$ which is trivial on K. The local ring $\mathcal{O}_{X,x}$ is the corresponding valuation ring. Let K' be the residue field of x and R' the integral closure of R in K'. Then R' is a complete discrete valuation ring, purely ramified over R. Since X/R is projective, the morphism $\operatorname{Spec} K' \to X$ corresponding to x extends uniquely to a morphism $\operatorname{Spec} R' \to X$ ([14] II, Th. 4.7 and Th. 4.9). Let $\bar{x}$ be the image of the special point of $\operatorname{Spec} R'$ on X. The point $\bar{x}$ is the unique closed point of the special fiber $\bar{X}$ contained in the Zariski closure of x on X. We will call $\bar{x}$ the **specialization** of x and write $x \rightsquigarrow \bar{x}$.

Let $\bar{x} \in \bar{X} \subset X$ be a closed point of the special fiber. The local ring $\mathcal{O}_{X,\bar{x}}$ is a regular local ring of dimension 2 with quotient field $K(X)$. Let $S_{\bar{x}}$ be the set of discrete valuations of $K(X)$ dominating $\mathcal{O}_{X,\bar{x}}$. A valuation $v \in S_{\bar{x}}$ corresponds to a point on X of codimension 1 whose Zariski closure contains $\bar{x}$. Hence we can identify $S_{\bar{x}}$ with the set $\{x \in X_K \mid x \rightsquigarrow \bar{x}\} \cup \{\eta\}$, where η is the generic point of the unique irreducible component of $\bar{X}$ containing $\bar{x}$. Therefore (54) becomes

$$\mathcal{O}_{X,\bar{x}} = (\bigcap_{x \rightsquigarrow \bar{x}} \mathcal{O}_{X,x}) \cap \mathcal{O}_{X,\eta}. \tag{56}$$

Let $D \subset X$ be a mark on X (see Section 2.2). In particular, $D = \operatorname{Spec} R'$ for a finite étale R-algebra R'. But since R is strictly henselian we have $R' = R \oplus \ldots \oplus R$. Therefore D is the Zariski closure of a finite set $\{x_1, \ldots, x_r\} \subset X_K$ of K-rational points whose specializations $\bar{x}_1, \ldots, \bar{x}_r \in \bar{X}$ are pairwise distinct. We will identify D with the set $\{x_1, \ldots, x_r\}$ and write $X(K)$ for the set of K-rational points of $X(K)$.

Lemma 4.2.3 *Let X be a smooth projective curve over R and let $D = \{x_1, \dots, x_r\}$ be a mark on X.*

(i) *For every closed point $\bar{x} \in \bar{X}$ there exists a point $x \in X(K)$ with $x \rightsquigarrow \bar{x}$.*

(ii) *Let $\rho : Y \to (X, D)$ be a tamely ramified cover and $a \in X(K)$ such that its specialization $\bar{a}$ is not an element of the reduction $\{\bar{x}_1, \dots, \bar{x}_r\}$ of D. Then for any point $\bar{b} \in \rho^{-1}(\bar{a})$ there is a unique point $b \in Y(K)$ with $b \rightsquigarrow \bar{b}$ and $\rho(b) = a$.*

Proof: Let $\bar{a} \in \bar{X}$. By Proposition 2.2.3 we have $\mathcal{O}_{X,\bar{a}} \xrightarrow{\sim} R[[z]]$ for some local coordinate z. Hence we can embed $K(X)$ (which is the quotient field of $\mathcal{O}_{X,\bar{a}}$) into $K((z))$. This defines a discrete valuation v_x with residue field K on $K(X)$. It is clear that the point $x \in X(K)$ corresponding to v_x specializes to $\bar{x}$. This proves the first claim. In the situation of (ii), ρ is étale at $\bar{b}$. Therefore $\mathcal{O}_{X,\bar{a}} \xrightarrow{\sim} \mathcal{O}_{Y,\bar{b}}$, proving the second claim. ∎

4.2.3   Reduction Now we assume in addition that the geometric fibers $X_{\bar{K}}$ and $\bar{X}$ are connected. For any finite extension K'/K, let R' be the integral closure of R in K'. Then $X' := X \times_R R'$ is a smooth projective curve over R'. The scheme X' is still connected, has special fiber $\bar{X}$ and function field $K'(X') = K(X) \otimes_K K'$.

Proposition 4.2.4 *Let $\rho_{\bar{K}} : Y_{\bar{K}} \to (X_{\bar{K}}, D_{\bar{K}})$ be a tamely ramified G-Galois cover. Assume that the order of G is prime to the characteristic of k. Then there exists a tamely ramified cover $\rho : Y \to (X, D)$ with $\rho_{\bar{K}} = \rho \times_R \bar{K}$. Its reduction $\bar{\rho} : \bar{Y} \to (\bar{X}, \bar{D})$ is a G-Galois cover.*

Proof: By assumption $Y_{\bar{K}}$ is connected. The G-Galois cover $\rho_{\bar{K}}$ is already defined over some finite extension K' of K, i.e. there is a tamely ramified G-Galois cover $\rho_{K'} : Y_{K'} \to (X_{K'}, D_{K'})$ with $\rho_{\bar{K}} = \rho_{K'} \times_{K'} \bar{K}$. Note that the corresponding extension of function fields $K'(Y_{K'})/K'(X_{K'})$ is a G-Galois extension. Our assumption on G allows us to apply Abhyankar's Lemma. More precisely, after a tamely ramified extension of K' we may assume that the discrete valuation of $K'(X')$ corresponding to the generic point of $\bar{X}$ is unramified in $K'(Y_{K'})$.

Let Y' be the normalization of X' in $K'(Y_{K'})$ and $\rho' : Y' \to X'$ the natural map ([14] II Ex. 3.8). By construction Y' is a normal integral scheme with function field $K'(Y') = K'(Y_{K'})$. We claim that ρ' is a tamely ramified cover of (X', D').

Let $\bar{x} \in \bar{X}$ be a closed point and let B be the integral closure of $\mathcal{O}_{X',\bar{x}}$ in $K'(Y_{K'})$. By Lemma 4.2.1 B is finite over $\mathcal{O}_{X',\bar{x}}$. Therefore the fiber $(\rho')^{-1}(\bar{x}) = \{\bar{y}_1, \dots, \bar{y}_s\}$ is a finite set and B is its semi-local ring (i.e. its localizations are exactly $\mathcal{O}_{Y',\bar{y}_j}$). In particular, ρ' is a finite morphism.

Let $S_{\bar{x}} := \{x \in X_{K'} \mid x \rightsquigarrow \bar{x}\} \cup \{\eta\}$ be the set of discrete valuations of $K'(X_{K'})$ specializing to $\bar{x}$. If $\bar{x} \notin \{\bar{x}_1, \ldots, \bar{x}_r\}$ then $K'(Y_{K'})/K'(X_{K'})$ is unramified over $S_{\bar{x}}$. If $\bar{x} = \bar{x}_i$ then it is tamely ramified over x_i and unramified over $S_{\bar{x}} - \{x_i\}$. Using Lemma 4.2.1 and Lemma 4.2.2, we conclude that ρ' is a tamely ramified cover of (X', D').

Let $\bar{\rho} := \rho' \times_{R'} k$ be the reduction to the special fiber; it is a tamely ramified cover of $(\bar{X}, \bar{D})$. By Corollary 3.1.3 it lifts uniquely to a tamely ramified cover $\rho : Y \to (X, D)$. The uniqueness of lifting implies $\rho' = \rho \times_R R'$. Therefore $\rho_K = \rho \times_R \bar{K}$.

Note that there is a natural G-action on Y' inducing a G-action on Y and hence on $\bar{Y}$. To prove that $\bar{\rho} : \bar{Y} \to \bar{X}$ is in fact a G-Galois cover, it remains to show that $\bar{Y}$ is connected. Assume that $\bar{Y}$ is the disjoint union of two open subset $\bar{Y}_1, \bar{Y}_2$. Then the natural maps $\bar{Y}_i \to (\bar{X}, \bar{D})$ are still tamely ramified covers, for $i = 1, 2$. Lift them to tamely ramified covers $Y_i \to (X, D)$, $i = 1, 2$. By the uniqueness of lifting, we conclude that Y is the disjoint union of Y_1 and Y_2. But Y is an integral normal scheme and hence connected ([14] II Proposition 3.1). Therefore one of the Y_i must be empty, proving that $\bar{Y}$ is connected. ∎

4.3 The specialization morphism

Let G be a profinite group. We define the **prime to p part** of G as the inverse limit over those finite quotients of G which are of order prime to p and denote it by $G^{p'}$. It is a profinite quotient of G. If $\phi : G \to H$ is a continuous morphisms of profinite groups, ϕ factors to a continuous morphism $\phi^{p'} : G^{p'} \to H^{p'}$. Note that surjectivity of ϕ implies surjectivity of $\phi^{p'}$. Remember that $\Gamma_{g,r}$ denotes the tame fundamental group of a smooth projective curve of genus g with r marked points in characteristic zero (Remark 4.1.2).

Theorem 4.3.1 (Grothendieck) *Let $\bar{X}$ be a smooth projective curve of genus g over an algebraically closed field k of characteristic $p > 0$ and $a_0, \bar{x}_1, \ldots, \bar{x}_r$ be pairwise distinct closed points. Let $\bar{D} := \{\bar{x}_1, \ldots, \bar{x}_r\}$. Then there is a surjective homomorphism of profinite groups*

$$\Gamma_{g,r} \longrightarrow \pi_1^{\bar{D}}(\bar{X}, a_0)$$

inducing an isomorphism on the prime to p parts, i.e. $\Gamma_{g,r}^{p'} \xrightarrow{\sim} \pi_1^{\bar{D}}(\bar{X}, a_0)^{p'}$

The proof of this theorem will be given in the rest of this section. In 4.3.1 we construct a lift of $\bar{X}$ to characteristic 0. In 4.3.2 we construct the specialization morphism between tame fundamental groups. In 4.3.3 it is shown that this specialization morphism is injective on the prime to p parts.

4.3.1   Let $\bar{X}$ be a smooth projective curve over an algebraically closed field of characteristic $p > 0$ and let $\bar{D} := \{\bar{x}_1, \ldots, \bar{x}_r\}$ be a mark on $\bar{X}$. Let

$R := W(k)$ be the ring of Witt vectors over k; it is a complete discrete valuation ring of characteristic 0. We adopt the notations K, $\bar{K}$ etc. from Section 4.2. By Proposition 4.3.2 below we can lift $\bar{X}$ to a smooth projective curve X over R; its geometric generic fiber $X_{\bar{K}}$ is connected. By Lemma 4.2.3 we can lift $\bar{x}_1, \ldots, \bar{x}_r \in \bar{X}$ to K-rational points $x_1, \ldots, x_r \in X(K)$. This defines a mark $D := \{x_1, \ldots, x_r\}$ on X.

Proposition 4.3.2 *Let $\bar{X}$ be a smooth projective curve over k.*

(i) *There exists a smooth projective curve X over R with $\bar{X} = X \times_R k$.*

(ii) *If $\bar{X}$ is connected, then for any X as in (i) the generic geometric fiber $X_{\bar{K}} := X \times_R \bar{K}$ is connected.*

Proof: We will give a proof of (i) only in characteristic different from 2. For the general case, see e.g. [8] III. If $\mathrm{char}(k) \neq 2$ then there exists a *simple morphism* $\bar{f} : \bar{X} \to \mathbb{P}^1_k$ (see [5] Proposition 8.1). In particular, $\bar{f}$ is tamely ramified, having only ramification of order 2. Therefore we can lift $\bar{f}$ to a tamely ramified cover $f : X \to \mathbb{P}^1_R$ (Corollary 3.1.3). This proves (i) in this special case.

Assume that $\bar{X}$ is connected and that X is a lift of $\bar{X}$ as in (i). Let $\bar{R}$ be the integral closure of R in $\bar{K}$. Then $X_{\bar{R}} := X \times_R \bar{R}$ is a smooth projective curve over the valuation ring $\bar{R}$ with special fiber $\bar{X}$. Since the natural morphism $X_{\bar{R}} \to \mathrm{Spec}\, \bar{R}$ is closed ([14] II Theorem 4.9), every closed subset of $X_{\bar{R}}$ must intersect the special fiber $\bar{X}$. Therefore $X_{\bar{R}}$ is connected. A connected regular scheme is integral. Therefore $X_{\bar{K}} \subset X_{\bar{R}}$ is integral and in particular connected. ∎

4.3.2   Let (X, D) be as constructed in 4.3.1. Choose a closed point $a_0 \in \bar{X} - \{\bar{x}_1, \ldots, \bar{x}_r\}$. By Lemma 4.2.3 (i) we can choose a point $a_1 \in X(K) - \{x_1, \ldots, x_r\}$ which specializes to a_0. We will regard a_0 (resp. a_1) as base points on $\bar{X}$ (resp. $X_{\bar{K}}$). As in 4.1.2, choose an algebraic closure $k(\bar{X}) \hookrightarrow \Omega_0$ and let Ω_0 be the maximal subextension of Ω_0 which is tamely ramified over $\bar{x}_1, \ldots, \bar{x}_r$. Extend the valuation on $k(\bar{X})$ corresponding to a_0 to a valuation $\tilde{a}_0$ on Ω_0. Similarly for $X_{\bar{K}}$: let Ω_1 be an algebraic closure of $\bar{K}(X)$, Ω_1 the maximal subextension tamely ramified over $x_1, \ldots, x_r$ and $\tilde{a}_1$ a valuation of Ω_1 lying over a_1. As in 4.1.2, we define the fundamental groups

$$\pi_1^{D_{\bar{K}}}(X_{\bar{K}}, a_1) := \mathrm{Gal}(\Omega_1/\bar{K}(X)), \quad \pi_1^{\bar{D}}(\bar{X}, a_0) := \mathrm{Gal}(\Omega_0/k(\bar{X})) \qquad (57)$$

We can interpret Lemma 4.2.3 as follows. The process of specialization $a_1 \rightsquigarrow a_0$ is a 'path' in $X - D$ connecting a_0 with a_1, and tamely ramified covers of (X, D) have a unique lifting property with respect to such paths. In close analogy to topological covering space theory, this lifting property defines a morphism $\pi_1^{D_{\bar{K}}}(X_{\bar{K}}, a_1) \longrightarrow \pi_1^{\bar{D}}(\bar{X}, a_0)$ of fundamental groups.

Proposition 4.3.3 *There exists a natural surjective morphism of profinite groups*

$$\phi : \pi_1^{D_{\bar{K}}}(X_{\bar{K}}, a_1) \longrightarrow \pi_1^{\bar{D}}(\bar{X}, a_0).$$

Proof: Let $G = \pi_1^{\bar{D}}(\bar{X}, a_0)/H$ be a finite quotient. By Proposition 4.1.1 it corresponds to a pointed G-Galois cover $\bar{\rho} : \bar{Y} \to (\bar{X}, \bar{D})$. Let $b_0 \in \bar{\rho}^{-1}(a_0)$ be the distinguished point.

Corollary 3.1.3 states that we can lift $\bar{\rho}$ to a tamely ramified cover $\rho : Y \to (X, D)$ with automorphism group G. In particular, G acts on $Y_{\bar{K}}$. Lemma 4.2.3 (ii) shows that specialization defines a G-equivariant bijection $\rho_{\bar{K}}^{-1}(a_1) \overset{\sim}{\to} \bar{\rho}^{-1}(a_0)$. Hence G acts transitively and fixed point free on $\rho_{\bar{K}}^{-1}(a_1)$. By Proposition 4.3.2 (ii) $Y_{\bar{K}}$ is connected. Hence $\rho_{\bar{K}} : Y_{\bar{K}} \to X_{\bar{K}}$ is a G-Galois cover. Let $b_1 \in \rho_{\bar{K}}^{-1}(a_1)$ be the unique lift of b_0 over a_1 (Lemma 4.2.3 (ii)). Consider $\rho_{\bar{K}}$ as a pointed G-Galois cover with distinguished point b_1. By Proposition 4.1.1, $\rho_{\bar{K}}$ corresponds to a surjective continuous homomorphism $\dot{\phi}_H : \pi_1^{D_{\bar{K}}}(X_{\bar{K}}, a_1) \to G$.

We claim that the maps $\dot{\phi}_H$ are compatible (where H runs over all normal subgroups of finite index) and lift to the desired surjective morphism ϕ. To prove this we have to show the following: if $G' = \pi_1^{\bar{D}}(\bar{X}, a_0)/H'$ is a smaller quotient than G, i.e. $H \subset H'$, and $\psi : G \to G'$ is the natural map, then $\psi \circ \dot{\phi}_H = \dot{\phi}_{H'}$.

Let $\bar{\rho}' : \bar{Y}' \to \bar{X}$ be the pointed G'-Galois cover corresponding to the natural map $\pi_1^{\bar{D}}(\bar{X}, a_0) \to G'$ (with distinguished point b_0'). Let $\rho' : Y' \to X$ be its lift to R and b_1' be the lift of b_0' in $(\rho')^{-1}(a_1)$. Let $\lambda_{b_1'} : \bar{K}(Y') \hookrightarrow \Omega_1$ be the unique embedding such that $\lambda_{b_1'}^{-1}(\tilde{a}_1)$ is the valuation on $\bar{K}(Y')$ corresponding to the point b_1'. The map $\dot{\phi}_{H'}$ is the restriction map of Galois groups corresponding to $\lambda_{b_1'}$.

By the definition of G and G' we have a natural inclusion $k(\bar{Y}') \subset k(\bar{Y}) \subset \Omega_0$ of Galois extensions, such that the restriction map from $G = \mathrm{Gal}(k(\bar{Y})/k(\bar{X}))$ to $G' = \mathrm{Gal}(k(\bar{Y}')/k(\bar{X}))$ is the natural map. It induces an $\bar{X}$-morphism $\bar{f} : \bar{Y} \to \bar{Y}'$ such that $\bar{f}(b_0) = b_0'$. By Corollary 3.1.3 $\bar{f}$ lifts uniquely to an X-morphism $f : Y \to Y'$. The uniqueness statement of Lemma 4.2.3 (ii) implies that $f(b_1) = b_1'$. Let $\mu : \bar{K}(Y') \hookrightarrow \bar{K}(Y)$ be the inclusion of function fields induced by f. We have

$$\lambda_{b_1'} = \lambda_{b_1} \circ \mu, \tag{58}$$

because both embeddings restrict $\tilde{a}_1$ to the valuation corresponding to b_1'. If we translate (58) into the corresponding equation for maps between Galois groups, we obtain $\psi \circ \dot{\phi}_H = \dot{\phi}_{H'}$. This proves the proposition.   ∎

4.3.3   We are now going to finish the proof of Theorem 4.3.1. Remember that $\bar{K}$ is an algebraically closed field of characteristic 0. By Remark 4.1.2

we can identify $\pi_1^{D_{\bar{K}}}(X_{\bar{K}}, a_1)$ with $\Gamma_{g,r}$. Thus, the proof of Theorem 4.3.1 is complete if we are able to show that the morphism ϕ given in Proposition 4.3.3 is injective on the prime to p parts.

Reconsidering the construction of the specialization morphism, we see that we have to prove the following. Let G be a finite quotient of $\pi_1^{D_{\bar{K}}}(X_{\bar{K}}, a_1)$ of order prime to p. Then the natural map $\psi : \pi_1^{D_{\bar{K}}}(X_{\bar{K}}, a_1) \to G$ factors over ϕ.

ψ corresponds to a punctured G-Galois cover $\rho_{\bar{K}} : Y_{\bar{K}} \to X_{\bar{K}}$ (with distinguished point $b_1 \in \rho_{\bar{K}}^{-1}(a_1)$. By Proposition 4.2.4 we can find a tamely ramified cover $\rho : Y \to X$ with $\rho_{\bar{K}} = \rho \times_R \bar{K}$. Its reduction $\bar{\rho} := \rho \times_R k$ is a G-Galois cover, which we consider as punctured by the specialization b_0 of b_1. Therefore $\bar{\rho}$ corresponds to a surjective map $\chi : \pi_1^{\bar{D}}(\bar{X}, a_0) \to G$. By the construction of ϕ we have $\psi = \chi \circ \phi$. ∎

5 Appendix

For the convenience of the reader, this appendix contains some results about étale ring extensions and henselian rings which are used in this paper. Our main reference is [21]. See also [18] Chapter I, §3.

5.1   Let A, B be rings, $\varphi : A \to B$ a morphism and $\mathfrak{q} \in \operatorname{Spec} B$. $A \to B$ is called **unramified** at $\mathfrak{q}$, if $\mathfrak{p}B = \mathfrak{q}$ and the residue field extension $A/\mathfrak{p}A \to B/\mathfrak{q}B$ is separable (where $\mathfrak{p} := \varphi^{-1}(\mathfrak{q}) \in \operatorname{Spec} A$). We say that B/A is unramified, if it is unramified for all $\mathfrak{q} \in \operatorname{Spec} B$. Assume for a moment that B is finitely generated as an A-algebra. Then $A \to B$ is unramified at $\mathfrak{q}$ iff $A \to B$ is 'net' at $\mathfrak{q}$ (in Raynaud's terminology, see [21] I, Def. 4) iff $(\Omega_{B/A}) = 0$ (see [21] III, §4, Prop. 9 u. Ex. 1). Since $\Omega_{B/A}$ is a finitely generated B-module, being unramified is an open property on $\operatorname{Spec} B$ ($\operatorname{supp} \Omega_{B/A} \subset \operatorname{Spec} B$ is closed).

$A \to B$ is called **étale**, if it is of finite presentation, flat and unramified. $A \to B$ is called **formally étale** if for every A-algebra C and ideal $I \triangleleft C$ with $I^2 = 0$ the canonical map

$$\operatorname{Hom}_A(B, C) \longrightarrow \operatorname{Hom}_A(B, C/I) \qquad (59)$$

is a bijection. By [21], Chapitre I, Definition 2 and Chapitre V, Corollaire 1, $A \to B$ is étale if and only if it is of finite presentation and formally étale.

Lemma 5.1.1 *Let A be a ring with a nilpotent ideal I; let $\bar{A} := A/I$. Then every finite étale $\bar{A}$-algebra $\bar{B}$ lifts uniquely to a finite étale A-algebra B.*

Proof: This follows from [21] V, Thm. 4, and from Nakayama's Lemma. ∎

Lemma 5.1.2 *Let A be a ring and B, C be A-algebras.*

(i) *If $A \to B$ is étale, then so is $C \to B \otimes_A C$.*

(ii) *Let $B \to C$ be an A-algebra morphism. If $A \to B$ and $A \to C$ are étale, then $B \to C$ is étale.*

(iii) *If $A \to B$ and $B \to C$ are étale, then $A \to C$ is étale.*

(iv) *Assume that B is a finite free A-algebra and let $(\delta_{B/A}) \lhd A$ be the discriminant ideal. Then $A \to B$ is étale if and only if $\delta_{B/A} \in A^\times$.*

Proof: As remarked above, a ring extension is étale if and only if it is formally étale and of finite presentation. This implies the Claims (i), (ii) and (iii), because the analogous statements hold both for formally étale and for finitely presented morphisms. To prove (iv), note that $A \to B$ is étale if and only if it is unramified. By [21] III Proposition 10 and 11, $A \to B$ is unramified if and only if $B \otimes_A (A / \mathfrak{p} A)$ is a separable $A / \mathfrak{p} A$-algebra, for all $\mathfrak{p} \in \operatorname{Spec} A$. It is well known that the latter holds if and only if $\delta_{B/A} \notin \mathfrak{p}$ (see e.g. [18] III Proposition 3.1). ∎

5.2   Let A be a ring and $A \to k$ a morphism to some field k. Denote by $\mathsf{Et}(A \to k)$ the category of étale A-algebras B equipped with an A-algebra morphism $B \to k$. Morphisms between objects $B \to k$ and $C \to k$ are A-algebra morphisms $B \to C$ compatible with the maps to k. If such a morphism exists, then we say that $C \to k$ is **smaller** than $B \to k$. A ring is called **indecomposable** if it is not the direct sum of two rings (equivalently, $\operatorname{Spec} A$ is connected, [14] II Ex. 2.19). Let $\mathsf{Et}'(A \to k)$ be the full subcategory of indecomposable objects of $\mathsf{Et}(A \to k)$.

Lemma 5.2.1 *Let $A \to k$ be as above and $B \to k$, $C \to k \in \mathsf{Et}(A \to k)$. If C is indecomposable, then there is at most one morphism from $B \to k$ to $C \to k$. In any case, there exists an indecomposable $D \to k \in \mathsf{Et}'(A \to k)$, smaller than $B \to k$ and $C \to k$.*

Proof: See [21] VIII §2, Prop. 2. ∎

The Lemmas 5.1.2 and 5.2.1 allow us to define the direct limit

$$\tilde{A} := \varinjlim A', \qquad A' \to k \in \mathsf{Et}'(A \to k). \tag{60}$$

The ring $\tilde{A}$ is called the **henselization** of A with respect to $A \to k$. There is a natural A-algebra morphism $\tilde{A} \to k$. If k is separably closed, then $\tilde{A}$ is called a **strict henselization** of A at $\operatorname{Ker}(A \to k)$.

A local ring A is called **henselian**, if every finite A-algebra is the direct sum of local factors (see [21], I, Def. 1). Or, equivalently, if A verifies Hensel's Lemma ([21] I, Prop. 5). The ring A is called **strictly henselian** if it is henselian and its residue field is separably closed. From [21] Chap. VIII, we get

Remark 5.2.2 $\tilde{A}$ is a local henselian ring and its residue field $\tilde{A}/\tilde{\mathfrak{m}}$ is the separable closure of $A/\mathfrak{p}$ inside k. $A \to \tilde{A}$ is flat and unramified. If A is noetherian, $\tilde{A}$ will be noetherian, too. $\tilde{A} \to k$ satisfies the following universal property: for every henselian local A-algebra B, equipped with an A-algebra morphism $B \to k$, and whose residue field contains the separable closure of $A/\mathfrak{p}$ inside k, there is a unique A-algebra morphism $\tilde{A} \to B$ compatible with the maps to k.

We define the **completion** $\hat{A}$ of A with respect to $A \to k$ as the completion of $\tilde{A}$ at its maximal ideal. If k is separably closed, $\hat{A}$ is called the **strict completion** of A at $\mathrm{Ker}(A \to k)$.

Proposition 5.2.3 *Let A be a noetherian ring, $A \to B$ be a finitely presented morphism and $B \to k$ a morphism to a field k. Then:*

   (i) *There are natural local morphisms $\tilde{A} \to \check{B}$ (resp. $\hat{A} \to \hat{B}$) between the henselizations (resp. completions) with respect to k.*

   (ii) *If $A \to B$ is flat, $\tilde{A} \to \check{B}$ and $\hat{A} \to \hat{B}$ are faithfully flat.*

   (iii) *If $B = A/I$ for some ideal $I \triangleleft A$, then $\check{B} = \tilde{A}/I\tilde{A}$ and $\hat{B} = \hat{A}/I\hat{A}$.*

   (iv) *$A \to B$ is unramified at $\mathfrak{q} := \mathrm{ker}(B \to k)$ iff $\tilde{A} \to \check{B}$ is surjective iff $\hat{A} \to \hat{B}$ is surjective.*

   (v) *$A \to B$ is étale at $\mathfrak{q}$ iff $\tilde{A} \xrightarrow{\sim} \check{B}$ is an isomorphism iff $\hat{A} \xrightarrow{\sim} \hat{B}$ is an isomorphism.*

   (vi) *$\tilde{A}$ is integrally closed in $\hat{A}$.*

Proof:
   (i): $\tilde{A} \to \check{B}$ exists by the universal property of henselization (see Remark 5.2.2). The existence of $\hat{A} \to \hat{B}$ is a direct consequence.

   (ii): If $A \to B$ is flat, its base change $\tilde{A} \to B \otimes_A \tilde{A}$ is flat, too. But $B \otimes_A \tilde{A}$ is the limit of the $B \otimes_A A' \to k \in \mathsf{Et}(B \to k)$, where $A' \to k$ runs over $\mathsf{Et}'(A \to k)$. Conclude with [21] VIII, §2, Prop. 5 that $\check{B}$ is also the henselization of $B \otimes_A \tilde{A}$ w.r.t. its natural morphism to k, i.e. $B \otimes_A \tilde{A} \to \check{B}$ is flat. Now transitivity of flatness and [17] 4.A shows that $\tilde{A} \to \check{B}$ is faithfully flat. As A is noetherian, all other rings we use are noetherian, too, and $\tilde{A} \hookrightarrow \hat{A}$, $\check{B} \hookrightarrow \hat{B}$ are faithfully flat ([17] 23.L Cor. 1). Look at the factorization $\hat{A} \to \hat{A} \otimes_{\tilde{A}} \check{B} \to \hat{B}$. The first arrow is faithfully flat by the base change rule, the second one is completion at the maximal ideal of $\hat{A} \otimes_{\tilde{A}} \check{B}$, hence faithfully flat, too. Again, transitivity shows that $\hat{A} \to \hat{B}$ is faithfully flat.

   (iii): We have a natural morphism $\tilde{A}/I\tilde{A} \to \check{B}$. Since $\tilde{A}$ is henselian and $\tilde{A} \to \tilde{A}/I\tilde{A}$ a finite local morphism, $\tilde{A}/I\tilde{A}$ must be henselian. By the universal property of henselization, there is a morphism $\check{B} \to \tilde{A}/I\tilde{A}$, which is easily seen to be the inverse of the morphism of the first sentence.

(iv): If $A \to B$ is unramified at $\mathfrak{q}$, then it is net in a neighborhood of $\mathfrak{q}$. By [21] V, §1 Thm. 1, we can find $A' \to k \in \mathsf{Et}'(A \to k)$ and $B' \to k \in \mathsf{Et}'(B \to k)$ with $B' = A'/I'$. Now (iii) implies $\tilde{B} = \tilde{A}/I\tilde{A}$. The surjectivity of $\tilde{A} \to \tilde{B}$ immediately implies the surjectivity of $\hat{A} \to \hat{B}$. Now assume the latter. Then $A \to \hat{A}$ and $\hat{A} \to \hat{B}$ are unramified, hence there composition is unramified, too. Thus, $A / \mathfrak{p} \to B / \mathfrak{q} \hookrightarrow \hat{B}/\hat{\mathfrak{m}}_B$ is separable and $\mathfrak{p}\hat{B} = \hat{\mathfrak{m}}_B$ (where $\hat{\mathfrak{m}}_A$, $\hat{\mathfrak{m}}_B$ are the maximal ideals of $\hat{A}$, $\hat{B}$ and $\mathfrak{p} := \ker(A \to k)$). Now the faithful flatness of $B \to \hat{B}$ and $\hat{\mathfrak{m}}_B = \mathfrak{q}\hat{B}$ implies $\mathfrak{p}B_\mathfrak{q} = \mathfrak{q}$, hence $A \to B$ is unramified at $\mathfrak{q}$.

(v): If $A \to B$ is étale at (and then also around) $\mathfrak{q}$, the definition of the henselization immediately shows that $\tilde{A} \cong \tilde{B}$, which trivially implies $\hat{A} \cong \hat{B}$. Assuming the latter, we conclude as in the proof of (iv) that $A \to B$ is unramified at $\mathfrak{q}$. Now the faithful flatness of $A \to \hat{A}$ and $B \to \hat{B}$ and [17] 4.B show that $A \to B$ is flat, therefore $A \to B$ is étale in a neighborhood of $\mathfrak{q}$.

(vi): Let $a \in \hat{A}$ be an element which is integral over $\tilde{A}$. Then $A' := \tilde{A}[a] \subset \hat{A}$ is a finite local extension of $\hat{A}$. But the completion of $\tilde{A}$ and A' is $\hat{A}$ for both rings. By (v), $\tilde{A} \to A'$ must be étale. Therefore $A' = \tilde{A}$ and $a \in \tilde{A}$. ∎

References

[1] S. Bosch, W. Lütkebohmert, and M. Raynaud. *Néron Models*. Number 21 in Ergebnisse der Mathematik und ihrer Grenzgebiete. Springer-Verlag, 1990.

[2] N. Bourbaki. *Algèbre Commutative*. Élément de Mathématique. Masson, 1983.

[3] P. Deligne and N. Katz. *Groupes de Monodromie en Géométrie Algebrique II (SGA 7)*. Number 340 in LNM. Springer-Verlag, 1973.

[4] S. J. Edixhoven. The modular curves $X_0(N)$. Preliminary lecture notes.

[5] W. Fulton. Hurwitz schemes and the irreducibility of the moduli of algebraic curves. *Ann. Math.*, 90:542–575, 1969.

[6] L. Gerritzen, F. Herrlich, and M. van der Put. Stable n-pointed trees of projective lines. *Proceedings of the Koninklijke Nederlandse Akademie van Wetenschappen*, 91:131–163, 1988.

[7] A. Grothendieck. Éléments de Géometrie Algébrique III. *Publ. Math. IHES*, (11), 1961. Avec la collaboration de J. Dieudonné.

[8] A. Grothendieck. *Revêtement Étales et Groupe Fondamental (SGA I)*. Number 224 in LNM. Springer-Verlag, 1971.

[9] D. Harbater. Deformation theory and the tame fundamental group. *Trans. of the AMS*, 262(2):399–415, 1980.

[10] D. Harbater. Galois coverings of the arithmetic line. In *Number Theory: New York, 1984-85*, number 1240 in LNM, pages 165–195. Springer-Verlag, 1987.

[11] D. Harbater. Abhyankar's conjecture on Galois groups over curves. *Invent. Math.*, 117:1–25, 1994.

[12] D. Harbater and K. F. Stevenson. Patching and thickening problems. To appear in J. of Algebra.

[13] J. Harris and D. Mumford. On the Kodaira dimension of the moduli space of curves. *Invent. Math.*, 67:23–86, 1982.

[14] R. Hartshorne. *Algebraic Geometry*. Number 52 in GTM. Springer-Verlag, 1977.

[15] Y. Ihara and H. Nakamura. On deformation of maximally degenerate stable curves and Oda's problem. *J. Reine Angew. Math.*, (487):125–151, 1997.

[16] S. Lang. *Algebra*. Addison Wesley, 3. edition, 1993.

[17] H. Matsumura. *Commutative Algebra*. Benjamin, 1980.

[18] J. S. Milne. *Étale Cohomology*. Princeton Univ. Press, 1980.

[19] S. Mochizuki. The geometry of the compactification of the Hurwitz scheme. *Publ. RIMS, Kyoto University*, 31(3):355–441, 1995.

[20] J. P. Murre. *Lectures on an Introduction to Grothendieck's Theory of the Fundamental Group*. Number 40 in Lectures on Mathematics and Physics. Tata Institut, Bombay, 1967.

[21] M. Raynaud. *Anneaux Locaux Henséliens*. Number 169 in LNM. Springer-Verlag, 1970.

[22] M. Raynaud. Revêtement de la droite affine en caractéristique $p > 0$ et conjecture d'Abhyankar. *Invent. Math.*, 116:425–462, 1994.

[23] M. Saïdi. Revêtements modérés et groupe fondamental de graphe de groupes. *Comp. Math.*, 107:321–340, 1997.

[24] H. Völklein. *Groups as Galois Groups*. Number 53 in Cambridge Studies in Adv. Math. Cambridge Univ. Press, 1996.

[25] S. Wewers. Construction of Hurwitz spaces. Preprint No. 21 of the IEM, Essen, 1998.